DIAGRAMMATICS

Lectures on Selected Problems in Condensed Matter Theory

2nd Edition

DIAGRAMMATICS

Lectures on Selected Problems in Condensed Matter Theory

2nd Edition

MICHAEL V. SADOVSKII

Russian Academy of Sciences, Russia

World Scientific

NEW JERSEY · LONDON · SINGAPORE · BEIJING · SHANGHAI · HONG KONG · TAIPEI · CHENNAI · TOKYO

Published by

World Scientific Publishing Co. Pte. Ltd.

5 Toh Tuck Link, Singapore 596224

USA office: 27 Warren Street, Suite 401-402, Hackensack, NJ 07601

UK office: 57 Shelton Street, Covent Garden, London WC2H 9HE

Library of Congress Control Number: 2019039110

British Library Cataloguing-in-Publication Data
A catalogue record for this book is available from the British Library.

DIAGRAMMATICS
Lectures on Selected Problems in Condensed Matter Theory
2nd Edition

Copyright © 2020 by World Scientific Publishing Co. Pte. Ltd.

ISBN 978-981-121-220-8

For any available supplementary material, please visit
https://www.worldscientific.com/worldscibooks/10.1142/11605#t=suppl

It is necessary to overcome mysticism with respect to technics...

Chairman Mao Zedong

Oh? The man works and doesn't tell his assistant what he is doing...?
He will never give that seminar.

W. Pauli to R.P. Feynman

L.P. Gorkov, A.A. Abrikosov, I.E. Dzyaloshinskii, Argonne, USA, 1998

Preface

At the end of the fifties and in early sixties of the last century there was a kind of "revolution" in the theory of condensed matter (at that time called mostly the theory of solid state and quantum liquids) which was due to the use of methods, developed a decade earlier in quantum field theory, mainly the method of Feynman diagrams. Since that time diagrammatic methods became the foundation of this section of theoretical physics, and the knowledge of these is absolutely necessary for any professional working in this field.

A number of good books are devoted to a rather detailed exposition of the general aspects of these methods, such as the introduction of diagrammatic formalism for different types of interactions [Abrikosov A.A., Gorkov L.P., Dzyaloshinskii I.E. (1963); Lifshits E.M., Pitaevskii L.P. (1980)]. Of course, most of these books contain also the discussion of some specific applications of these methods to concrete physical problems. At the same time, up to now there are almost no books, where the reader can find the detailed description of calculations and methodical "know how" for specific problems, at the beginner level (like graduate or postgraduate students).[1] During the last decades a great number of problems were solved (or analyzed) using Feynman diagram technique and the results are scattered in the numerous original papers, reviews and books.

---

[1]The author knows only one such attempt [Levitov L.S., Shitov A.V. (2003)], which remained unpublished for a long time, and finally was published only in Russian. A comprehensive review of the applications of field theory methods to different problems of solid state theory and the theory of quantum liquids is contained in [Mahan G.D. (1981)], but it is in fact a review for a professional, not a textbook.

The aim of these lectures is precisely the demonstration of the power of diagram technique as applied to the solution of different problems of condensed matter theory, most of which a long time ago became a kind of "gold reserve" of this theory, while different concepts and methodical developments constitute a part of a working "folklore" of modern theorists. Our choice of problems is based both on their importance and personal interests of the author. Some of these problems are not "finally" solved up to now, so that further development of the results of almost any section of this book may be the starting point of a serious theoretical study. Actually, we limit ourselves only to the selected problems of electronic theory of solids, dropping any discussion of Bose-liquids, most problems of the theory of magnetism, as well as the theory of critical phenomena, where diagrammatic methods are also quite important. It should be clearly understood that the material discussed in every chapter of this book can be a part of a separate lecture course, and we do not pretend to give a self-contained review of any of these parts of the modern theory.

It is obvious, that the application of quantum field theory methods to the theory of condensed matter is not limited to diagrammatic methods only. In particular, there was a great temptation to pay some attention to the functional integrals or renormalization group. But finally a decision was made to limit discussion only to diagrammatic approaches and problems, which can be solved within more or less standard perturbation theory, dropping almost all modern aspects of the theory of strongly correlated systems. This was due to a wish to make these lectures more or less "compact", as well as to demonstrate the "richness" of results, which can be obtained in this way.

To understand these lectures it is necessary to know the basic notions of Feynman diagram technique, approximately within the limits of chapters II and III of the notorious "AGD" book [Abrikosov A.A., Gorkov L.P., Dzyaloshinskii I.E. (1963)], where anybody can find a presentation, which remains unsurpassed up to now.[2]

---

[2]In fact the material presented in this book was used by the author as a *second* part of the lecture course, taught at the Ural State University in Ekaterinburg. The first part of this course is actually based on these chapters of "AGD".

The author is grateful to Dr. K.K. Phua of World Scientific for the invitation to publish an English version of this book. This new edition is revised and expanded by introduction of some additional material in Chapters 3, 4 and 5.

M.V. Sadovskii, Ekaterinburg, 2019

Note to the Reader:
The parts of the main text in smaller font represent either some technical remarks or some material that requires a slightly higher theoretical background.

Contents

Chapter 1

Introduction

The concept of quasiparticles is of major importance in the theory of condensed matter. This concept can be rigorously justified within the Green's function formalism, which a long time ago became the main working tool of all modern approaches to many particle systems. The method of Green's functions allows to formulate criteria for the existence of quasiparticles in specific models of interacting particles, as well as constitutes the universal method of practical calculations of arbitrary physical properties of many particle systems with the account of different types of interactions. This method originated in quantum field theory, where quite effective and convenient approach, based on the use of *Feynman diagrams* appeared for the first time. The following transfer of these methods to the theory of many particle systems, in fact, lead to the formulation of the modern theory of condensed matter [Abrikosov A.A., Gorkov L.P., Dzyaloshinskii I.E. (1963); Lifshits E.M., Pitaevskii L.P. (1980)].

In this lecture course we do not present step by step derivation of the Green's function formalism itself, our aim is to teach how to use this method for solution of concrete physical problems. It is assumed that the basic principles of construction of Feynman diagrams for different types of interactions are already known, both for the case of zero temperature $T = 0$, as well as for finite temperatures (Matsubara formalism) [Abrikosov A.A., Gorkov L.P., Dzyaloshinskii I.E. (1963)]. The structure of the course is clear from the Contents. Separate chapters are devoted to the analysis of different types of interactions, which are studied within the electronic theory of the solid state, and also to a

number of major electronic instabilities (phase transitions). At first, in
each chapter we formulate the rules of diagram technique, appropriate
for the interaction under study, then we analyze different problems, in
most cases presenting all the details of calculations, or at least giving all
the information necessary to reproduce the results. Practically everywhere in these lectures we tried to adhere to the rules and major notations used in "AGD" book [Abrikosov A.A., Gorkov L.P., Dzyaloshinskii
I.E. (1963)], though due to rather "informal" style of our presentation,
we can not guarantee the absence of some "randomness" in notations
between different sections. In fact, each chapter can be used as the
introduction to the problems of the appropriate part of the solid state
theory. In this sense the chapters can be read independently of each
other, but it should be noted that all the problems under discussion
has much in common and are, in fact, deeply connected to each other.
Bibliography is in no sense complete, we quote only the sources, from
which we have taken the material used in our presentation, limiting
ourselves mainly to textbooks or reviews. Accordingly, there are practically no references to original papers and no discussion of priorities,
in most important cases we just quote the name of the author (with an
approximate year, when the result was obtained). Some of the material
of these lectures is based on personal exercises by the author, no specific
references are given in most of such cases.

The main idea of diagrammatic approach in the theory of condensed matter reduces in fact to the summation of an infinite series of Feynman diagrams for the single-particle or many-particle (in
most cases two-particle) Green's functions (and (or) appropriate vertex parts, describing multi-particle interactions). Usually it is possible to perform a certain *partial* summation of some classes (types)
of diagrams of perturbation series, which are "dominating" over some
physical parameter (e.g. dimensionless coupling constant, density of
particles, or some other combination of parameters, characteristic
for the problem under discussion). In most cases, such dominating
classes of diagrams were determined already during the initial stages
of the development of diagrammatic approaches to different kinds of
interactions [Abrikosov A.A., Gorkov L.P., Dzyaloshinskii I.E. (1963);
Lifshits E.M., Pitaevskii L.P. (1980)], and we shall consider a number

of such typical cases and physical results obtained. In some (very) rare cases and for (mostly) oversimplified model cases, it is possible to perform a *complete* summation of *all* Feynman diagrams. These cases (problems) are much less known, but mostly are quite important and instructive. We shall consider a number of such problems, both to illustrate technical aspects and also to analyze nontrivial conclusions, such as the "destruction" of the concept of quasiparticles itself, which being quite useful certainly has its limits. Here we shall move closer to most modern aspects of the theory.

Practically everywhere in these lectures we use the natural units with $\hbar = c = 1$, "restoring" $\hbar$ and c only in some final expressions and estimates. Boltzmann's constant is always taken as $k_B = 1$.

1.1 Quasiparticles and Green's functions

Though we shall not be presenting any systematic derivation of diagrammatic approach to many-particle systems, let us start with some short introduction of some elementary concepts and definitions, just for coherence of presentation and to remind a reader basic physical ideas behind the application of Green's functions in condensed matter theory.

Consider first the case of temperature $T = 0$, i.e. the system at its ground state. Let us start from the elementary problem of a *single* quantum particle moving in some time-independent external potential (or field), and described by the usual (time-dependent) Schroedinger equation with appropriate Hamiltonian H:

$$i\frac{\partial \psi(\mathbf{r}, t)}{\partial t} - H\psi(\mathbf{r}, t) = 0 \tag{1.1}$$

Instead of solving this equation directly (with some initial condition for the wave-function) we introduce the Schroedinger-like equation for the *Green's function* $G(\mathbf{r}, t; \mathbf{r}', t')$, depending on *two* values of time and coordinate:

$$i\frac{\partial G}{\partial t} - HG = i\delta(\mathbf{r} - \mathbf{r}')\delta(t - t') \tag{1.2}$$

with initial condition $G(\mathbf{r}, t + 0; \mathbf{r}', t) = \delta(\mathbf{r} - \mathbf{r}')$. Physically, Green's function represents the *probability amplitude* for a particle transition from (initial) point $\mathbf{r}'$ at the moment of time t' to the some point $\mathbf{r}$ at the moment t. Squared modulus of this amplitude gives the probability of such transition. This is easily checked expressing ψ-function at the moment $t + \tau$ via ψ-function at the moment t as:

$$\psi(\mathbf{r}, t + \tau) = \int d\mathbf{r}' G(\mathbf{r}, t + \tau; \mathbf{r}' t) \psi(\mathbf{r}', t) \qquad (1.3)$$

It is easily seen that this expression for $\psi(\mathbf{r}, t + \tau)$ satisfies the Schroedinger equation (1.1), and for $\tau \to 0$ it coincides with $\psi(\mathbf{r}, t)$ due to the initial condition $G(\mathbf{r}, t + 0; \mathbf{r}', t) = \delta(\mathbf{r} - \mathbf{r}')$. Obviously, we have to assume $G = 0$ for $\tau < 0$ to guarantee causality.

Let us now introduce some set of eigenfunctions of the stationary Schroedinger equation:

$$H\varphi_\lambda(\mathbf{r}) = \varepsilon_\lambda \varphi_\lambda(\mathbf{r}) \qquad (1.4)$$

Depending on the problem at hand, the quantum numbers λ can have different physical meaning. If our problem (Hamiltonian) is translation invariant $\lambda \to \mathbf{p}$, e.g. the momentum of a free particle, for the system in an external magnetic field λ represents the set of Landau quantum numbers, for a particle moving in some arbitrary (or random) potential, these may be some (in general unknown to us) quantum numbers of the states diagonalizing the Hamiltonian.

Consider the simple case of a particle moving in some potential:

$$H = \frac{p^2}{2m} + V(\mathbf{r}) \qquad (1.5)$$

Any solution of the Schroedinger equation (1.1) can be expanded using the complete system of eigenfunctions of (1.4):

$$\psi(\mathbf{r}, t) = \sum_\lambda c_\lambda(t) \varphi_\lambda(\mathbf{r}) \qquad (1.6)$$

Then we can write (1.3) as an equation for the coefficients of this expansion:

$$c_\lambda(t + \tau) = \sum_{\lambda'} G_{\lambda\lambda'}(\tau) c_{\lambda'}(t) \qquad (1.7)$$

and obtain:

$$G_{\lambda\lambda'}(\tau) = \int d^3r d^3r' G(\mathbf{r}, \mathbf{r}'\tau) \varphi_\lambda^\star(\mathbf{r}) \varphi_{\lambda'}(\mathbf{r}') \qquad (1.8)$$

— the Green's function in the representation of quantum numbers λ. As φ_λ is an exact stationary state of the (time-independent)

Hamiltonian H, there are no transitions to another states, so that $c_\lambda(t+\tau) = e^{-i\varepsilon_\lambda \tau} c_\lambda(t)$, i.e.

$$G_{\lambda\lambda'}(\tau) = G_\lambda(\tau)\delta_{\lambda\lambda'} = e^{-i\varepsilon_\lambda \tau}\theta(\tau) \qquad (1.9)$$

where $\theta(\tau) = 1$ for $\tau \geq 0$ and $\theta(\tau) = 0$ for $\tau < 0$. Consider now the Fourier transform:[1]

$$G_\lambda(\varepsilon) = \frac{1}{i}\int_{-\infty}^{\infty} d\tau e^{i\varepsilon\tau} G_\lambda(\tau) \qquad (1.10)$$

$$G_\lambda(\tau) = i\int_{-\infty}^{\infty} \frac{d\varepsilon}{2\pi} e^{-i\varepsilon\tau} G_\lambda(\varepsilon) \qquad (1.11)$$

Then, after elementary integration we get:

$$G_\lambda(\varepsilon) = \frac{1}{\varepsilon - \varepsilon_\lambda + i\delta}, \quad \delta \to +0 \qquad (1.12)$$

The sign of $\delta \to 0$ is chosen to guarantee $G_\lambda(\tau) = 0$ for $\tau < 0$. In fact we have:

$$G_\lambda(\tau) = i\int_{-\infty}^{\infty} \frac{d\varepsilon}{2\pi} \frac{e^{-i\varepsilon\tau}}{\varepsilon - \varepsilon_\lambda + i\delta}$$

$$= \begin{cases} e^{-i\varepsilon_\lambda\tau} & \text{for} \quad \tau > 0 \\ 0 & \text{for} \quad \tau < 0 \end{cases} \qquad (1.13)$$

To convince yourself note, that the integrand here has a pole at $\varepsilon = \varepsilon_\lambda - i\delta$. Then for $\tau > 0$ we can close the integration contour in the lower half-plane of complex variable ε (as the factor $e^{-i\varepsilon\tau}$ guarantees the exponential damping of the integrand at the semicircle at infinity in the lower half-plane), then the pole of the integrand is inside the contour of integration and using Cauchy theorem we obtain the result given in Eq. (1.13). For $\tau < 0$, to guarantee the zero contribution from the semicircle, we have to close integration contour in the upper half-plane of ε. Then there is no pole inside the contour and the integral reduces to zero.

---

[1]Note the additional factor i which we introduced in (1.10), (1.11) and below in (1.19) which guarantees correspondence with standard notations of "AGD". Usually this factor is just added in the definition of Green's function [Abrikosov A.A., Gorkov L.P., Dzyaloshinskii I.E. (1963)].

In a mixed $(\mathbf{r}, \varepsilon)$ representation we obtain:

$$G(\mathbf{r}, \mathbf{r}', \varepsilon) = \sum_{\lambda, \lambda'} G_{\lambda\lambda'}(\varepsilon)\varphi_\lambda(\mathbf{r})\varphi_{\lambda'}^\star(\mathbf{r}')$$

$$= \sum_\lambda \frac{\varphi_\lambda(\mathbf{r})\varphi_\lambda^\star(\mathbf{r}')}{\varepsilon - \varepsilon_\lambda + i\delta} \tag{1.14}$$

Here the sum over λ includes summation over all bound states and integration over the continuous part of the spectrum. We can see that $G(\mathbf{r}, \mathbf{r}', \varepsilon)$ possesses poles at the values of ε equal to ε_λ, i.e. at the energies of bound states, and the cut (continuum of the poles) on the part of the real axis of ε, corresponding to the continuous part of the spectrum.

Consider now the many-particle system. Let us limit discussion only to the case of (many) Fermions. Similar analysis can be given for the system of Bose particles, but we skip it referring the reader to the general courses [Abrikosov A.A., Gorkov L.P., Dzyaloshinskii I.E. (1963); Lifshits E.M., Pitaevskii L.P. (1980)]. Consider first the case of non-interacting Fermions (Fermi-gas). Elementary excitations in this case are pairs of excited particles (above the Fermi surface) and holes (below the Fermi surface).

Let us determine Green's function for a particle excitation $G_{\lambda\lambda'}(\tau)$, i.e. the transition amplitude of a particle from some state λ to a state λ' (for the case of non-interacting Fermions). We have to take into account limitations due to Pauli principle, i.e. exclude transitions to occupied states. This can be achieved by an additional factor $(1 - n_\lambda)$ in the definition of the Green's function, where

$$n_\lambda = \begin{cases} 1 & \text{for} \quad \varepsilon_\lambda \leq \varepsilon_F \\ 0 & \text{for} \quad \varepsilon_\lambda > \varepsilon_F \end{cases} \tag{1.15}$$

is just the particle number in a state λ (Fermi distribution for $T = 0$). Thus we obtain:

$$G_{\lambda\lambda'}^+(\tau) = (1 - n_\lambda)\delta_{\lambda\lambda'} \begin{cases} e^{-i\varepsilon_\lambda\tau} & \text{for} \quad \tau > 0 \\ 0 & \text{for} \quad \tau < 0 \end{cases} \tag{1.16}$$

Let us now find similar expression for holes. As the number of available states for holes at the state λ is just n_λ, we get:

$$G_{\lambda\lambda'}^-(\tau) = n_\lambda\delta_{\lambda\lambda'} \begin{cases} e^{i\varepsilon_\lambda\tau} & \text{for} \quad \tau > 0 \\ 0 & \text{for} \quad \tau < 0 \end{cases} \tag{1.17}$$

where we have taken into account also that the hole energy (with respect to the Fermi level) is opposite in sign to the particle energy.

It is convenient to introduce Green's function $G_\lambda(\tau)$, defined both for $\tau > 0$ and $\tau < 0$:

$$G_\lambda(\tau) = \begin{cases} G_\lambda^+(\tau) & \text{for } \tau > 0 \\ -G_\lambda^-(-\tau) & \text{for } \tau < 0 \end{cases} \tag{1.18}$$

Fourier transform of this function is easily calculated as:

$$G_\lambda(\varepsilon) = -i(1 - n_\lambda) \int_0^\infty d\tau e^{-i\varepsilon_\lambda \tau + i\varepsilon\tau} + in_\lambda \int_{-\infty}^0 d\tau e^{i\varepsilon_\lambda \tau + i\varepsilon\tau}$$

$$= \frac{1 - n_\lambda}{\varepsilon - \varepsilon_\lambda + i\delta} + \frac{n_\lambda}{\varepsilon - \varepsilon_\lambda - i\delta} \tag{1.19}$$

where it is necessary to introduce $\delta \to +0$ to guarantee convergence. It is convenient to rewrite this expression as:

$$G_\lambda(\varepsilon) = \frac{1}{\varepsilon - \varepsilon_\lambda + i\delta \, sign\varepsilon_\lambda}$$

$$= \begin{cases} \frac{1}{\varepsilon - \varepsilon_\lambda + i\delta} & \text{for } \varepsilon_\lambda > \varepsilon_F \\ \frac{1}{\varepsilon - \varepsilon_\lambda - i\delta} & \text{for } \varepsilon_\lambda < \varepsilon_F \end{cases} \tag{1.20}$$

where we have introduced sign-function: $sign(x) = 1$ for $x > 0$ and $sign(x) = -1$ for $x < 0$. Note that the Fourier transform of this Green's function has a pole at ε equal to a particle (hole) energy.

Consider now the system of *interacting* Fermions. Single-particle Green's function in a system of interacting Fermions can be defined as:

$$G^+(\mathbf{r}t; \mathbf{r}'t')_{t>t'} = <0|\hat{\psi}(\mathbf{r}t)\hat{\psi}^+(\mathbf{r}'t')|0> \tag{1.21}$$

where $|0>$ is an exact ground state ("vacuum") of our system, corresponding to the filled Fermi-sphere, $\hat{\psi}(\mathbf{r}t)$ is second quantized operator of a Fermi field in Heisenberg representation [Abrikosov A.A., Gorkov L.P., Dzyaloshinskii I.E. (1963)]:

$$\hat{\psi}(\mathbf{r}t) = e^{iHt}\hat{\psi}(\mathbf{r})e^{-iHt} \tag{1.22}$$

with H — the Hamiltonian of our many-particle (interacting) system. Operator $\hat{\psi}(\mathbf{r})$ can be expressed in a standard way via annihilation operators a_λ of a particle in λ-states (while $\hat{\psi}^+$ is similarly expressed via creation operators a_λ^+):

$$\hat{\psi}(\mathbf{r}) = \sum_\lambda a_\lambda \varphi_\lambda(\mathbf{r}) \tag{1.23}$$

Eq. (1.21) obviously gives us the transition amplitude for a particle propagating from $(\mathbf{r}'t')$ to $(\mathbf{r}t)$.

Similar expression can be written for propagating hole:

$$G^-(\mathbf{r}t;\mathbf{r}'t')_{t>t'} = <0|\hat{\psi}^+(\mathbf{r}t)\hat{\psi}(\mathbf{r}'t')|0> \tag{1.24}$$

where it is taken into account that annihilation of a particle in a given point is equivalent to creation of a hole.

Both expressions (1.21) and (1.24) are defined for $t > t'$. It is convenient to can write down a single expression, which for $t > t'$ describes propagating particle, while for $t < t'$ – propagating hole (similarly to Eq. (1.18)):

$$G(\mathbf{r}t;\mathbf{r}'t') = \begin{cases} G^+(\mathbf{r}t;\mathbf{r}'t') & \text{for } t > t' \\ -G^-(\mathbf{r}'t';\mathbf{r}t) & \text{for } t < t' \end{cases} \tag{1.25}$$

Another way to write this is:[2]

$$G(x,x') = <0|T\hat{\psi}(x)\hat{\psi}^+(x')|0> \tag{1.26}$$

where we have denoted $x = (\mathbf{r}t)$, and the symbol of T-ordering means that all the operators standing to the right of T are placed in order over time arguments, with those corresponding to later moments standing to the left from those corresponding to earlier times, with the account of a sign change due to permutations of Fermion operators (necessary to place operators in the "right" order in time arguments). Formal definition of T-ordering taken from the quantum field theory is given by:

$$T\{F_1(t_1)F_2(t_2)\} = \begin{cases} F_1(t_1)F_2(t_2) & \text{for } t_1 > t_2 \\ -F_2(t_2)F_1(t_1) & \text{for } t_1 < t_2 \end{cases} \tag{1.27}$$

for Fermion operators and

$$T\{B_1(t_1)B_2(t_2)\} = \begin{cases} B_1(t_1)B_2(t_2) & \text{for } t_1 > t_2 \\ B_2(t_2)B_1(t_1) & \text{for } t_1 < t_2 \end{cases} \tag{1.28}$$

---

[2]Standard definition of "AGD" differs by an additional factor of $-i$, which we have taken into account in Fourier transforms above.

for Boson operators. Green's function defined by Eq. (1.26) is usually called Feynman or causal (T-ordered).[3]

If we deal with an infinite homogeneous (translation invariant) system we have $G(\mathbf{r}t; \mathbf{r}'t') = G(\mathbf{r} - \mathbf{r}', t - t')$ and it is convenient to perform Fourier transformation both in $t - t'$ and $\mathbf{r} - \mathbf{r}'$:

$$G(\mathbf{p}\tau) = \int d^3 r G(\mathbf{r}\tau) e^{-i\mathbf{p}\mathbf{r}} \tag{1.29}$$

where

$$G(\mathbf{p}\tau) = \begin{cases} < 0|a_{\mathbf{p}} e^{-iH\tau} a_{\mathbf{p}}^+|0 > e^{iE_0\tau} & \tau > 0 \\ - < 0|a_{\mathbf{p}}^+ e^{iH\tau} a_{\mathbf{p}}|0 > e^{-iE_0\tau} & \tau < 0 \end{cases} \tag{1.30}$$

where E_0 is the ground state energy (in our case just equal to Fermi energy E_F).

Quasiparticles can be a viable concept if the single-particle Green's function of a system under consideration can be expressed as ($\tau > 0$):

$$G(\mathbf{p}\tau) \approx Z e^{-i\xi(\mathbf{p})\tau - \gamma(\mathbf{p})\tau} + \dots \quad \text{and} \quad \gamma(\mathbf{p}) \ll \xi(\mathbf{p}) \tag{1.31}$$

where $\xi(\mathbf{p}) = \varepsilon(\mathbf{p}) - E_F$, i.e. it contains a contribution of the form similar to that of the Green's function of the free (non-interacting) Fermi gas. Eq. (1.31) means the presence (with amplitude Z in the ground state $|0 >$) of a wave-packet, corresponding to a quasiparticle with energy $\xi(\mathbf{p})$ and *damping* $\gamma(\mathbf{p})$. We have to require that $\gamma(\mathbf{p}) \ll \xi(\mathbf{p})$, i.e. the weakness of damping to make quasiparticles "well defined".[4] In a similar way, for $\tau < 0$ we can define the Green's function for a quasihole. Finally, in a system with well defined quasiparticles the

---

[3]Note that this definition does not coincide with that of the so-called two-time Green's function introduced by Bogoliubov and Tyablikov and used in the theory of linear response [Zubarev D.N. (1974)], even if we go there to the limit of zero temperature. The advantage of introducing Feynman's functions is in the availability of diagram technique, giving the universal method to calculate these Green's functions via perturbation theory. There is no (convenient) diagram technique for Green's functions of Bogoliubov and Tyablikov. There are a number of exact relations and methods, allowing to express the Green's functions of linear response theory via Feynman's functions for $T = 0$ and appropriate generalizations for the case of finite temperatures (Matsubara formalism) which we shall use below [Abrikosov A.A., Gorkov L.P., Dzyaloshinskii I.E. (1963); Lifshits E.M., Pitaevskii L.P. (1980)].

[4]This condition, as we shall see below, is satisfied in Landau theory of Fermi liquids, where close to the Fermi surface we have: $\xi(\mathbf{p}) \approx v_F(|\mathbf{p} - p_F|)$, while $\gamma(\mathbf{p}) \sim (|\mathbf{p}| - p_F)^2$ (v_F is Fermi velocity).

Fourier transform of the Green's function (1.26) can be written as:

$$G(\mathbf{p}\varepsilon) = Z \left\{ \frac{1 - n_{\mathbf{p}}}{\varepsilon - \xi(\mathbf{p}) + i\gamma(\mathbf{p})} + \frac{n_{\mathbf{p}}}{\varepsilon - \xi(\mathbf{p}) - i\gamma(\mathbf{p})} \right\} + G_{reg}(\mathbf{p}\varepsilon)$$

$$= \frac{Z}{\varepsilon - \xi(\mathbf{p}) + i\gamma(\mathbf{p})sign(p - p_F)} + G_{reg}(\mathbf{p}\varepsilon) \qquad (1.32)$$

We see that the poles of this expression define the quasiparticle spectrum and damping. This is a general property of Green's functions, allowing to determine the quasiparticle spectrum in many-particle system. The value of G_{reg} in (1.32) is determined by the contribution of many-particle excitations, and in most cases is of no special importance. However, in systems with strong correlations (interactions) we may meet with situation, when there is actually no quasiparticle poles in the Green's function, so that there is no single-particle excitations at all and everything is actually determined by G_{reg}, making the studies of the properties of such systems much more complicated.

Why do we need Green's functions at all? First they give us the general method to study the spectrum of excitations in many-particle (interacting) systems. It happens also, that the knowledge of Green's functions allows to calculate ground state ($T = 0$) averages of arbitrary physical characteristics of many-particle systems. Let us consider simple examples. Using the introduced *single-particle* Green's function we may calculate the ground state averages of operators which can be written as a sum of single-particle contributions (one-particle operators) [Bogoliubov N.N. (1991a); Sadovskii M.V. (2019a)]:

$$\hat{A} = \sum_i \hat{A}_i(x_i, \mathbf{p}_i) \qquad (1.33)$$

where x_i represents e.g. both spatial and spin variables, while $\mathbf{p}_i$ are the momenta (operators!) of separate particles of our system. Typical examples are:

$$n(\mathbf{r}) = \sum_i \delta(\mathbf{r} - \mathbf{r}_i) \qquad (1.34)$$

— operator of the particle density at the point $\mathbf{r}$,

$$\mathbf{j}(\mathbf{r}) = \frac{e}{m} \sum_i \mathbf{p}_i \delta(\mathbf{r} - \mathbf{r}_i) \qquad (1.35)$$

— current density at $\mathbf{r}$ etc.

Operator $\hat{A}$ in second quantized form can be written as:

$$\hat{A} = \int dx \psi^+(x) A(x, \mathbf{p}) \psi(x) \tag{1.36}$$

Consider Green's function (1.25), (1.26) at $t = t' - 0$:

$$G(x, x', \tau)|_{\tau \to -0} = - <0|\psi^+(x')\psi(x)|0> \tag{1.37}$$

Then the average value of $\hat{A}$ in the ground state can be written as:

$$<A> = \int dx A(x, \mathbf{p}) G(x, x', \tau = -0)|_{x=x'} = -Sp AG|_{\tau=-0} \tag{1.38}$$

Thus, the value of $G|_{\tau=-0}$ up to a sign coincides with single-particle density matrix at $T = 0$:

$$\rho(x', x) = <0|\psi^+(x')\psi(x)|0> = -G|_{\tau=-0} \tag{1.39}$$

To determine the average values of two-particle operators:

$$\hat{B} = \sum_{ik} B_{ik}(x_i \mathbf{p}_i; x_k \mathbf{p}_k) \tag{1.40}$$

we have to calculate *two-particle* Green's function, defined usually as:

$$G_2(1, 2; 3, 4) = <0|T\psi(1)\psi(2)\psi^+(3)\psi^+(4)|0> \tag{1.41}$$

etc. Thus, the problem of finding the average values of multi-particle operators, requires the knowledge of appropriate density matrices [Bogoliubov N.N. (1991a)], which can be expressed via corresponding multi-particle Green's functions.

1.2 Diagram technique. Dyson equation

Feynman diagrams give an elegant graphical representation of arbitrary contributions to perturbation series for Green's functions. The standard way to obtain specific diagram rules for a given interacting system reduces to the study of (scattering) S-matrix perturbation expansion and the use of the Wick's theorem [Abrikosov A.A., Gorkov L.P., Dzyaloshinskii I.E. (1963); Lifshits E.M., Pitaevskii L.P. (1980)]. Typical graphic elements of any diagram technique are Green's functions *lines* and interaction *vertices*, which are combined into Feynman diagrams of a certain "topology", depending on the nature of interaction under consideration. Below we shall formulate these rules explicitly

[Abrikosov A.A., Gorkov L.P., Dzyaloshinskii I.E. (1963)] for different kinds of interactions, which will be studied in these lectures.

Wonderful aspect of Feynman diagram technique is the possibility to perform *graphical summation* of infinite (sub)series of diagrams. Consider the simplest (and actually most important!) example of such summation, leading to the derivation of the so-called Dyson's equation [Abrikosov A.A., Gorkov L.P., Dzyaloshinskii I.E. (1963); Lifshits E.M., Pitaevskii L.P. (1980)]. Let us denote an *exact* Green's function by a "fat" (or "dressed" line), and a free-particle Green's function via "thin" line. Full transition amplitude (Green's function) of a transition from point 2 to point 1 is obviously equal to the sum of all possible transition amplitudes, appearing at all orders of perturbation theory, i.e. to the sum of all possible Feynman diagrams for the Green's function. Let us classify diagrams in the following way. First of all separate the only graph (line), corresponding to the propagation of a free particle. The remaining diagrams has the following form: up to some point the particle is propagating freely, then some scattering occurs, resulting in creation and annihilation of a number of particles and holes (or the particle is just scattered by the other particles, belonging to the Fermi "sea", below the Fermi level), then again we have a freely propagating particle, then scattering processes (interactions) are repeated etc. Let us denote as Σ the sum of all diagrams, which can not be separated in two parts by *cutting a single Fermion line*, this "block" Σ is called the irreducible *self-energy* of a particle (Fermion). Now we can easily convince ourselves that the full Green's function is determined by the Dyson equation, graphically shown in Fig. 1.1. In analytic form it is an integral equation:

$$G(1,2) = G_0(1,2) + \int d\tau_3 d\tau_4 G_0(1,3)\Sigma(3,4)G(4,2) \qquad (1.42)$$

Iterating this equation we obtain the full perturbation series for the Green's function. After Fourier transformation Dyson equation is reduced to the algebraic one:

$$G(\mathbf{p}\varepsilon) = G_0(\mathbf{p}\varepsilon) + G_0(\mathbf{p}\varepsilon)\Sigma(\mathbf{p}\varepsilon)G(\mathbf{p}\varepsilon), \qquad (1.43)$$

which is easily solved:

$$G(\mathbf{p}\varepsilon) = \frac{1}{\varepsilon - \varepsilon(\mathbf{p}) - \Sigma(\mathbf{p}\varepsilon)} \qquad (1.44)$$

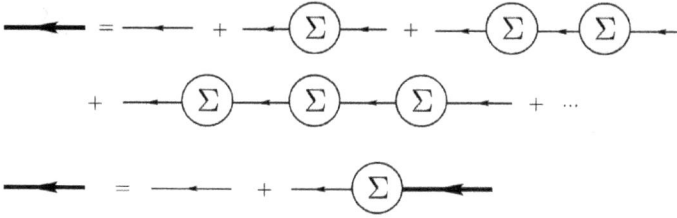

Fig. 1.1 Diagrammatic derivation of the Dyson equation.

where we have taken into account the explicit form of $G_0(\mathbf{p}\varepsilon)$. It is clear that the self-energy $\Sigma(\mathbf{p}\varepsilon)$ represents in a compact form all changes in a particle motion as a result of its interaction with other particles of a system. In general case, self-energy is complex, i.e. consists of real and imaginary parts (that is the reason why in (1.44) we have dropped an infinitesimal imaginary term of the free particle $i\delta sign(\varepsilon - \varepsilon_F)$). Solving Dyson equation in some approximation (or, in rare cases, exactly) allows us to analyze the energy (excitation) spectrum of many-particle interacting systems.

1.3 Green's functions at finite temperatures

Green's functions formalism is almost directly generalized for the case of finite temperatures T [Abrikosov A.A., Gorkov L.P., Dzyaloshinskii I.E. (1963)]. To remind the reader the essence of this (Matsubara) formalism we again restrict ourselves mainly to the case of Fermions. So-called thermodynamic (or Matsubara) Green's function is defined as:

$$\mathcal{G}(\mathbf{p}, \tau_2 - \tau_1) = -i < T_\tau a_{\mathbf{p}}(\tau_2)a_{\mathbf{p}}^+(\tau_1) > \qquad (1.45)$$

where we use "interaction" representation for operators in the following form:

$$a_{\mathbf{p}}(\tau) = e^{(H-\mu N)\tau}a_{\mathbf{p}}e^{-(H-\mu N)\tau} \qquad (1.46)$$

where Matsubara "time" $0 < \tau_1, \tau_2 < \beta = \frac{1}{T}$ and μ is the chemical potential, while angular brackets denote the averaging over grand canonical Gibbs ensemble, which is conveniently written as:

$$< A > = \frac{Sp\rho A}{Sp\rho} \quad \text{where} \quad \rho = e^{-\beta(H-\mu N)} \qquad (1.47)$$

with $Z = Sp\rho$.

The reason why Matsubara Green's functions $\mathcal{G}$ can be expanded in (almost) the same diagrammatic expansion, as quantum mechanical Green's functions G in the case of $T = 0$, is as follows. We have seen that diagrammatic expansion for G directly follows from the time dependent Schroedinger equation. Statistical operator ρ, written as in (1.47), satisfies the so-called Bloch equation:

$$\frac{\partial \rho}{\partial \beta} = -(H - \mu N)\rho \tag{1.48}$$

as is easily checked just by differentiation. There is direct correspondence between Schroedinger equation (1.1) and Bloch equation (1.48):

$$\psi \leftrightarrow \rho \quad H \leftrightarrow H - \mu N \quad it \leftrightarrow \beta \tag{1.49}$$

Thus, after the simple substitution of

$$H \to H - \mu N \quad it \to \tau \tag{1.50}$$

in the expressions of the previous section, we can obtain Matsubara Green's function formalism for $\mathcal{G}$ of almost the same form as in the case of $T = 0$ for quantum mechanical G. Substitution $H \to H - \mu N$ leads only to the appropriate change of the single particle energy by μ:

$$H_0 - \mu N = \sum_{\mathbf{p}} (\varepsilon(\mathbf{p}) - \mu)a_{\mathbf{p}}^+ a_{\mathbf{p}} \tag{1.51}$$

Though Matsubara Green's functions $\mathcal{G}$ depend on "imaginary time" τ,[5] we may always return to the real time via substitution (in final expressions) $\tau \to it$, or, strictly speaking, via analytical continuation of Matsubara expressions from imaginary to real time axis.

Above we have already noted that the values of τ_1 and τ_2 in (1.45) are limited to the interval from 0 to β. Accordingly, to perform Fourier transformation over τ we have to introduce $\mathcal{G}$ periodically continued on the interval from $-\infty$ to ∞. Then we can write down the Fourier expansion as:

$$\mathcal{G}(\mathbf{p}\tau) = \frac{1}{\beta} \sum_{n=-\infty}^{\infty} e^{-i\omega_n \tau} \mathcal{G}(\mathbf{p}\omega_n) \tag{1.52}$$

---

[5]The value τ was taken real, but Green's function $\mathcal{G}$ can be obtained from G by a substitution $it \to \tau$, so that in thermodynamic formalism we are dealing with a transition to $t = -i\tau$, i.e. "imaginary time".

where summation is over discrete (Matsubara) frequencies $\omega_n = \pi n T$. Then:

$$\mathcal{G}(\mathbf{p}\omega_n) = \frac{1}{2} \int_{-\beta}^{\beta} d\tau\, e^{i\omega_n \tau} \mathcal{G}(\mathbf{p}\tau) \tag{1.53}$$

"Time" difference $\tau = \tau_2 - \tau_1$ belongs to the interval $(-\beta, \beta)$, as both τ_1 and τ_2 vary on the interval $(0, \beta)$. The function $\mathcal{G}(\mathbf{p}\tau)$ periodically reproduces itself on intervals $(-\beta, \beta), (\beta, 3\beta), (3\beta, 5\beta), ..., (-3\beta, -\beta), ...$ For the system consisting of Fermions, the even values of n drop out from the series for $\mathcal{G}(\mathbf{p}\tau)$ due to "antiperiodic" boundary condition:

$$\mathcal{G}(\mathbf{p}, \tau) = -\mathcal{G}(\mathbf{p}, \tau + \beta) \quad \text{for} \quad \tau < 0 \tag{1.54}$$

Validity of this expression is checked using the property of the trace: $Sp AB = Sp BA$. For $\tau' - \tau > 0$ we have:

$$\mathcal{G}(\mathbf{p}, \tau - \tau') = \frac{i}{Z} Sp e^{-\beta(H-\mu N)} a_{\mathbf{p}}^{+}(\tau') a_{\mathbf{p}}(\tau)$$

$$= \frac{i}{Z} Sp a_{\mathbf{p}}(\tau) e^{-\beta(H-\mu N)} a_{\mathbf{p}}^{+}(\tau') e$$

$$= \frac{i}{Z} Sp e^{-\beta(H-\mu N)} e^{\beta(H-\mu N)} a_{\mathbf{p}}(\tau) e^{-\beta(H-\mu N)} a_{\mathbf{p}}^{+}(\tau')$$

$$= \frac{i}{Z} Sp e^{-\beta(H-\mu N)} a_{\mathbf{p}}(\tau + \beta) a_{\mathbf{p}}^{+}(\tau') \tag{1.55}$$

or

$$\mathcal{G}(\mathbf{p}, \tau - \tau') = -\mathcal{G}(\mathbf{p}, \tau - \tau' + \beta) \tag{1.56}$$

which for $\tau' = 0$ gives us (1.54). The minus sign appears here due to anticommutation of Fermi operators. Substituting (1.54) into (1.52) we see, that all contributions with even n are just zero. Thus for the Fermion case we are always dealing with the odd Matsubara frequencies:

$$\omega_n = \frac{(2n + 1)\pi}{\beta} = (2n + 1)\pi T \tag{1.57}$$

For Bosons, in a similar way, only contributions from even Matsubara frequencies

$$\omega_n = \frac{2n\pi}{\beta} = 2n\pi T \tag{1.58}$$

survive in the Fourier series for the Green's function.

Returning to Eqs. (1.16), (1.17) and (1.18) for Green's functions of the free particles at $T = 0$, we can easily write down the free-particle Matsubara Green's function as:

$$\mathcal{G}_0(\mathbf{p}, \tau_2 - \tau_1) = -i\{\theta(\tau_2 - \tau_1)(1 - n(\mathbf{p})) - \theta(\tau_1 - \tau_2)n(\mathbf{p})\}e^{-(\varepsilon(\mathbf{p}) - \mu)(\tau_2 - \tau_1)}$$
(1.59)

where $n(\mathbf{p}) = [e^{\beta(\varepsilon(\mathbf{p}) - \mu)} + 1]^{-1}$ is the Fermi distribution for finite T. Thus, the step functions entering the definition of G_0 at $T = 0$ are smeared by finite temperatures, leading to the simultaneous appearance of both particles and holes in a state with a given $\mathbf{p}$.

Substituting (1.59) into (1.53) we find:[6]

$$\mathcal{G}_0(\mathbf{p}\omega_n) = \frac{i}{i\omega_n - \varepsilon(\mathbf{p}) + \mu}, \quad \omega_n = (2n + 1)\pi T \qquad (1.60)$$

With only the major change to discrete frequencies, Matsubara diagram technique at finite T is practically identical with quantum mechanical diagram technique at $T = 0$. The full Green's function is determined from Dyson equation:

$$\mathcal{G}(\mathbf{p}\omega_n) = \frac{i}{i\omega_n - \varepsilon(\mathbf{p}) + \mu - \Sigma(\mathbf{p}\omega_n)}, \quad \omega_n = (2n + 1)\pi T \qquad (1.61)$$

Let us stress, however, that Matsubara Green's functions do not have the meaning of any kind of "transition amplitudes" (propagators) of the quantum (field) theory.

Calculating Matsubara Green's functions we can, in principle, find any thermodynamic characteristic of any many-particle system at equilibrium. Description of general non-equilibrium processes can be based on the more general formalism of Keldysh Green's functions [Lifshits E.M., Pitaevskii L.P. (1980)] and appropriate diagram technique. However, this formalism is outside the scope of our lectures.

---

[6]Here again we have an extra factor of i in comparison with standard notations of "AGD", which actually appeared in our Eq. (1.45).

Chapter 2

Electron–Electron Interaction

2.1 Diagram rules

Consider the system of interacting (nonrelativistic) Fermions. In the following we speak mainly about electrons in a metal. Interaction Hamiltonian can be written as:

$$H_{int} = \frac{1}{2} \int d\mathbf{r}_1 d\mathbf{r}_2 \psi_\alpha^+(\mathbf{r}_1)\psi_\beta^+(\mathbf{r}_2)V(\mathbf{r}_1 - \mathbf{r}_2)\psi_\beta(\mathbf{r}_2)\psi_\alpha(\mathbf{r}_1) \qquad (2.1)$$

where $V(\mathbf{r})$ — is the (static) interaction potential. $\psi_\alpha^+(\mathbf{r}), \psi_\alpha(\mathbf{r})$ — creation and annihilation operators of Fermions at the point $\mathbf{r}$, α — spin index.

General rules of diagram technique to calculate interaction corrections to *single-particle* Green's function in momentum representation $G(p)$ are given in [Abrikosov A.A., Gorkov L.P., Dzyaloshinskii I.E. (1963)]. Let us formulate the summary of these rules for the case of zero temperature $T = 0$:

- Diagram of n-th order in interaction contains $2n$ vertices, $2n+1$ full (electronic) lines and n wave-like (interaction) lines. To all lines we attribute definite 4-momenta, conserving at the interaction vertices.
- Full line denotes Green's function of a free electron (Fermion):

$$G_0(p) = \frac{\delta_{\alpha\beta}}{\varepsilon - \xi(\mathbf{p}) + i\delta \, sign\xi(\mathbf{p})} \qquad \text{where} \qquad \delta \to +0 \quad (2.2)$$

where

$$\xi(\mathbf{p}) = \frac{p^2}{2m} - \mu \approx v_F(|\mathbf{p}| - p_F) \qquad (2.3)$$

17

is the energy spectrum of free electrons, with energy calculated from the Fermi level (chemical potential μ), p_F and v_F — are Fermi momentum and velocity at the Fermi surface.

- Wave-like line denotes the Fourier transform of the potential $U(\mathbf{q})$.
- We must integrate over n independent momenta and frequencies (4-momenta).
- Final expression is multiplied by $(i)^n (2\pi)^{-4n} (2s+1)^F (-1)^F$, where F — is the number of closed Fermionic loops and s — Fermion spin (for electrons $s = 1/2$, so that we always have $2s + 1 = 2$).

For the case of finite temperatures, in Matsubara formalism [Abrikosov A.A., Gorkov L.P., Dzyaloshinskii I.E. (1963)], diagram rules for calculation of k-th order correction to $G(\varepsilon_n \mathbf{p})$ are formulated as follows:

- Diagram of k-th order possesses $2k$ vertices, $2k + 1$ full (electronic) lines and k wave-like (interaction) lines. To all lines we attribute momenta and (Matsubara) frequencies, satisfying conservation laws in each vertex. Frequencies at Bose lines are always even ($\omega_m = 2\pi m T$), while frequencies of Fermi lines are odd ($\varepsilon_n = (2n + 1)\pi T$).
- We must integrate over all independent momenta and sum over independent Matsubara frequencies.
- Each full line with momentum $\mathbf{p}$ and frequency ε_n denotes free electron Green's function in Matsubara representation:

$$G_0(\varepsilon_n \mathbf{p}) = \frac{\delta_{\alpha\beta}}{i\varepsilon_n - \xi(\mathbf{p})} \qquad (2.4)$$

while each wave-like line with momentum $\mathbf{q}$ and frequency ω_m denotes $V(\mathbf{q})$.

- Final expression is multiplied by $(-1)^k \frac{T^k}{(2\pi)^{3k}} (2s+1)^F (-1)^F$, where F again denotes the number of Fermion loops in a given diagram, while s is Fermion (electron) spin.

2.2 Electron gas with Coulomb interaction

If we try to perform direct calculations of interaction corrections to the Green's function of an electron in a normal metal using diagram rules given above, we immediately discover that appropriate analytic expressions just diverge due to the singularity of Coulomb interaction at small momentum transfers $\mathbf{q}$:

$$V(\mathbf{q}) = \frac{4\pi e^2}{\mathbf{q}^2} \tag{2.5}$$

reflecting the long-range nature of Coulomb interaction. This problem can be solved performing summation of an infinite series of diagrams, describing the *screening* of Coulomb potential by free electrons.

Let us introduce an effective interaction ("dressed" wave-like line) defined by diagrams shown in Fig. 2.1, where *polarization operator* is determined via the sum of diagrams, shown in Fig. 2.2. It is important to stress that an expansion shown on Fig. 2.2 contains no diagrams,

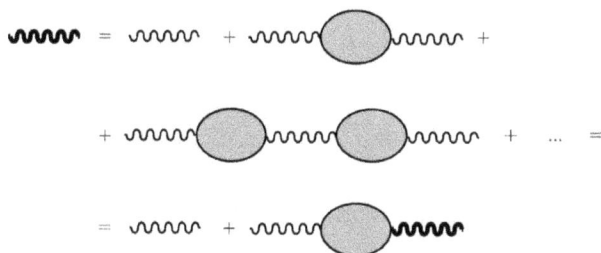

Fig. 2.1 Diagrammatic definition of an effective interaction.

Fig. 2.2 Diagrams for irreducible polarization operator.

Fig. 2.3 An example of reducible diagram.

which can be "cut" through one interaction line, of the type shown in Fig. 2.3, defining *irreducible* polarization operator. Thus, an expansion shown in Fig. 2.1 is an analogue of Dyson equation. Analytically the effective interaction can be written as:

$$\mathcal{V}(\mathbf{q}\omega) = V(\mathbf{q}) + V(\mathbf{q})\Pi(\mathbf{q}\omega)\mathcal{V}(\mathbf{q}\omega) \tag{2.6}$$

Effective interaction $\mathcal{V}(\mathbf{q}\omega)$ is in general dependent on frequency ω, corresponding to the account of retardation effects due to characteristic time of electron response to instantaneous Coulomb interaction.

Solving Eq. (2.6) we obtain:

$$\mathcal{V}(\mathbf{q}\omega) = \frac{V(\mathbf{q})}{1 - V(\mathbf{q})\Pi(\mathbf{q}\omega)} \equiv \frac{V(\mathbf{q})}{\epsilon(\mathbf{q}\omega)} \tag{2.7}$$

where we introduced dielectric function (permeability):

$$\epsilon(\mathbf{q}\omega) = 1 - V(\mathbf{q})\Pi(\mathbf{q}\omega) \tag{2.8}$$

So-called random phase approximation (RPA)[1] corresponds to the simplest approximation of polarization operator by the loop of two free-electron Green's functions, as shown by the diagram of Fig. 2.4(a):[2]

$$\Pi_0(\mathbf{q}\omega) = -2i \int \frac{d^4p}{(2\pi)^4} G_0(p+q) G_0(p) \tag{2.9}$$

Effective interaction is defined now by diagrams shown in Fig. 2.4(b). Equation for the effective interaction can also be written in another form, shown in Fig. 2.5(a), where we introduced *reducible* polarization operator $\tilde{\Pi}(\mathbf{q}\omega)$, defined by diagrams of Fig. 2.5(b):

$$\mathcal{V} = V + V\tilde{\Pi}V \tag{2.10}$$

From Fig. 2.5(b) it is clear that:

$$\tilde{\Pi} = \frac{\Pi_0}{1 - V\Pi_0} \tag{2.11}$$

From (2.10) using (2.11) we get:

$$\mathcal{V} = V(1 + \tilde{\Pi}V) = V\left(1 + \frac{\Pi_0 V}{1 - V\Pi_0}\right)$$

$$= \frac{V}{1 - V\Pi_0} = \frac{V}{\epsilon} \tag{2.12}$$

---

[1]This term has purely historic meaning.
[2]Note that in many books and papers the definition of $\Pi(\mathbf{q}\omega)$ is taken with different sign (e.g. see [Schrieffer J.R. (1964)]), here we use notations of [Abrikosov A.A., Gorkov L.P., Dzyaloshinskii I.E. (1963)]

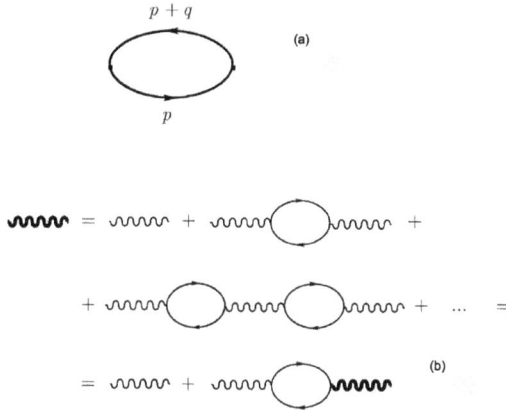

Fig. 2.4 Random phase approximation (RPA) for polarization operator and effective (screened) Coulomb interaction.

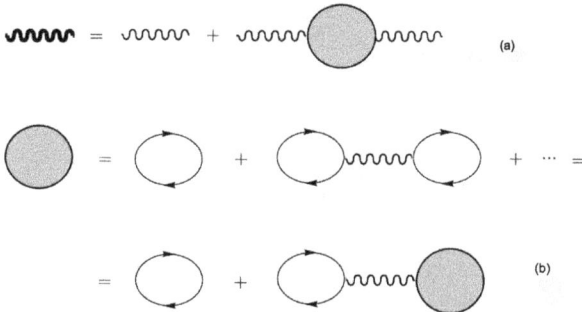

Fig. 2.5 Effective interaction expressed via reducible polarization operator $\tilde{\Pi}$.

which coincides with (2.7), with dielectric function taken in RPA approximation. Expression (2.11) defines full polarization of the systems.

Similarly we can obtain RPA expression for magnetic susceptibility. In this case we have to analyze the response of a system to infinitesimal magnetic field, flipping electronic spin. Dropping technical details, we just note that here it is sufficient to consider diagrams shown in Fig. 2.6 [Khomskii D.I. (2010); Levitov L.S., Shitov A.V. (2003)]. If we consider

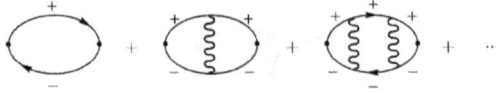

Fig. 2.6   Diagrams for magnetic susceptibility. $\pm$ denote spin projections.

popular model with (point-like) Hubbard interaction U of electrons, i.e.

$$\mathcal{H}_{int} = \sum_i U n_{i\uparrow} n_{i\downarrow}, \tag{2.13}$$

where $n_{i\uparrow}$ and $n_{i\downarrow}$ are operators of electronic density at a given lattice site i with opposite spin projections, these diagrams are easily summed and we obtain [Khomskii D.I. (2010)]:

$$\chi(\mathbf{q}\omega) = \frac{\chi_0(\mathbf{q}\omega)}{1 + U\Pi_0(\mathbf{q}\omega)} \tag{2.14}$$

where $\chi_0(\mathbf{q}\omega)$ is proportional to $\Pi_0(\mathbf{q}\omega)$ defined by (2.9):

$$\chi_0(\mathbf{q}\omega) = -\frac{1}{4}g^2\mu_B^2\Pi_0(\mathbf{q}\omega) \qquad \Pi_0(\mathbf{q}\omega) = -\frac{4}{g^2\mu_B^2}\chi_0(\mathbf{q}\omega) \tag{2.15}$$

where μ_B is the Bohr magneton and g is so-called g-factor (for free electrons $g = 2$). Note the sign change in the denominator of (2.14) in comparison with (2.8). This is due to the fact that during the derivation of $\epsilon(\mathbf{q}\omega)$ we have dealt with the response function of the density–density type and summed electronic loops contributing extra factors of -1. Here, calculating the linear response we sum "ladder" diagrams (cf. Fig. (2.6)), while loops are prohibited due to spin conservation (lines of particles and holes in Fig. (2.6) correspond to different spin projections). However, both expressions for $\epsilon(\mathbf{q}\omega)$ and $\chi(\mathbf{q}\omega)$ are quite similar and defined, in fact, by the same expression for $\Pi_0(\mathbf{q}\omega)$ defined in (2.9).

2.3   Polarization operator of free electron gas at $T = 0$

Let us start now with calculation of $\Pi_0(\mathbf{q}\omega)$, defined by Eq. (2.9). Equivalently we can write it as:

$$\Pi_0(\mathbf{q}\omega) = -2i \int \frac{d^3p}{(2\pi)^3} \int_{-\infty}^{\infty} \frac{d\varepsilon}{2\pi} G_0(\varepsilon_+ \mathbf{p}_+) G_0(\varepsilon_- \mathbf{p}_-) \tag{2.16}$$

where $\varepsilon_\pm = \varepsilon \pm \frac{\omega}{2}$, $\mathbf{p}_\pm = \mathbf{p} \pm \frac{1}{2}\mathbf{q}$. In the integral appearing here the main contribution comes from the vicinity of the Fermi surface, thus for $q \ll p_F$ we may write $|\mathbf{p}_\pm| = p \pm \frac{1}{2}q\cos\theta$, with θ — an angle between vectors $\mathbf{p}$ and $\mathbf{q}$, and take:

$$G_0(\varepsilon_\pm \mathbf{p}_\pm) = \frac{1}{\varepsilon_\pm - \xi_\pm(\mathbf{p}) + i\delta\, sign\xi_\pm(\mathbf{p})} \qquad (2.17)$$

where

$$\xi_\pm(\mathbf{p}) = \xi(\mathbf{p}_\pm) = \xi(\mathbf{p}) \pm \frac{1}{2}v_F q\cos\theta \qquad (2.18)$$

Integration over ε in (2.16) we can perform closing integration contour in the upper half-plane of the complex variable ε and expanding the product of two G_0's via simple fractions. The integral is different from zero only if poles of both Green's functions G_0 are in different half-planes. Finally we get:

$$\int_{-\infty}^{\infty} \frac{d\varepsilon}{(\varepsilon + \frac{\omega}{2} - \xi_+ + i\delta\, sign\xi_+)(\varepsilon - \frac{\omega}{2} - \xi_- + i\delta\, sign\xi_-)}$$
$$= \frac{2\pi i(n(\xi_-) - n(\xi_+))}{\omega - v_F q\cos\theta + i\delta(sign\xi_+ - sign\xi_-)} \qquad (2.19)$$

where:

$$n(\xi) = \begin{cases} 1 & \text{for} \quad \xi \leq 0 \\ 0 & \text{for} \quad \xi > 0 \end{cases} \qquad (2.20)$$

is the Fermi distribution at $T = 0$. As we are interested in small q, the difference $n(\xi_-) - n(\xi_+)$ is nonzero only in rather thin layer close to the Fermi surface. Thus instead of performing full p-integration, we can just integrate over the *linearized* spectrum ξ, using simple integration rule:

$$\int \frac{d^3 p}{(2\pi)^3} ... \approx \frac{\nu_F}{2} \int_{-\infty}^{\infty} d\xi \int_{-1}^{1} d(\cos\theta)... \qquad (2.21)$$

where

$$\nu_F = \frac{m p_F}{2\pi^2 \hbar^3} \qquad (2.22)$$

is the density of states at the Fermi level (for a single spin projection). Depending on the sign of $\cos\theta$ we have to consider two cases:

(1) $\cos\theta > 0$ — so that (2.19) is nonzero for $-\frac{v_F q}{2}\cos\theta < \xi < \frac{v_F q}{2}\cos\theta$, and $n(\xi_-) - n(\xi_+) = 1$;

(2) $\cos\theta < 0$ — in this case (2.19) is nonzero for $\frac{v_F q}{2}\cos\theta < \xi < -\frac{v_F q}{2}\cos\theta$, and we have $n(\xi_-) - n(\xi_+) = -1$.

Now we have only to take the following integral over the angle θ:

$$\Pi_0(\mathbf{q}\omega) = \nu_F \int_{-1}^{1} d\cos\theta \frac{v_F q\cos\theta}{\omega - v_F q\cos\theta + i\delta sign\omega} \qquad (2.23)$$

This integral is calculated directly using

$$\int_{-1}^{1} \frac{x\,dx}{x_0 - x + i\delta signx_0} = A + iB \qquad (2.24)$$

$$A = -2 + x_0\ln\left|\frac{x_0 + 1}{x_0 - 1}\right| \qquad B = \begin{cases} 0 & \text{for} \quad |x_0| > 1 \\ -\pi x_0 & \text{for} \quad 0 < x_0 < 1 \\ \pi x_0 & \text{for} \quad -1 < x_0 < 0 \end{cases}$$

Finally we get:

$$\Pi_0(\mathbf{q}\omega) = -2\nu_F\left\{1 - \frac{\omega}{2v_F q}\ln\left|\frac{\omega + v_F q}{\omega - v_F q}\right| + \frac{i\pi}{2}\frac{|\omega|}{v_F q}\theta\left(1 - \frac{|\omega|}{v_F q}\right)\right\} \qquad (2.25)$$

For $\omega = 0$ we obtain:

$$\Pi_0(\mathbf{q}\omega = 0) = -2\nu_F = -N(E_F) \qquad (2.26)$$

where we have introduced:

$$N(E_F) = 2\nu_F = \frac{mp_F}{\pi^2\hbar^3} \qquad (2.27)$$

— the density of states at the Fermi level for both spin projections. For $\omega \gg v_F q$ we have:

$$\Pi_0(\mathbf{q}\omega) \approx N(E_F)\frac{1}{3}\frac{v_F^2 q^2}{\omega^2}\left(1 + \frac{3}{5}\frac{v_F^2 q^2}{\omega^2}\right) \qquad (2.28)$$

These expressions will be often used in the future.

2.4 Dielectric function of an electron gas

Using (2.26) in (2.8) we obtain dielectric function describing the usual (Debye or Thomas–Fermi) screening:

$$\epsilon(\mathbf{q}, 0)|_{q\to 0} = 1 + \frac{\kappa_D^2}{q^2} \qquad (2.29)$$

where the inverse square of screening length is:

$$\kappa_D^2 = 4\pi e^2 N(E_F) = \frac{4e^2 m p_F}{\pi} = \frac{6\pi n e^2}{E_F} \tag{2.30}$$

where $n = \frac{p_F^3}{3\pi^2}$ is the density of electron gas. Then the Fourier transform of effective interaction is:

$$\mathcal{V}(\mathbf{q}, 0) = \frac{4\pi e^2}{q^2 + \kappa_D^2} \tag{2.31}$$

In coordinate space this corresponds to the screened potential:

$$\mathcal{V}(r) = \frac{e^2}{r} e^{-\kappa_D r} \tag{2.32}$$

Using in (2.8) the asymptotic behavior given in (2.28), in the limit of $q \to 0$ we get:

$$\epsilon(\omega) = 1 - \frac{\omega_p^2}{\omega^2} \tag{2.33}$$

where for the square of plasma frequency we have the usual expression:

$$\omega_p^2 = \frac{4\pi n e^2}{m} \tag{2.34}$$

Taking into account the second term in (2.28), from the condition $\epsilon(q\omega) = 0$ we obtain the spectrum of plasmons:

$$\omega^2(q) = \omega_p^2 + \frac{3}{5} v_F^2 q^2 \tag{2.35}$$

More accurate analysis, taking into account the imaginary part of polarization operator, allows to study plasmon damping [Schrieffer J.R. (1964); Nozieres P., Pines D. (1966)]. With the growth of q the spectrum (2.35) enters the region of single-particle excitations (electron-hole pairs) as shown in Fig. 2.7(a), where strong damping appears and plasmons cease to exist as well defined collective excitations.

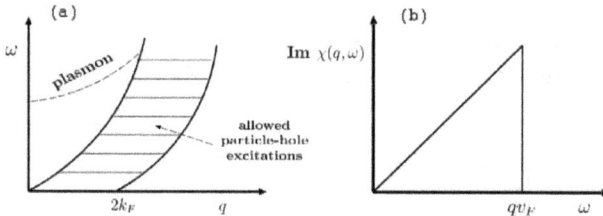

Fig. 2.7 (a) Dashed region are the allowed values of the energy of electron-hole pair excitations in the Fermi system, corresponding to the region of strong plasmon damping. (b) Imaginary part of generalized density–density susceptibility in electron-hole channel.

Energy of electron-hole excitation in the system of free electrons is:

$$\omega_{\mathbf{pq}}^0 = \xi_{\mathbf{p+q}} - \xi_{\mathbf{p}} = \frac{(\mathbf{p+q})^2}{2m} - \frac{p^2}{2m} = \frac{\mathbf{qp}}{m} + \frac{q^2}{2m} \tag{2.36}$$

The spectrum of these excitations with momentum $\mathbf{q}$ forms the continuum belonging to:

$$0 \le \omega_{\mathbf{pq}}^0 \le \frac{qp_F}{m} + \frac{q^2}{2m} \quad \text{for} \quad q < 2p_F$$

$$-\frac{qp_F}{m} + \frac{q^2}{2m} \le \omega_{\mathbf{pq}}^0 \le \frac{qp_F}{m} + \frac{q^2}{2m} \quad \text{for} \quad q > 2p_F \tag{2.37}$$

This region is shown as dashed in Fig. 2.7(a). Below we shall show that the imaginary part of polarization operator given by (2.25), for $\omega > 0$ coincides (up to a sign) with the imaginary part of density–density response function (generalized susceptibility), as shown in Fig. 2.7(b).

It is clear that our calculations leading to (2.25) are valid only for small ω and q. In fact, polarization operator can be found for arbitrary q and ω (J. Lindhardt, 1954). Let us quote some of the results [Schrieffer J.R. (1964)]. Static dielectric function is given by the following expression:

$$\epsilon(q,0) = 1 + \frac{4me^2 p_F}{\pi q^2} u\left(\frac{q}{2p_F}\right)$$

$$= 1 + \left(\frac{4}{9\pi^4}\right)^{1/3} \frac{r_s}{x^2} u(x) = 1 + 0.66 r_s \left(\frac{p_F}{q}\right)^2 u\left(\frac{q}{2p_F}\right) \tag{2.38}$$

where

$$u(x) = \frac{1}{2}\left\{1 + \frac{1-x^2}{2x} \ln\left|\frac{1+x}{1-x}\right|\right\} \tag{2.39}$$

In (2.38) we have introduced the standard notations of the theory of electron gas, where r_s is determined by the relation: $\frac{4\pi r_s^3 a_0^3}{3} = \frac{1}{n}$, where n is the density of electrons, and $a_0 = \frac{\hbar^2}{me^2}$ is the Bohr radius. We see that r_s is just the mean distance between electrons in units of Bohr radius.

Small parameter for perturbation theory in our model (RPA) is the ratio of characteristic Coulomb (interaction) energy and Fermi energy:

$$\frac{V_C}{E_F} \sim \frac{e^2 p_F}{p_F^2} m \sim \frac{e^2}{\hbar v_F} \sim \frac{\hbar}{p_F a_0} \sim \frac{a}{a_0} \sim r_s \tag{2.40}$$

In real metals we have $1 < r_s < 5$, so that RPA is obviously rather bad approximation. It works well for the case of highly compressed electron gas and is usually called "high-density approximation".

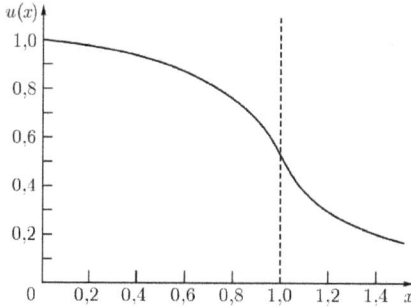

Fig. 2.8   Plot of $u(x)$. The derivative of this function is logarithmically divergent at $x = 1$.

The plot of $u(x)$ is shown in Fig. 2.8. For $q \to 0$ we obviously again get the simple result (2.29). It is important to discuss the region of $q \sim 2p_F$. From (2.38) and (2.39) it is seen that the derivative $\frac{\partial \epsilon(q,0)}{\partial q} \to \infty$ for $q \to 2p_F$. This leads to a number of anomalies of physical properties. For example, the spatial dependence of the screened interaction potential is not as simple as given by Eq. (2.32). In fact, asymptotic behavior of the Fourier integral $\int dq e^{iqr} f(q)$ is determined by singularities of $f(q)$ and its derivatives, within integration interval. Consider the case of $f(q) \to \infty$ at $q = q_0$, e.g. $f(q) \sim \delta(q - q_0)$. Then for $f(r)$ we obviously get the oscillating contribution $\sim e^{iq_0 r}$. Similarly, singularity of the derivative $\frac{\partial \epsilon}{\partial q}$ at $q = 2p_F$ leads to the appearance of *long-range and oscillating* contribution to interaction potential:

$$\mathcal{V}(r)|_{r \to \infty} \sim \frac{\cos(2p_F r + \phi)}{r^3} \tag{2.41}$$

Then the screening charge around e.g. charged impurity in a metal also oscillates according to (2.41) (Friedel oscillations).

Even more important is the similar effect in the theory of magnetic interactions in metals. We have already noted that paramagnetic susceptibility of electron gas in fact is determined by the same "polarization loop" (cf. (2.15)). Then:

$$\chi_0(q\omega = 0) = \frac{3g^2 \mu_B^2 n}{8E_F} u\left(\frac{q}{2p_F}\right) \tag{2.42}$$

Then the spin density $\mathbf{s}(\mathbf{r})$ on some distance r from the magnetic impurity with spin $\mathbf{S}_a$, determined as

$$\mathbf{s}(\mathbf{r}) = \frac{J}{g^2\mu_B^2}\sum_{\mathbf{q}}\chi_0(\mathbf{q})e^{i\mathbf{qr}}\mathbf{S}_a \qquad (2.43)$$

will also be oscillating function similar to (2.41). Here J determines contact exchange interaction of impurity with conduction electrons: $-J\mathbf{S}_a\mathbf{s}$. Now, if we place another magnetic impurity $\mathbf{S}_b$, it will interact with conduction electrons in a similar way, and we obtain an effective exchange interaction of two impurity spins via conduction electrons (Ruderman–Kittel–Kasuya–Yosida). This so-called RKKY interaction can be written as:

$$J_{RKKY}(\mathbf{r}_a - \mathbf{r}_b) = -\frac{J^2}{g^2\mu_B^2}\sum_{\mathbf{q}}\chi_0(\mathbf{q})e^{i\mathbf{q}(\mathbf{r}_a-\mathbf{r}_b)} \sim \frac{J^2}{E_F}\frac{\cos(2p_F r_{ab} + \phi)}{r_{ab}^3}$$

$$(2.44)$$

It is seen that this interaction oscillates as a function of the distance between impurities $r_{ab} = |\mathbf{r}_a - \mathbf{r}_b|$. This oscillating nature of exchange interaction of localized spins in metals leads to a number of important physical effects. According to (2.44) in coordinate space appear regions with different signs of exchange interaction (i.e. where interaction is either of ferromagnetic or antiferromagnetic nature) leading to the formation of complicated magnetic structures, e.g. in metallic compounds with regular sublattices of rare-earth elements (magnetic spirals or helicoidal structures) [Khomskii D.I. (2010)]. In case of randomly placed magnetic impurities in non magnetic metal, Eq. (2.44) produces random signs of exchange interaction between spins at different sites, which leads to the formation of quite unusual magnetic state — spin glass [Ginzburg S.L. (1989)].

2.5   Electron self-energy, effective mass and damping of quasiparticles

Our final aim is to calculate single-particle Green's function in a system with Coulomb (or also some other) interaction. This Green's function can always be written in Dyson's form:

$$G(\mathbf{p}\varepsilon) = \frac{1}{\varepsilon - \varepsilon_p - \Sigma(p\varepsilon)} \qquad (2.45)$$

where self-energy $\Sigma(p\varepsilon)$ is taken in some approximation, obtained e.g. via some partial summation of diagram series. What physical information can be obtained in this way? We know that the Green's function of free electrons $G_0(\mathbf{p}\varepsilon)$ has a pole at $\varepsilon_p = \frac{p^2}{2m} - \mu$. Let us *assume* that in the interacting system Green's function also has a pole:

$$G(\mathbf{p}\varepsilon) \approx \frac{1}{\varepsilon - \tilde{\varepsilon}_p} \qquad (2.46)$$

where $\tilde{\varepsilon}_p$ represents the spectrum of "renormalized" quasiparticles. Comparing with (2.45) we see, that the spectrum $\tilde{\varepsilon}_p$ is defined by the equation:

$$\varepsilon - \varepsilon_p - Re\Sigma(p\varepsilon) = 0 \qquad \text{or} \qquad \tilde{\varepsilon}_p - \varepsilon_p - Re\Sigma(p\tilde{\varepsilon}_p) = 0 \qquad (2.47)$$

where, just to simplify our calculations (and only for a time!), we have neglected $Im\Sigma$, which (as we shall see later) determines quasiparticle damping. Let us expand (2.45) in the vicinity of the pole:

$$G(\mathbf{p}\varepsilon) = \frac{1}{\varepsilon - \varepsilon_p - \Sigma(\mathbf{p}\varepsilon)} = \frac{1}{\varepsilon - \varepsilon_p - \Sigma(\mathbf{p}\tilde{\varepsilon}_p) - \frac{\partial \Sigma}{\partial \varepsilon}\big|_{\varepsilon=\tilde{\varepsilon}_p}(\varepsilon - \tilde{\varepsilon}_p)}$$
$$(2.48)$$

Taking into account (2.47) we can rewrite (2.48) in the following form:

$$G(\mathbf{p}\varepsilon) = \frac{1}{\varepsilon - \tilde{\varepsilon}_p - \frac{\partial \Sigma}{\partial \varepsilon}\big|_{\varepsilon=\tilde{\varepsilon}_p}(\varepsilon - \tilde{\varepsilon}_p)} = \frac{\frac{1}{1-\frac{\partial \Sigma}{\partial \varepsilon}|_{\varepsilon=\varepsilon_p}}}{\varepsilon - \tilde{\varepsilon}_p} \equiv \frac{Z_p}{\varepsilon - \tilde{\varepsilon}_p} \qquad (2.49)$$

where we have introduced "residue" at the quasiparticle pole as:

$$Z_p = \frac{1}{1 - \frac{\partial \Sigma}{\partial \varepsilon}\big|_{\varepsilon=\varepsilon_p}} \qquad (2.50)$$

Sometimes Z_p is also called a factor of "wave function renormalization". From general grounds it is clear that $Z_p \leq 1$ and equality is only reached for the ideal (free) Fermi gas. Spectral density corresponding to the Green's function (2.49) is given by:

$$A(\mathbf{p}\varepsilon) = Z_p\delta(\varepsilon - \tilde{\varepsilon}_p) \qquad (2.51)$$

i.e. it is represented by δ-function peak at $\varepsilon = \tilde{\varepsilon}_p$ (quasiparticle energy) as in the case of free electron gas. In fact, inequality $Z_p < 1$ means that in a system with interactions the quasiparticle contribution to $A(\mathbf{p}\varepsilon)$ is slightly suppressed due to appearance of an additional "multi particle" (incoherent) contribution to the spectral density [Migdal A.B. (1967)], which we just dropped in this simplified analysis. Neglecting quasiparticle damping we obtain here the quasiparticle contribution to $A(\mathbf{p}\varepsilon)$ in the form infinitesimally narrow δ-function, finite damping (as we shall show below) leads to the appearance of the finite width of this peak.

Suppose now that the spectrum of "renormalized" quasiparticles can be described by an effective mass approximation:

$$\tilde{\varepsilon}_p = \frac{p^2}{2m^*} - \mu \tag{2.52}$$

Then we easily get:

$$\frac{1}{2m^*} = \frac{\partial \tilde{\varepsilon}_p}{\partial(p^2)} = \frac{\partial \varepsilon_p}{\partial(p^2)} + \left\{ \frac{\partial \Sigma}{\partial(p^2)} + \frac{\partial \Sigma}{\partial \tilde{\varepsilon}_p} \frac{\partial \tilde{\varepsilon}_p}{\partial(p^2)} \right\}$$

$$= \frac{1}{2m} + \frac{\partial \Sigma}{2m\partial\left(\frac{p^2}{2m}\right)} + \frac{\partial \Sigma}{\partial \varepsilon}|_{\varepsilon=\tilde{\varepsilon}_p} \frac{\partial \tilde{\varepsilon}_p}{\partial(p^2)} \tag{2.53}$$

or

$$\frac{1}{m^*}\left(1 - \frac{\partial \Sigma}{\partial \varepsilon}|_{\varepsilon=\tilde{\varepsilon}_p}\right) = \frac{1}{m}\left(1 + \frac{\partial \Sigma}{\partial \varepsilon_p}\right) \tag{2.54}$$

so that

$$\frac{m^*}{m} = \frac{1 - \frac{\partial \Sigma}{\partial \varepsilon}|_{\varepsilon=\tilde{\varepsilon}_p}}{1 + \frac{\partial \Sigma}{\partial \varepsilon_p}} = \frac{1}{Z_p} \frac{1}{1 + \frac{\partial \Sigma}{\partial \varepsilon_p}} \tag{2.55}$$

which gives us an important relation between "mass renormalization" m^*/m and residue at the pole of the green's function Z_p. In the simplest case, when the self-energy has no dependence on the momentum p (or, equivalently, on $\tilde{\varepsilon}_p$), this relation is especially simple:

$$\frac{m^*}{m} = \frac{1}{Z_p} \tag{2.56}$$

so that the effective mass in a system with interactions is enhanced in comparison with the case of an ideal gas.

General behavior of damping is connected with $Im\Sigma$ and will be discussed later in detail. However, even from this simplified analysis, it is clear that simple relations obtained above allow us to calculate effective parameters of many particle system (quasiparticles) from some approximate form of electron self-energy, obtained from specific diagrams of perturbation theory.

As an example, consider again high density approximation for electronic gas. Let us analyze simplest contributions to electron self-energy. In fact we can just drop Hartree-like diagrams, as they cancel with similar contributions due to electron interaction with spatially homogeneous positively charged "ion background", which is necessary to introduce to guarantee charge neutrality. This becomes clear if we consider the sum of simplest diagrams of this type shown in Fig. 2.9. In obvious notations we have:

$$2i \int \frac{dp'}{(2\pi)^4} V(0) G(p') - 2i \int \frac{dp'}{(2\pi)^4} V(0) G_i(p')$$

$$= 2i \int \frac{d\mathbf{p}'}{(2\pi)^3} \int \frac{d\varepsilon}{2\pi} V(0) G(\varepsilon \mathbf{p}') - 2i \int \frac{d\mathbf{p}'}{(2\pi)^3} \int \frac{d\varepsilon}{2\pi} V(0) G_i(\varepsilon \mathbf{p}')$$

$$= 2V(0) \left[\int \frac{d\mathbf{p}}{(2\pi)^3} n_{\mathbf{p}} - \int \frac{d\mathbf{p}}{(2\pi)^3} n_{\mathbf{p}}^i \right] = 2V(0)(n - n_i) = 0$$

$$(2.57)$$

so that we have total cancellation of these contributions (charge density of electrons n is equal to charge density of ions (positive "background") n^i).

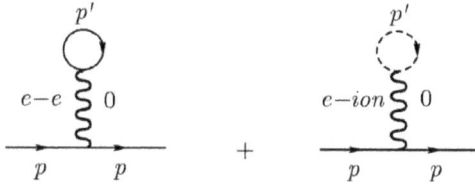

Fig. 2.9 Simplest Hartree-like diagrams describing electron–electron interaction and similar interaction with positive "background" of ions (Green's function of ions is shown by dashed line).

Thus in RPA approximation the problem is reduced to calculation of self-energy diagram shown in Fig. 2.10, where "dressed" wave-like line describes effectively screened Coulomb interaction of Fig. 2.4(b):

$$\Sigma(p) = i \int \frac{d^4 q}{(2\pi)^4} V(q) G_0(p + q) \qquad (2.58)$$

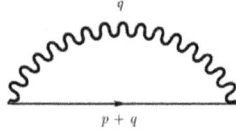

Fig. 2.10   Electron self-energy in RPA approximation.

Though both $G_0(p)$ and $\mathcal{V}(q)$ entering (2.58) are known exactly, integrations here are very complicated and we just quote the final results (J.J. Quinn, R.A. Ferrell, 1958) [Schrieffer J.R. (1964)]. Excitation energy of a quasiparticle (measured from the Fermi energy E_F) in this approximation can be written as:

$$\tilde{\varepsilon}_p = E_F \left\{ \frac{p^2}{p_F^2} - 0.166r_s \left[\frac{p}{p_F}(\ln r_s + 0.203) + \ln r_s - 1.80 \right] \right\} - E_F \tag{2.59}$$

Quasiparticle damping in RPA is given by:

$$|\gamma_p| = E_F(0.252r_s^{1/2}) \left(\frac{p}{p_F} - 1 \right)^2 \tag{2.60}$$

and it is small for $p \to p_F$, in a sense that we can guarantee $|\tilde{\varepsilon}_p| \gg |\gamma_p|$, in full accordance with general conclusions of (phenomenological) Landau theory of Fermi-liquids [Lifshits E.M., Pitaevskii L.P. (1980); Nozieres P. (1964); Nozieres P., Pines D. (1966)]. This allows us to speak about well-defined quasiparticles close to the Fermi surface. From (2.59) and (2.60) it is clear that in RPA quasiparticles are well-defined roughly for $|\tilde{\varepsilon}_p| < E_F/5$. From (2.59) we can easily get the expression for the effective mass of an electron as:

$$\frac{1}{m^*} = \frac{1}{p_F} \frac{\partial \tilde{\varepsilon}_p}{\partial p}|_{p=p_F} = \frac{1}{m}[1 - 0.083r_s(\ln r_s + 0.203)] \tag{2.61}$$

It is well known that electronic contribution to the specific heat is proportional to m^*. Then, from (2.61) we immediately obtain (M. Gell-Mann, 1957) [Schrieffer J.R. (1964)]:

$$\frac{c}{c_0} = 1 + 0.083r_s(\ln r_s + 0.203) \tag{2.62}$$

where c_0 is the specific heat of an ideal Fermi-gas.

Finally note, that all the results quoted are valid for an electronic gas of high enough density, when $r_s \ll 1$. In real metals we mostly have $1 < r_s < 5$, and you should be careful while using RPA for any estimates.

2.6 RKKY-oscillations

Let us return to more detailed discussions of RKKY-oscillations [Levitov L.S., Shitov A.V. (2003)]. Consider localized spin **S** surrounded by an ideal Fermi gas of (conduction) electrons (e.g. magnetic impurity in a normal metal) and interacting with local spin-density of these electrons via contact (point-like) exchange interaction:

$$H_{int} = -J \int d\mathbf{r} S^i \delta(\mathbf{r}) \psi^+(\mathbf{r}) \hat{\sigma}^i \psi(\mathbf{r}) \qquad (2.63)$$

Assuming that exchange coupling J is small enough, we can find (first-order) contribution to spin polarization of conduction electrons:

$$\sigma^i(\mathbf{r}) = < \psi^+(\mathbf{r}) \hat{\sigma}^i \psi(\mathbf{r}) > \qquad (2.64)$$

at the distance $|\mathbf{r}|$ from localized spin (impurity) **S**.

Consider first the case of zero temperature $T = 0$. Let us write down Green's function of a free electron in coordinate representation. It is useful (methodically) to make calculations in two ways. Start with angular integration:

$$
\begin{aligned}
G(\varepsilon, \mathbf{r}) &= \int_0^\infty \int_0^\pi \frac{dp p^2 \sin\theta d\theta}{2\pi^2} \frac{e^{ipr\cos\theta}}{\varepsilon - \xi(p) + i\delta \, signe} \\
&= \frac{1}{2\pi^2 r} \int_0^\infty \frac{dp p \sin pr}{\varepsilon - \xi(p) + i\delta \, signe}
\end{aligned}
\qquad (2.65)
$$

First and simplest way to proceed is to change integration variable from p to linearized (in the vicinity of the Fermi surface) spectrum ξ and perform contour integration in the complex plain:

$$G(\varepsilon, r) \approx \frac{1}{2\pi^2 r} \int_{-\infty}^\infty \frac{d\xi}{v_F} p_F \frac{\sin(p_F + \xi/v_F)r}{\varepsilon - \xi + i\delta \, signe} = -\frac{m}{2\pi r} e^{ir(signe \varepsilon p_F + |\varepsilon|/v_F)} \qquad (2.66)$$

We see that the Green's function is oscillating with the period determined by the Fermi wavelength $\lambda_F = \frac{2\pi\hbar}{p_F}$. The phase of these oscillations changes sign at the Fermi level (for $\varepsilon = 0$) due to the effects of Fermi statistics.

Another way is to perform integration over p exactly. As the integrand in (2.65) is even, we can make integration over the whole real axis of p, dividing the result by two:

$$G(\varepsilon, r) = \frac{1}{4\pi^2 r} \int_{-\infty}^{\infty} \frac{dp\, p \sin pr}{\varepsilon - \frac{p^2}{2m} + E_F + i\delta sign\varepsilon} \quad (2.67)$$

Expanding the integrand here into simple fractions and performing integration we get:

$$G(\varepsilon, r) = \frac{m}{4\pi^2 r} \int_{-\infty}^{\infty} dp \sin pr \left[\frac{1}{\kappa - p} - \frac{1}{\kappa + p} \right] = -\frac{m}{2\pi r} e^{isign\varepsilon\kappa r} \quad (2.68)$$

where $\kappa = \sqrt{2m(\varepsilon + E_F + i\delta sign\varepsilon)}$. Comparing this result with (2.66) we can see, that simplified calculation using ξ-integration gives rather good approximation of an exact result for $|\varepsilon| \ll E_F$, i.e. in the immediate vicinity of the Fermi surface.

Let us return now to calculation of spin polarization (2.64), writing this expression via an exact Green's function (accounting for electron interactions with localized spin):

$$\hat{\sigma}^i(\mathbf{r}) = -i \lim_{t' \to t+0 \, \mathbf{r} = \mathbf{r}'} Sp\hat{\sigma}^i G(\mathbf{r}t; \mathbf{r}'t') \quad (2.69)$$

Trace[3] here is calculated over spin indices of $\hat{\sigma}^i$ and G. Now, let us just take first order expression for Green's function correction due to interaction (2.63):

$$G_{\alpha\beta}^{(1)}(\varepsilon, \mathbf{r}, \mathbf{r}') = -\sigma_{\alpha\beta}^i G_0(\varepsilon\mathbf{r}) G_0(\varepsilon, -\mathbf{r}') JS^i \quad (2.70)$$

Substituting (2.70) into (2.69) and using $Sp\sigma^i\sigma^j = 2\delta^{ij}$ we obtain:

$$\sigma^i(\mathbf{r}) = 2iJS^i \int \frac{d\varepsilon}{2\pi} G_0^2(\varepsilon r) \quad (2.71)$$

which in fact coincides with (2.43) written above. Use now (2.66) and get (for $p_F r \gg 1$):

$$\int \frac{d\varepsilon}{2\pi} G_0^2(\varepsilon r) = \frac{1}{2\pi} \left(\frac{m}{2\pi r} \right)^2 \int_0^{\infty} d\varepsilon \left(e^{i2p_F r + \frac{2i\varepsilon}{v_F} r} + e^{-i2p_F r + \frac{2i\varepsilon}{v_F} r} \right)$$

$$= i \frac{mp_F}{(2\pi)^3} \frac{\cos 2p_F r}{r^3} \quad (2.72)$$

[3] According to Russian and German literature tradition we always use Sp-notation instead of English Tr.

Now for the spin density we immediately obtain slowly damping oscillations with period π/p_F:

$$\sigma^i(r) = -JS^i \frac{mp_F}{4\pi^3} \frac{\cos 2p_F r}{r^3} \tag{2.73}$$

More accurate expression for spin density at the point r can be obtained using exact r-dependence of the Green's function (2.68) and integrating its square in (2.71) over ε. Then we get:

$$\sigma^i(r) = -JS^i \frac{2mp_F^4}{\pi^3} \left(\frac{\cos 2p_F r}{(2p_F r)^3} - \frac{\sin 2p_F r}{(2p_F r)^4} \right) \tag{2.74}$$

In the limit of $p_F r \gg 1$ Eq. (2.74) actually goes to (2.73), in accordance with our "ideology" of ξ-integration. Note also that an exact Eq. (2.74) has only $1/r$ singularity at small r of. Thus, as we perform integration over d^3r no divergence of full polarization appears. Approximate Eq. (2.73) is more singular for $r \to 0$, but it is inappropriate at small r as ξ-integration guarantees correct answer only at large distances.

Let us now consider the case of finite temperatures. Again, we start with calculating free (Matsubara!) Green's function in coordinate representation (similarly to the way we obtained (2.66)):

$$
\begin{aligned}
G(\varepsilon_n \mathbf{r}) &= \int \frac{d^3 p}{(2\pi)^3} \frac{e^{i\mathbf{pr}}}{i\varepsilon_n - \xi(p)} = \frac{\nu_F}{pr} \int_{-\infty}^{\infty} d\xi \frac{\sin\left(p_F + \frac{\xi}{v_F}\right) r}{i\varepsilon_n - \xi} \\
&= \frac{\nu_F}{2ipr} \int_{-\infty}^{\infty} d\xi \frac{e^{i(p_F + \xi/v_F)r} - e^{-i(p_F + \xi/v_F)r}}{i\varepsilon_n - \xi} \\
&= -\frac{m}{2\pi r} e^{i(p_F + i\varepsilon_n/v_F)r \, \text{sign} \varepsilon_n} \tag{2.75}
\end{aligned}
$$

In the same way as above we can express spin density via an exact Green's function. All expressions differ now only by replacements of $-i$ by -1 here and there. Finally we get:

$$\sigma^i(\mathbf{r}) = -2JS^i T \sum_{\varepsilon_n} G_0^2(\varepsilon_n, r) \tag{2.76}$$

and:

$$\sigma^i(r) = -2JS^i T \left(\frac{m}{2\pi r}\right)^2 \left\{ \sum_{\varepsilon_n > 0} e^{2ip_F r - 2\varepsilon_n r/v_F} + \sum_{\varepsilon_n < 0} e^{-2ip_F r + 2\varepsilon_n r/v_F} \right\} \tag{2.77}$$

Sums over Matsubara frequencies are calculated in elementary way and we obtain:

$$\sigma^i(r) = -JS^i \frac{m^2 T}{2\pi^2 r^2} \frac{\cos 2p_F r}{sh\frac{2\pi T r}{v_F}} \qquad (2.78)$$

For $T \to 0$ this expression goes to our previous result (2.73). The length at which oscillations are damped is now equal to $\frac{\hbar v_F}{2\pi T}$ (just look at the argument of hyperbolic sine!). Thus, at finite temperatures RKKY oscillations persist at the distances smaller than the "thermal" length given by $l_T = \frac{\hbar v_F}{T}$ and are exponentially small for $r > l_T$.

2.7 Linear response

Calculation of the linear response of a many-particle system to some external perturbation (field) is one of the central tasks of the theory of condensed matter. Below we shall show how this problem is solved within the formalism of Matsubara Green's functions.

Let us return to the analysis of dielectric function. Strictly speaking, the permeability defined by Eq. (2.8), *does not* represent correct response function! In particular, it does not possess correct analytic properties as (cf. (2.25)) it has singularities in $Im\omega > 0$ — halfplane, which breaks Kramers–Kronig relations [Sadovskii M.V. (2019a)]. This is clear also from (2.19), where, depending on the signs of ξ_+ and ξ_- the pole in ω may lay in the upper halfplane, in lower halfplane, or on the real axis. This automatically leads to the breaking of causality (Kramers–Kronig relations), which is necessary for any correct response function [Sadovskii M.V. (2019a)]. The reason for this behavior is that during our calculation of the polarization "bubble" (at $T = 0$) we have used Feynman Green's functions (on which diagram technique at $T = 0$ is built), which lead to $\Pi(-\omega) = \Pi(\omega)$, while any generalized *susceptibility* (retarded response function) has to satisfy $\chi(-\omega) = \chi^*(\omega)$. So we need some special discussion, how to find a correct response function?

The standard approach to deal with susceptibilities uses Matsubara diagram technique with *analytic continuation* of the final result from discrete imaginary frequencies to the real one. This procedure gives the appropriate *retarded* susceptibility (Green's function) [Abrikosov A.A., Gorkov L.P., Dzyaloshinskii I.E. (1963)]. How to perform such calculations in general case we shall see later, but first we shall discuss a simple case of non-interacting particles [Levitov L.S., Shitov A.V. (2003)]

Generalized susceptibility of some quantum operator A with respect to another operator B is given by the famous Kubo expression [Sadovskii M.V. (2019a); Zubarev D.N. (1974)]:

$$\chi_{AB}(\omega) = i \int_0^\infty dt e^{i\omega t} < [\hat{A}(t), \hat{B}(0)] > \qquad (2.79)$$

Here we see the commutator (averaged over the ground state or Gibbs ensemble) which originates form the appropriate retarded two-time Green's function of Bogoliubov and Tyablikov [Sadovskii M.V. (2019a); Zubarev D.N. (1974)].[4] In *non-interacting* case we can write second quantized expressions for (single-particle) operators $\hat{A}$ and $\hat{B}$ using some full system of eigenfunctions ψ_m with eigenenergies E_m appropriate to our system:[5]

$$\hat{A}(t) = \sum_{mk} A_{mk} \hat{a}_m^+ \hat{a}_k e^{-i(E_k - E_m)t} \qquad (2.80)$$

$$\hat{B}(t) = \sum_{mk} B_{mk} \hat{a}_m^+ \hat{a}_k e^{-i(E_k - E_m)t} \qquad (2.81)$$

where A_{mk} and B_{mk} are the matrix elements of operators, calculated using eigenfunctions ψ_m and ψ_k, while $\hat{a}_m^+$, $\hat{a}_k$ are appropriate Fermion creation and annihilation operators for these eigenstates. Substituting these expressions to $< [\hat{A}(t), \hat{B}(0)] >$ and calculating this commutator directly, using Wick's theorem [Abrikosov A.A., Gorkov L.P., Dzyaloshinskii I.E. (1963)], we get:

$$\chi_{AB}(\omega) = \sum_{mk} A_{mk} B_{km} \frac{n(E_m) - n(E_k)}{E_k - E_m - \omega - i\delta} \qquad (2.82)$$

where $n(E_k) = < \hat{a}_k^+ \hat{a}_k >$ reduces to the usual Fermi distribution.

---

[4]$\theta(t)$-function, entering the definition of this Green's function, leads to the appearance in (2.79) of time integration from $t = 0$ to $t = \infty$.

[5]For non-interacting particles such eigenfunctions and eigenstates always can be found (at least in principle!) by solving the appropriate stationary Schroedinger equation similar to (1.4). As was already noted, these states may be just plane-waves for an ideal Fermi gas, Landau states for the same gas in an external magnetic field, or these may be some exact (but unknown to us!) states of an electron in a random potential field (if we are dealing with a disordered system). Note that notations here are slightly different from those of the previous Chapter, where, in particular, we denoted eigenstates as φ_λ, eigenergies ε_λ, etc.

In case when eigenstates of an electron are unknown (e.g. in case we are considering the levels of an electron in a specific realization of the random potential in a disordered system) it is useful to express susceptibility via Green's function. Let us remind definitions of retarded and advanced Green's functions $G^R(\varepsilon)$ and $G^A(\varepsilon)$. These functions are related to the causal (Feynman) Green's function as [Abrikosov A.A., Gorkov L.P., Dzyaloshinskii I.E. (1963)]:

$$G(t,t') = \begin{cases} G^R(t,t') & t > t' \\ G^A(t,t') & t < t' \end{cases} \tag{2.83}$$

After the Fourier transformation:

$$G^{R(A)}(\varepsilon\mathbf{p}) = \frac{1}{\varepsilon - \xi(p) \pm i\delta} \tag{2.84}$$

so that Feynman Green's function contains contributions from both electrons and holes (cf. Chapter I):

$$\begin{aligned} G(\varepsilon\mathbf{p}) &= (1 - n(p))G^R(\varepsilon\mathbf{p}) + n(p)G^A(\varepsilon\mathbf{p}) \\ &= \frac{1 - n(p)}{\varepsilon - \xi(p) + i\delta} + \frac{n(p)}{\varepsilon - \xi(p) - i\delta} \end{aligned} \tag{2.85}$$

where

$$n(p) = \begin{cases} 1 & p \leq p_F \\ 0 & p > p_F \end{cases} \tag{2.86}$$

is just Fermi distribution at $T = 0$.

To express susceptibility via $G^R(\varepsilon)$ and $G^A(\varepsilon)$, let us represent energy denominator in (2.82) as an integral over an additional energy variable:

$$\begin{aligned} \frac{1}{E_k - E_m - \omega - i\delta} &= -\frac{1}{2\pi i} \int_{-\infty}^{\infty} d\varepsilon \frac{1}{(\varepsilon - \omega - E_m - i\delta)(\varepsilon - E_k + i\delta)} \\ &= -\frac{1}{2\pi i} \int_{-\infty}^{\infty} d\varepsilon G_m^A(\varepsilon - \omega) G_k^R(\varepsilon) \end{aligned} \tag{2.87}$$

Substituting this expression to (2.82), we obtain the general operator expression for susceptibility:

$$\chi_{AB}(\omega) = \frac{1}{2\pi i} \int_{\infty}^{\infty} d\varepsilon Sp([\hat{G}^R(\varepsilon)\hat{B}, \hat{G}^A(\varepsilon - \omega)\hat{A}]\hat{\rho}) \tag{2.88}$$

where $\hat{\rho}$ is the density matrix (in diagonalizing representation we have $\rho_{mk} = n(E_m)\delta_{mk}$). The main advantage of Eq. (2.88) in comparison with (2.82) is its validity for an arbitrary representation, even when we do not know exact eigenstates.

Up to now we have used Green's function formalism for $T = 0$. For $T > 0$ we have to use Matsubara technique. It may seem that to analyze dynamics in real time t (necessary to calculate (2.79)) Matsubara formalism is useless, as it deals with *imaginary* time τ. However, as we shall see now, Matsubara technique allows rather simple approach to calculation of the linear response.

Let us introduce Matsubara susceptibility as ($\omega_m = 2\pi m T$):

$$\chi_{AB}^{(M)}(\omega_m) = \frac{1}{2}\int_{-\beta}^{\beta} d\tau < T_\tau \hat{A}(\tau)\hat{B}(0) > e^{i\omega_m \tau} \qquad (2.89)$$

Now we can use the following spectacular theorem [Abrikosov A.A., Gorkov L.P., Dzyaloshinskii I.E. (1963)]:

- Analytic continuation of $\chi_{AB}^{(M)}(\omega_m)$ from the discrete set of points at positive imaginary half-axis of frequency $\omega = i\omega_n$ $(n > 0)$ to the real axis $(Im\omega \to +0)$ precisely gives us the retarded susceptibility $\chi_{AB}(\omega)$.

This theorem allows us to determine $\chi_{AB}(\omega)$ using $\chi_{AB}^{(M)}(\omega_m)$ calculated using Matsubara diagram technique. In the absence of interactions, Matsubara susceptibility is given by a single "bubble" diagram ("polarization operator") with operators $\hat{A}$ and $\hat{B}$ standing at the vertices. In diagram technique for $T = 0$ susceptibility and polarization operator possess different analyticity properties (cf. above). However, according to just formulated statement, to find the correct susceptibility it is sufficient to calculate "polarization bubble" with Matsubara Green's functions, and then just continue it analytically to the real frequencies.

Let us now give a proof of our major statement (theorem). We have to calculate Kubo susceptibility:

$$\chi_{AB}(\omega) = i\int_0^\infty dt e^{i\omega t} < [\hat{A}(t), \hat{B}(0)] > \qquad (2.90)$$

where $< ... > = Sp(e^{-\beta H}...)/Sp(e^{-\beta H})$ is the usual Gibbs average, $\hat{A}(t) = e^{itH}\hat{A}e^{-itH}$-operator in Heisenberg representation. It is easily seen that (2.90) can be written as:

$$\chi_{AB}(\omega) = \frac{i}{Z}\int_0^\infty dt e^{i\omega t}\sum_{mn} e^{-\beta E_n}\left(e^{i\omega_{nm}t} <n|\hat{A}|m><m|\hat{B}|n>\right.$$
$$\left. - e^{-i\omega_{nm}t} <n|\hat{B}|m><m|\hat{A}|n>\right) \tag{2.91}$$

where $\omega_{nm} = E_n - E_m$, $Z = Spe^{-\beta H}$, and n, m numerate exact energy levels of *many-particle interacting* system. Changing m to n and vice versa in the second term in the sum and integrating over t, we obtain:

$$\chi_{AB}(\omega) = \frac{1}{Z}\sum_{mn}\frac{e^{-\beta E_n} - e^{-\beta E_m}}{\omega_{nm} - \omega - i\delta} <n|A|m><m|B|n> \tag{2.92}$$

Imaginary term $i\delta(\delta \to +0)$ appears here due to the factor of $e^{-\delta t}$, which has to be added into formally divergent integral over t to guarantee convergence.

Now calculate in a similar way Matsubara response function:

$$\chi_{AB}^{(M)}(\omega_m) = \frac{1}{2}\int_{-\beta}^\beta d\tau e^{i\omega_m\tau} <T_\tau\hat{A}(\tau)\hat{B}(0)> \tag{2.93}$$

where $\hat{A}(\tau) = e^{\tau H}\hat{A}e^{-\tau H}$. We have:

$$\chi_{AB}^{(M)}(\omega_m) = \frac{1}{2Z}\int_0^\beta d\tau e^{i\omega_m\tau}\sum_{mn}e^{-\beta E_n}e^{\omega_{nm}\tau} <n|A|m><m|B|n>$$
$$+\frac{1}{2Z}\int_{-\beta}^0 d\tau e^{i\omega_m\tau}\sum_{mn}e^{-\beta E_n}e^{-\omega_{nm}\tau} <n|B|m><m|A|n> \tag{2.94}$$

Again changing summation indices in the second sum, taking into account $\omega_m\beta = 2\pi m$, and performing integration over τ, we get:

$$\chi_{AB}^{(M)}(\omega_m) = \frac{1}{Z}\sum_{mn}\frac{e^{-\beta E_n} - e^{-\beta E_m}}{\omega_{mn} - i\omega_m} <n|A|m><m|B|n> \tag{2.95}$$

Now everything is ready! Susceptibility $\chi_{AB}(\omega)$ is an analytic function of ω in the upper half-plane of complex frequency. This is a general property of the Fourier transform of a function, which is different from zero only for $t > 0$ (retarded response!) [Sadovskii M.V. (2019a)]. Now, such a function can be analytically continued from the real axis to the positive imaginary half-axis. Obviously, at points $\omega_m = 2\pi m T$ this function just coincides with $\chi_{AB}^{(M)}$, it is seen by direct comparison of

(2.92) and (2.95). Suppose the existence of an analytical continuation of $\chi_{AB}^{(M)}$ from the positive imaginary half-axis to the whole upper half-plane of complex ω. Then, this analytically continued function have to coincide with $\chi_{AB}(\omega)$, as according to the well known theorem of the theory of complex variables, two functions analytic in some region of the complex plane and coinciding on the infinite subset of discrete points (possessing the limiting point at $m \to \infty$) just coincide in the whole complex plane.

Note that the case of $T = 0$ sometimes is also conveniently analyzed within Matsubara formalism. In this case we have just to transform summation over Matsubara frequencies to integration over continuous (imaginary) frequencies, as for $T \to 0$ discrete points $i\omega_m$ "fill" all the imaginary axis of complex plane of ω, so that $T \sum_m \dots \to \int \frac{d\omega}{2\pi} \dots$. During such calculations we do not have to care about the rules of overcircling the poles of Green's functions as the direction of integration is correct automatically.

Let us see how it works on a typical example of calculations of the polarization operator of the free electron gas, i.e. of the dielectric function (response function!) in RPA approximation. We have:

$$\Pi^{(M)}(\omega_m, \mathbf{q}) = 2T \sum_n \int \frac{d^3p}{(2\pi)^3} G(\varepsilon_n \mathbf{p}) G(\varepsilon_n + \omega_m, \mathbf{p} + \mathbf{q}) \qquad (2.96)$$

First we perform summation over ε_n. Let us write down the sum as follows:

$$T \sum_n \frac{1}{i\varepsilon_n + i\omega_m - \xi(\mathbf{p}+\mathbf{q})} \frac{1}{i\varepsilon_n - \xi(\mathbf{p})}$$

$$= T \sum_n \frac{1}{i\omega_m - \xi(\mathbf{p}+\mathbf{q}) + \xi(\mathbf{p})} \left(\frac{1}{i\varepsilon_n - \xi(\mathbf{p})} - \frac{1}{i\varepsilon_n + i\omega_m - \xi(\mathbf{p}+\mathbf{q})} \right)$$

$$\qquad (2.97)$$

In the second term we may change summation index from $n \to n-m$, so that ω_m just disappears. In both contributions the real part of the sum converges, while imaginary part is formally divergent. At the same time, this imaginary part is odd over n and cancels during the summation over n. Thus it is sufficient to calculate only the following sum:

$$S(\xi) = T \sum_n \frac{\xi}{\varepsilon_n^2 + \xi^2} \qquad (2.98)$$

It can be done using the identity:[6]

$$\sum_{n=-\infty}^{\infty} \frac{1}{(2n+1)^2\pi^2 + x^2} = \frac{1}{2x}th\frac{x}{2} \qquad (2.100)$$

Then:

$$S(\xi) = \frac{1}{2}th\frac{\xi}{2T} = \frac{1}{2} - n(\xi) \qquad (2.101)$$

where $n(\xi) = (e^{\frac{\xi}{T}} + 1)^{-1}$ is the Fermi distribution. Finally, we get the following expression, which is very useful in calculations of response functions of the Fermi gas:

$$T\sum_{n} \frac{1}{(i\varepsilon_n + i\omega_m - \xi(\mathbf{p+q}))(i\varepsilon_n - \xi(\mathbf{p}))} = -\frac{n(\xi(\mathbf{p+q})) - n(\xi(\mathbf{p}))}{i\omega_m - \xi(\mathbf{p+q}) + \xi(\mathbf{p})} \qquad (2.102)$$

Using this identity in (2.96), we obtain:

$$\Pi^{(M)}(\omega_m\mathbf{q}) = -2\int \frac{d^3p}{(2\pi)^3} \frac{n(\xi(\mathbf{p+q})) - n(\xi(\mathbf{p}))}{i\omega_m - \xi(\mathbf{p+q}) + \xi(\mathbf{p})} \qquad (2.103)$$

Consider now the limit of $T = 0$. In this case we have $n(\xi) = \theta(-\xi)$. Changing, in the usual way, to integration over ξ we limit ourselves to small $q \ll p_F$. Then we have:

$$\Pi^{(M)}(\omega_m\mathbf{q}) = -2\int \frac{d^3p}{(2\pi)^3} \frac{\partial n}{\partial\xi(p)} \frac{\mathbf{vq}}{i\omega_m - \mathbf{vq}} = 2\nu_F\int \frac{d\Omega}{4\pi} \frac{\mathbf{v}_F\mathbf{q}}{i\omega_m - \mathbf{v}_F\mathbf{q}} \qquad (2.104)$$

Performing angular integration as it was done in (2.23), we get:

$$\Pi^{(M)}(\omega_m q) = -2\nu_F\left\{1 + \frac{i\omega_m}{2v_Fq}\ln\frac{i\omega_m - v_Fq}{i\omega_m + v_Fq}\right\} \qquad (2.105)$$

To perform analytic continuation of this expression to the real axis of frequencies, we have only to make a substitution $i\omega_m \to \omega + i\delta$. Finally, we obtain:

$$\Pi^R(\omega + i\delta q) = -2\nu_F\left\{1 + \frac{\omega}{2v_Fq}\ln\frac{\omega - v_Fq + i\delta}{\omega + v_Fq + i\delta}\right\} \qquad (2.106)$$

---

[6]This result can be obtained as follows:

$$\frac{1}{(2n+1)^2\pi^2 + x^2} = \frac{1}{2x}\left\{\frac{1}{x + i\pi(2n+1)} + \frac{1}{x - i\pi(2n+1)}\right\}$$
$$= \frac{1}{2x}\int_0^\infty dze^{-xz}[e^{-i\pi(2n+1)z} + e^{i\pi(2n+1)z}] \qquad (2.99)$$

Now just sum the progression under the integral.

which determines the dielectric function of electron gas in RPA approximation (as a response function with correct analytical properties).

In fact, taking the real and imaginary parts of (2.106) and changing the sign, we obtain the density–density response function as:

$$Re\chi(\omega q) = 2\nu_F \left\{ 1 + \frac{\omega}{2v_F q} \ln \left| \frac{\omega - v_F q}{\omega + v_F q} \right| \right\} \qquad (2.107)$$

$$Im\chi(\omega q) = \pi\nu_F \frac{\omega}{v_F q} \theta(v_F q - |\omega|) \qquad (2.108)$$

Opposite to the case of a similar expression (2.25) (which appeared via summation of Feynman diagrams for $T = 0$), this result satisfies all analyticity requirements for response functions [Lifshits E.M., Pitaevskii L.P. (1980); Sadovskii M.V. (2019a)].[7] In particular, it satisfies the Kramers–Kronig relation:

$$\chi(\omega) = \frac{1}{\pi} \int_{-\infty}^{\infty} d\omega' \frac{Im\chi(\omega')}{\omega' - \omega - i\delta} \qquad (2.109)$$

Thus, dielectric permeability (response function) is defined as:

$$\epsilon(q\omega) = 1 + \frac{4\pi e^2}{q^2} \chi(q\omega) \qquad (2.110)$$

where $\chi(q\omega)$ is the retarded density–density response function [Nozieres P., Pines D. (1966)], which is obtained (up to a sign) via analytic continuation of Matsubara polarization operator.

Dielectric function is directly connected with electric conductivity of a system [Zubarev D.N. (1974)]:

$$\sigma(q\omega) = \frac{i\omega}{4\pi} (1 - \epsilon(q\omega)) = -\frac{ie^2}{q^2} \omega\chi(q\omega) \qquad (2.111)$$

It can be seen as follows. The current density induced by an external electric field $\mathbf{E} = -\nabla\varphi$ (where φ is the scalar potential) is given by:

$$\mathbf{j}(\mathbf{q}\omega) = \sigma(\mathbf{q}\omega)\mathbf{E}(\mathbf{q}\omega) = -i\sigma(\mathbf{q}\omega)\mathbf{q}\varphi(\mathbf{q}\omega) \qquad (2.112)$$

Charge conservation is expressed via continuity equation:

$$e\frac{\partial}{\partial t} n(\mathbf{r}t) + \nabla\mathbf{j}(\mathbf{r}t) = 0 \qquad (2.113)$$

or, in Fourier components:

$$- i\omega e\delta n(\mathbf{q}\omega) + i\mathbf{q}\mathbf{j}(\mathbf{q}\omega) = 0 \qquad (2.114)$$

---

[7]It is precisely Eq. (2.108), which is shown in Fig. 2.7(b).

where δn is some deviation of the density from spatially homogeneous (equilibrium) value n. This deviation is defined (in linear response theory) as [Nozieres P., Pines D. (1966)]:

$$\delta n(\mathbf{q}\omega) = e\chi(\mathbf{q}\omega)\varphi(\mathbf{q}\omega) \tag{2.115}$$

Combining (2.112)–(2.115) we immediately obtain (2.111).

In experiment we usually deal with the limit of $q \to 0$ (homogeneous external field). Then the conductivity is defined as:

$$\sigma(\omega) = -\lim_{q\to 0}\frac{ie^2}{q^2}\omega\chi(q\omega) \tag{2.116}$$

In the simplest case of free electron gas, to calculate the limit of $q \to 0$ for finite ω we use (2.28) and obtain ($\omega \to \omega + i\delta, \quad \delta \to +0$):

$$\sigma(\omega) = \lim_{q\to 0}\frac{ie^2}{q^2}\omega\frac{2\nu_F}{3}\frac{v_F^2 q^2}{\omega^2} = \frac{p_F^3}{3\pi^2 m}i\frac{e^2}{\omega} = \frac{ne^2}{m}\frac{i}{\omega + i\delta} \tag{2.117}$$

i.e. the usual Drude relation for conductivity of electron gas without any scatterings (ideal conductor!). For the real part of conductivity we get:

$$Re\sigma(\omega) = \frac{\omega}{4\pi}Im\epsilon(\omega) = \frac{ne^2}{m}\pi\delta(\omega) \tag{2.118}$$

where we have used $\frac{1}{\omega+i\delta} = \frac{1}{\omega} - i\pi\delta(\omega)$. Phenomenologically, we may take scattering into account replacing $\delta \to \gamma = \frac{1}{\tau}$, where γ is some scattering rate and τ is the mean-free time.

2.8 Microscopic foundations of Landau–Silin theory of Fermi-liquids

In real metals with $r_s \sim 2 - 3$ electron interaction is not weak and we cannot limit ourselves by the sum of any specific diagram subseries (like in RPA, which is valid for $r_s \ll 1$). At the same time, phenomenological theory of Fermi-liquids, introduced by Landau and Silin [Nozieres P., Pines D. (1966)], is quite successful even in the case of Fermi-systems with pretty strong interactions. Let us consider the basics of its microscopic justification [Lifshits E.M., Pitaevskii L.P. (1980); Migdal A.B. (1967); Nozieres P. (1964)].

In fact, Landau just assumed that the ground state of the Fermi-liquid is *qualitatively* the same as that of a Fermi-gas, while the low energy excitations can be described as *quasiparticles*, similar to particles and holes in a Fermi-gas, despite the strong interactions present in a real system of Fermions (like electrons in metals, atoms of He^3, protons and neutrons in an atomic nuclei etc.). The basic assumption here is that of an existence of well defined Fermi surface with the Fermi momentum p_F, define by the usual "gaseous" relation:

$$n = \frac{N}{V} = \frac{p_F^3}{3\pi^2 \hbar^3} \qquad (2.119)$$

relating it to the full particle density. This statement can be, in fact, *proven* in any order of perturbation theory over interaction, using the general properties of Green's functions, and is known as *Luttinger theorem* (J.M. Luttinger, 1960) [Abrikosov A.A., Gorkov L.P., Dzyaloshinskii I.E. (1963); Lifshits E.M., Pitaevskii L.P. (1980)]. This proof is rather complicated and technical, we just drop it.[8] It should be clearly understood that the ground state of a *normal* Fermi-liquid is not the only possible ground state of the system of interacting Fermions. For example we know, that the system may be in the superconducting (superfluid) state, when Luttinger theorem does not apply and there is *no* Fermi surface in the usual sense — it is "closed" by the energy gap. Presently, much attention is being paid to *strongly correlated* electronic systems with many "scenarios" of the formation of *non* Fermi-liquid state. However, below we shall mainly concentrate on the analysis of microscopic foundations of the theory of normal Fermi-liquids [Migdal A.B. (1967); Nozieres P. (1964)].

Basic physical reason for an interacting system of Fermions to has much in common with free Fermion case is due to restrictions introduced by Pauli principle. As we shall see shortly, mainly Pauli "correlations" allow us to observe well defined quasiparticle excitations close to the Fermi surface. In an infinite and homogeneous system the Green's function $G_{\alpha\beta}(p)$ is diagonal in spin indices and same for both[9] spin projections (in the absence of an external magnetic field or spontaneous magnetization), so we just drop all these indices. Now introduce

---

[8]In Appendix A we give some general topological arguments, justifying the stability of the Fermi surface towards adiabatic "switching on" of interparticle interactions (G.E. Volovik, 1991).

[9]We shall deal mostly with the case of spin $s = 1/2$.

Fermion self-energy as usual and write down the Dyson equation:

$$G^{-1}(\varepsilon \mathbf{p}) = \varepsilon - \frac{p^2}{2m} + \mu - \Sigma(\varepsilon \mathbf{p}) \qquad (2.120)$$

What can be said for the "general" case of interacting system? Let us estimate the contribution to the imaginary part of Σ from the process of creation of three quasiparticles (see Fig. 2.11) — the simplest process leading to the finite lifetime of a quasiparticle. This process reduces to the excitation of another particle (e.g. electron) from below of the Fermi surface, i.e. to the creation of an electron-hole pair. Then we have the usual conservation laws:

$$\mathbf{p}_1 + \mathbf{p}_2 = \mathbf{p}_3 + \mathbf{p}_4 \qquad \varepsilon_1 + \varepsilon_2 = \varepsilon_3 + \varepsilon_4 \qquad (2.121)$$

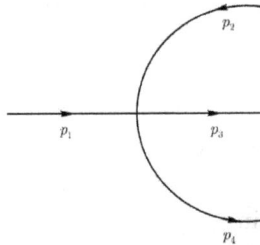

Fig. 2.11   Creation of three quasiparticles in Fermi-liquid.

and in our case

$$|\mathbf{p}_1|, |\mathbf{p}_3|, |\mathbf{p}_4| \geq p_F, \qquad |\mathbf{p}_2| \leq p_F$$
$$\varepsilon_1, \varepsilon_2, \varepsilon_4 \geq 0, \qquad \varepsilon_2 \leq 0 \qquad (2.122)$$

Now it is clear that for the case of $|\mathbf{p}_1| \to p_F$, for all the other excitations we also have $|\mathbf{p}_2|, |\mathbf{p}_3|, |\mathbf{p}_4| \to p_F$, while for $\varepsilon_1 \to +0$ leads also to $\varepsilon_\alpha(\alpha = 2, 3, 4) \to 0$. When $\mathbf{p}_1$ is somewhere "above" p_F, remaining values of $(|\mathbf{p}_\alpha| - p_F)$ are of the same order as $(|\mathbf{p}_1| - p_F)$. Then, the probability amplitude for the process shown by the diagram of Fig. 2.11 is proportional to:

$$W = \frac{1}{\tau} \sim \int \delta(\varepsilon_1 + \varepsilon_2 - \varepsilon_3 - \varepsilon_4) d\mathbf{p}_2 d\mathbf{p}_3 \qquad (2.123)$$

It is clear, as the momentum $\mathbf{p}_1$ is fixed, while $\mathbf{p}_4 = \mathbf{p}_1 + \mathbf{p}_2 - \mathbf{p}_3$, so that we have only two independent momenta for integration, as is written in

(2.123). As both $\mathbf{p}_2$ and $\mathbf{p}_3$ are close to p_F, we have $(|\mathbf{p}_{2,3}| - p_F) \sim (|\mathbf{p}_1| - p_F)$. Allowed values for the modules of $\mathbf{p}_2$ and $\mathbf{p}_3$ belong to the intervals: $p_F < p_3 < p_1 + p_2 - p_F$ and $2p_F - p_1 < p_2 < p_F$. The angle between $\mathbf{p}_1$ and $\mathbf{p}_3$ can be arbitrary, while an angle between $\mathbf{p}_3$ and $\mathbf{p}_1 + \mathbf{p}_2$ is determined by energy conservation and integration over this angle "cancels" δ-function in (2.123). Accordingly, the integration over $dp_2 dp_3$ is done for $p_2 \approx p_3 \approx p_F$ which leads to (2.123) being of the order of $\sim (|\mathbf{p}_1| - p_F)^2$. Finally, we have for the inverse lifetime of a electron (Fermion) with momentum p:

$$\frac{1}{\tau} \sim Im\Sigma \sim (p - p_F)^2 \sim \varepsilon^2 \qquad (2.124)$$

We can easily convince ourselves that the statistical weight of the processes with larger number of excited quasiparticles is proportional to higher powers of ε. For example, $Im\Sigma_5 \sim |\varepsilon^3|\varepsilon$ [Migdal A.B. (1967)]. For finite temperatures, "smearing" of Fermi distribution $\sim T$ leads to the appropriate contribution to damping due to thermally excited quasiparticles $\sim \frac{T^2}{E_F}$. Then, taking (2.124) into account, we can write down the following general estimate:

$$\frac{1}{\tau} = A\left(\frac{\varepsilon^2}{E_F} + \frac{T^2}{E_F}\right) \approx Max\left(\frac{\varepsilon^2}{E_F}, \frac{T^2}{E_F}\right) \qquad (2.125)$$

where $A \sim const$. Using the simple Drude-like expression for conductivity:

$$\sigma = \frac{ne^2}{m}\tau \qquad (2.126)$$

and taking $\tau^{-1} = A\frac{T^2}{E_F}$ we obtain (use also (2.119)) the following estimate of resistivity:

$$R = \frac{1}{\sigma} \sim \frac{T^2 m}{E_F p_F^3 e^2} = \frac{1}{e^2 p_F}\left(\frac{T}{E_F}\right)^2 \qquad (2.127)$$

This gives characteristic temperature dependence of resistivity due to electron–electron scattering $R \sim T^2$ (L.D. Landau, I.Ya. Pomeranchuk, 1937). Typically in metals we have $p_F \sim \frac{\hbar}{a}$ (where a is interatomic spacing) and $R \sim \frac{\hbar a}{e^2}\left(\frac{T}{E_F}\right)^2 \sim 10^{-3}$ Ohm cm $\left(\frac{T}{E_F}\right)^2$, which corresponds to a very small contribution to resistivity for most typical values of T, which is usually completely "masked" by other scattering mechanisms (e.g. due to phonons). Experimental observation of $\sim T^2$ contribution to resistivity of metals is possible in very pure samples and usually for temperatures $T < 1K$, when we can neglect scattering of electrons by phonons. These simple findings are often forgotten in modern literature.

Thus, close to the pole $\varepsilon = \varepsilon(p)$ of the Green's function (2.120) we always have $Re\varepsilon(p) \approx v_F(p - p_F) \gg Im\varepsilon(p) \sim \tau^{-1} \sim (p - p_F)^2$, which corresponds to "well defined" quasiparticles close to the Fermi level. More accurately (than was done deriving Eqs. (2.46)–(2.50)) we have to proceed in the following way. In homogeneous and isotropic system (Fermi-liquid) the value of $Re\Sigma(\varepsilon\mathbf{p})$ depends only on modulus of the momentum $p = |\mathbf{p}|$. Let us define the Fermi momentum p_F for the interacting system by the following relation:

$$\frac{p_F^2}{2m} + \Sigma(p_F, 0) = \mu \qquad (2.128)$$

Expanding $\Sigma(p\varepsilon)$ in powers of $p - p_F$ and ε, we obtain the expression for $G(p\varepsilon)$ valid close to the Fermi surface ($\varepsilon \to 0$, $p \to p_F$) as:

$$G^{-1}(\varepsilon p) \approx \varepsilon - \frac{p^2}{2m} + \mu - \Sigma(p_F, 0) - \left(\frac{\partial\Sigma}{\partial p}\right)_F (p - p_F) - \left(\frac{\partial\Sigma}{\partial\varepsilon}\right)_F \varepsilon$$

$$+ i\alpha'|\varepsilon|\varepsilon = \left[1 - \left(\frac{\partial\Sigma}{\partial\varepsilon}\right)_F\right]\varepsilon - \left[\frac{p_F}{m} + \left(\frac{\partial\Sigma}{\partial p}\right)_F\right](p - p_F) + i\alpha'|\varepsilon|\varepsilon$$

$$(2.129)$$

where we have taken into account Eq. (2.124) and guaranteed the correct sign change of the imaginary part of (Feynman) Green's function at $\varepsilon = 0$. Thus we find that the Green's function for (presumably) arbitrary system interacting Fermions can be written close to the Fermi surface as [Migdal A.B. (1967)]:

$$G(\varepsilon p) = \frac{Z}{\varepsilon - v_F(p - p_F) + i\alpha|\varepsilon|\varepsilon} + G^{reg}(\varepsilon p) \qquad (2.130)$$

where $G^{reg}(\varepsilon p)$ is some regular (non-singular) part with no poles close to the Fermi surface (and due to multi-particle excitations of the systems [Migdal A.B. (1967)]). Here in (2.130) we have introduced the following notations:

$$\frac{1}{Z} = 1 - \left(\frac{\partial\Sigma}{\partial\varepsilon}\right)_F = \left(\frac{\partial G^{-1}}{\partial\varepsilon}\right)_F \qquad (2.131)$$

for the residue at the pole of the Green's function and

$$v_F = \frac{\frac{p_F}{m} + \left(\frac{\partial\Sigma}{\partial p}\right)_F}{\left(\frac{\partial G^{-1}}{\partial\varepsilon}\right)_F} = -\frac{\left(\frac{\partial G^{-1}}{\partial p}\right)_F}{\left(\frac{\partial G^{-1}}{\partial\varepsilon}\right)_F}; \qquad \alpha = Z\alpha' \qquad (2.132)$$

for the velocity at the Fermi surface. Eq. (2.130) defines the general form of the single-particle Green's function in a system of interacting

Fermions (Fermi-liquid). It is easily seen that specific expressions, obtained above within RPA, are precisely of this form.

Now we can easily show, that Eq. (2.130) directly leads to the existence (at $T = 0$) of a *discontinuity* in *particle* distribution in momentum space even in the case of interacting Fermions (A.B. Migdal, 1957). To see this we have to calculate the difference of the values of particle distribution function $n(p)$ at both sides of the Fermi surface, i.e. the limit of $n(p_F + q) - n(p_F - q)$ for $q \to +0$. Momentum distribution of particles in Green's function formalism is expressed as (cf. (1.34), (1.38)) [Abrikosov A.A., Gorkov L.P., Dzyaloshinskii I.E. (1963)]:

$$n(p) = -i \lim_{t \to -0} \int_{-\infty}^{\infty} \frac{d\varepsilon}{2\pi} e^{-i\varepsilon t} G(\varepsilon p) \qquad (2.133)$$

Now use here (2.130). As $G^{reg}(\varepsilon p)$ is regular, it is clear that its contribution to the difference of integrals will tend to zero with $q \to 0$. Thus, it is sufficient to analyze only the difference of integrals from the poles of the Green's function (2.130). Then we get:

$$n(p_F - q) - n(p_F + q) = -i \int_{-\infty}^{\infty} \frac{d\varepsilon}{2\pi} \left\{ \frac{Z}{\varepsilon + v_F q - i\delta} - \frac{Z}{\varepsilon - v_F q + i\delta} \right\} \qquad (2.134)$$

where we have taken into account that close to the pole $sign\varepsilon = sign(p - p_F)$, and dropped the factor of $e^{-i\varepsilon t}$ (with $t \to 0$) due to convergence of the integral. Closing the integration contour at infinity (no matter in the lower or in the upper half-plane), we obtain:

$$n(p_F - 0) - n(p_F + 0) = Z \qquad (2.135)$$

As we obviously have $n(p) \le 1$, it follows that:

$$0 < Z \le 1 \qquad (2.136)$$

and the limiting value of $Z = 1$ is reached only in the case of and ideal Fermi-gas. Thus we see that the momentum distribution of particles in the Fermi-liquid at $T = 0$ has (similarly to the case of Fermi-gas) a finite discontinuity at the Fermi surface, as is qualitatively shown in Fig. 2.12. Two major differences with the case of an ideal gas are that discontinuity is less than unity, while distribution function $n(p)$ itself is finite also in the region of $p > p_F$ (particles are "pushed" to this region by interaction!). In fact, the existence of discontinuity in particle

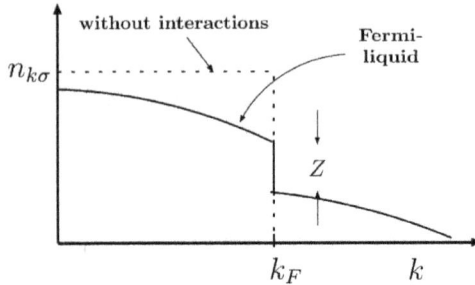

Fig. 2.12   Qualitative form of particle distribution function in the Fermi-liquid at $T = 0$.

distribution allows strict definition of the Fermi surface in the system of interacting Fermions.

Spectral density $A(\mathbf{p}\varepsilon) = -sign\varepsilon\frac{1}{\pi}ImG(\varepsilon p)$, corresponding to the Green's function (2.130) has a typical form of a smeared quasiparticle (Lorentzian) peak at $\varepsilon = \varepsilon_p$ (quasiparticle energy), on the smooth background due to multi-particle excitations, as shown in Fig. 2.13(b), while in an ideal gas of Fermions it reduces to δ-function, shown in Fig. 2.13(a). Note, that spectral densities of electrons in interacting systems can, in fact, be measured experimentally via photoemission with angular resolution (ARPES),[10] which allows also to study the form of real Fermi surfaces, even for very complicated compounds.[11] These measurements, performed in recent years, confirmed qualitative predictions of the theory of Fermi-liquids for majority of "metallic" systems. Deviations from Fermi-liquid behavior, observed in some systems are usually attributed to the effects of *strong correlations* [Varma C.N., Nussinov Z., Wim van Saarloos (2002)].

---

[10] J.C. Campuzano, M.R. Norman, M. Randeria. Photoemission in the High T_c Superconductors. ArXiv: cond-mat/0209476.

[11] A. Damascelli, D.H. Lu, Z.-X. Shen. From Mott insulator to overdoped superconductor: Evolution of the electronic structure of cuprates studied by ARPES. Rev. Mod. Phys. **75**, 473 (2003).

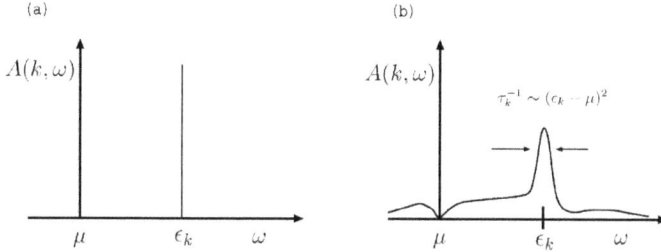

Fig. 2.13 Spectral density in Fermi-gas (a) and in Fermi-liquid (b).

2.9 Interaction of quasiparticles in Fermi-liquid

Interactions of quasiparticles in Fermi-liquid is described by the *two-particle* Green's function [Migdal A.B. (1967)]:

$$K =< T\psi(1)\psi(2)\psi^+(3)\psi^+(4) > \qquad (2.137)$$

which is determined by the sum of all diagrams, describing the propagation of two particles from the points $(1, 2)$ to $(3, 4)$. First of all, we can separate diagrams with no interactions between these two particles, but with all possible interactions of each of the particles with the "background", of the type shown in Fig. 2.14. It is obvious that here we are dealing with two independent sum of diagrams, each reducing to the full single-particle Green's function G. Accordingly, we have:

$$K_0 = G(1, 3)G(2, 4) - G(1, 4)G(2, 3) \qquad (2.138)$$

Minus sign before the second (exchange) term here is due to the antisymmetry of Fermions under permutations.

All the remaining diagrams for K describe interactions of the particles with each other. Let us denote as V all the graphs of this type, which can not be separated into two parts, connected by two electronic lines, as shown in Fig. 2.15. Then for the two-particle Green's function K we can write down the following equation:

$$K = K_0 - GGVK \qquad (2.139)$$

as the sum of all diagrams following after V again reduces to K. Of course, Eq. (2.139) is in fact an integral equation and "operator" multiplication of Green's functions denote here (and in similar cases below) appropriate integrations.

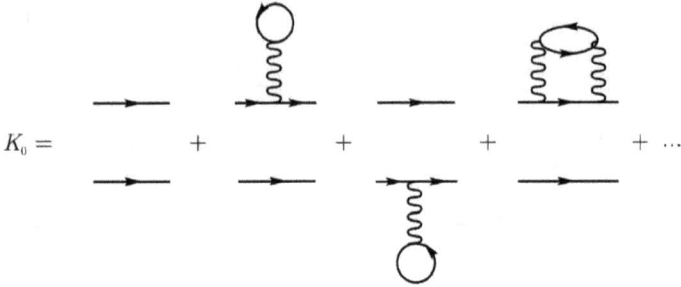

Fig. 2.14   Independent propagation of two particles in Fermi-liquid.

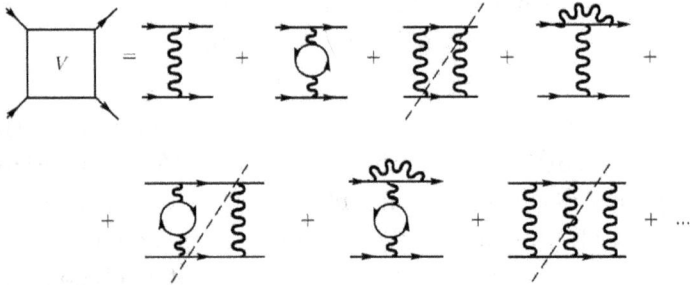

Fig. 2.15   Diagrammatic definition of the block V. Crossed out are diagrams, which can be cut by two particle lines.

To describe interaction itself it is convenient to introduce the vertex part (scattering amplitude) Γ defined by the following expression:

$$K - K_0 = -GG\Gamma GG \qquad (2.140)$$

This vertex Γ is represented by the sum of all diagrams, starting and ending with interaction lines, it does not contain the lines of particles, entering or leaving the whole block (i.e. the "external" lines are just cut off). Substituting (2.140) into (2.139) we get:

$$K - K_0 = -GG\Gamma GG = -GGVK = -GGVK_0 + GGVGG\Gamma GG \qquad (2.141)$$

Introduce the obvious relation:

$$GGVK_0 = GG(V - \tilde{V})GG \qquad (2.142)$$

where $\tilde{V}$ denotes V with exchange of outgoing external lines, Then from (2.141), after the multiplication from the left and right by $(GG)^{-1}$, we obtain:

$$\Gamma = V - \tilde{V} - VGG\Gamma \qquad (2.143)$$

From the definitions of Γ and K it follows that:

$$\Gamma(1, 2; 3, 4) = -\Gamma(2, 1; 3, 4) = -\Gamma(1, 2; 4, 3)$$
$$\Gamma(1, 2; 3, 4) = \Gamma(3, 4; 1, 2) \qquad (2.144)$$

reflecting the antisymmetry of wave functions in the system of Fermions.

Eq. (2.143) for Γ can be obtained also directly from the equation for K. Separating the block V, it is simple to obtain the diagrammatic equation, shown in Fig. 2.16, which (after symmetrization) reduces to (2.143).

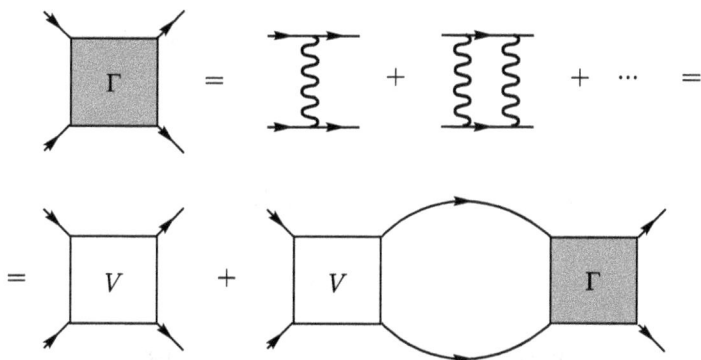

Fig. 2.16 Diagrammatic equation for the vertex Γ in the particle–particle channel.

To derive phenomenological equations of Fermi-liquid theory it is convenient to rewrite equation for Γ in another form. Above we introduced the block V, which could not be cut by two lines in the particle–particle channel (called also irreducible vertex in this channel). We may act also in another way and separate from all diagrams for the vertex Γ, those representing the block (vertex) U, irreducible in the particle–hole channel, consisting of diagrams, which can not be separated in two parts connected by two lines, representing a particle and a hole. Appropriate

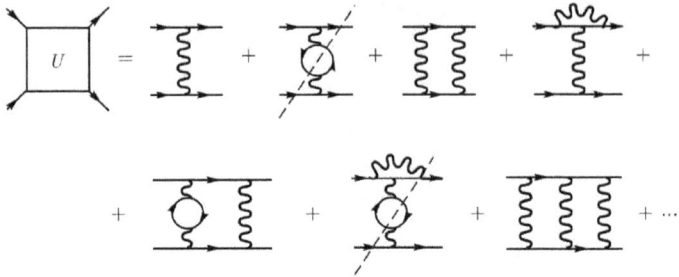

Fig. 2.17 Diagrammatic definition of the block (irreducible vertex) U. Crossed out are diagrams which can be cut by two lines of a particle and a hole.

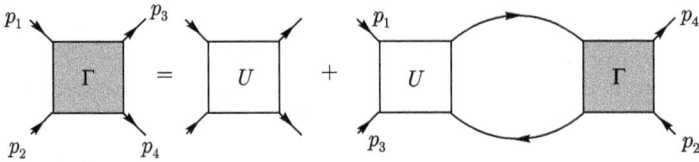

Fig. 2.18 Diagrammatic equation for the vertex Γ in particle–hole channel.

diagrams are shown in Fig. 2.17. Then for Γ we can write diagrammatic equation, shown in Fig. 2.18, or analytically:

$$\Gamma = U + UGG\Gamma \qquad (2.145)$$

In momentum representation, the difference of 4-momenta, entering Green's functions G (and satisfying the conservation law $p_1 + p_2 = p_3 + p_4$), is equal to the transferred momentum $q = p_1 - p_3 = p_4 - p_2$, which is equal to the sum of momenta in the particle–hole channel and is the same in each "cross-section" in this channel. In equation shown in Fig. 2.16, the sum of momenta entering into Γ, is equal to the full momentum of the system of two particles $q' = p_1 + p_2 = p_3 + p_4$, which is the same in each "cross-section" of particle–particle channel. Introduction of different blocks (irreducible vertices), of the type we used above, becomes convenient in the case, when such a block happens to be a smooth function of its variables (momenta), as in such a case it can be replaced by some effective constant.

Remarks on "parquet".

In some cases it is convenient to introduce block (vertex) W, which can not be cut by two lines (irreducible) both in particle–particle and particle–hole channels. Then, besides Eqs. (2.16), (2.18) we have to write down also equations, establishing connection of vertices U and V with vertex W, as shown in Fig. 2.19. Together with equations, shown in Fig. 2.16 and Fig. 2.18, equations shown in Fig. 2.19 form a system of the so-called "parquet" equations. Here we understand that W should be taken in symmetrized form, i.e. for Fermions we should take the difference $W_{\alpha\beta\gamma\delta}(p_1, p_2; p_3, p_4) - W_{\alpha,\beta,\delta,\gamma}(p_1, p_2; p_4, p_3)$. In analytic form we have:

$$V - \tilde{V} = W + UGG\Gamma \tag{2.146}$$

$$U = W - VGG\Gamma \tag{2.147}$$

Using for Γ the equation shown in Fig. 2.18, we obtain from (2.146):

$$U + V - \tilde{V} = W + \Gamma \tag{2.148}$$

The same result follows from (2.147) if we use for Γ the equation, shown in Fig. 2.16. These equations allow us to express Γ via the irreducible vertex W. As a result we obtain the following nonlinear (integral) equation:

$$\Gamma = W + \frac{1}{2}WGG\Gamma - \frac{1}{2}\Gamma GG(\Gamma + W)GG\Gamma \tag{2.149}$$

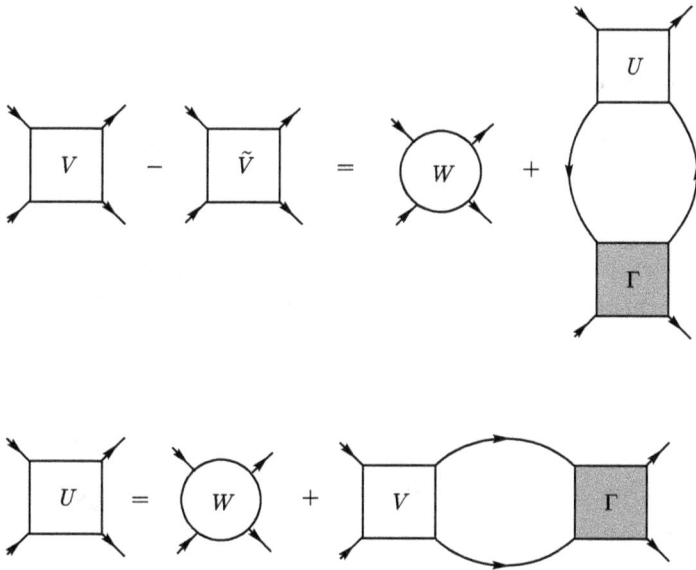

Fig. 2.19   "Parquet" equations for the vertex parts.

Usually the irreducible vertex W, taken in momentum representation, is rather weakly dependent on incoming and outgoing momenta, while the vertex U, according to (2.147), possess a significant dependence on the (small) sum of these momenta. Similarly, the vertex V, according to (2.146), has an important dependence on (small) transferred momentum $q \to 0$.

Introduction of the block (irreducible vertex) U is useful to study the properties of the full vertex Γ at small transferred momenta, while block V (and the use of equations, shown in Fig. 2.16) is conveniently used in case of small sum of incoming (outgoing) momenta (cf. below the analysis of Cooper instability!).

In Landau theory of normal Fermi-liquid the equation, shown in Fig. 2.18, is of special significance. In general, the second term in the r.h.s. of this equation contains integrations both in the vicinity and far away from the Fermi surface. However, Landau has shown that at small momentum transfers this equation may be transformed to another ("renormalized") equation for Γ with momenta close to the Fermi surface and with all integrations performed also in the close vicinity of the Fermi surface (i.e. we can obtain the closed equation for Γ at the Fermi surface).

Let us write down the equation, shown in Fig. 2.18, explicitly (in momentum representation):

$$\Gamma(p, p', q) = U(p, p', q) - i \int \frac{d^4 p''}{(2\pi)^4} U(p, p'', q) G\left(p'' + \frac{q}{2}\right)$$
$$\times G\left(p'' - \frac{q}{2}\right) \Gamma(p'', p', q) \qquad (2.150)$$

where integration is supposed to include summation over the spin indices of internal lines (which we just drop for the shortness of presentation). Here we introduced the following notations:

$$\Gamma(p_1, p_2, p_3, p_4) = \Gamma(p, p', q)(2\pi)^4 \delta(p_1 + p_2 + p_3 + p_4)$$
$$U(p_1, p_2, p_3, p_4) = U(p, p', q)(2\pi)^4 \delta(p_1 + p_2 + p_3 + p_4) \qquad (2.151)$$

where incoming (p_1, p_2) and outgoing (p_3, p_4) 4-momenta are connected with p and p' as:

$$p_1 = p + \frac{q}{2} \qquad p_2 = p' - \frac{q}{2}$$
$$p_3 = p - \frac{q}{2} \qquad p_4 = p' + \frac{q}{2} \qquad (2.152)$$

so that the transferred momentum is $q = (\mathbf{q}, \omega)$ and we have a conservation law: $p_1 + p_2 = p_3 + p_4$. In the first order of perturbation theory we have:

$$U(p, p', q) = \int d(\mathbf{r}_1 - \mathbf{r}_2) e^{-i\mathbf{q}(\mathbf{r}_1 - \mathbf{r}_2)} V(\mathbf{r}_1 - \mathbf{r}_2) \qquad (2.153)$$

where $V(\mathbf{r}_1 - \mathbf{r}_2)$ is the potential of interparticle interaction.

In Eq. (2.130) we have written down the general form of the single-particle Green's function in the Fermi-liquid:

$$G(p) = \frac{Z}{\varepsilon - \varepsilon(p) + i\gamma(\varepsilon)} + G^{reg}(p) \qquad (2.154)$$

where $\varepsilon(p) = v_F(p - p_F)$,

$$Z^{-1} = \left(\frac{\partial G^{-1}}{\partial \varepsilon} \right)_F; \qquad \gamma(\varepsilon) \sim \varepsilon^2 sign\varepsilon \qquad (2.155)$$

When $q \to 0$, the poles of both Green's functions in (2.150) move to each other and effectively we obtain a δ-like maximum close to the Fermi surface. Accordingly, we can write down the following representation of this product of Green' functions (considered in a sense of the kernel of the integral equation) in Eq. (2.150) (all energies are calculated with respect to $\varepsilon = E_F = 0$):

$$G\left(p + \frac{q}{2}\right) G\left(p - \frac{q}{2}\right) \approx Z^2 \delta(\varepsilon) \int_{-\infty}^{\infty} d\varepsilon G_0\left(p + \frac{q}{2}\right) G_0\left(p - \frac{q}{2}\right) + B(p, q)$$
$$(2.156)$$

where

$$G_0(p) = \frac{1}{\varepsilon - \varepsilon(p) + i\gamma sign(\varepsilon)}; \qquad \gamma \to +0 \qquad (2.157)$$

is just a free Green's function. The integral entering (2.156) was, in fact, already calculated above in (2.19). Using the result of this calculation, we have:

$$\int_{-\infty}^{\infty} d\varepsilon G_0\left(p + \frac{q}{2}\right) G_0\left(p - \frac{q}{2}\right)$$
$$= -2\pi i \frac{n(\mathbf{p} + \mathbf{q}/2) - n(\mathbf{p} - \mathbf{q}/2)}{\omega - \varepsilon(\mathbf{p} + \mathbf{q}/2) + \varepsilon(\mathbf{p} - \mathbf{q}/2) + i\gamma sign\omega} \qquad (2.158)$$

where

$$n(\mathbf{p}) = \begin{cases} 1 & \text{for} \quad |\mathbf{p}| \leq p_F \\ 0 & \text{for} \quad |\mathbf{p}| > p_F \end{cases} \qquad (2.159)$$

From here we easily obtain (cf. (2.19), (2.23) etc.):

$$G_0 \left(p + \frac{q}{2} \right) G_0 \left(p - \frac{q}{2} \right)_{q \to 0}$$

$$= iZ^2 \delta(\varepsilon) 2\pi \frac{\mathbf{qv}}{\omega - \mathbf{qv} + i\gamma \, sign\omega} \frac{\delta(|\mathbf{p}| - p_F)}{p_F} m^* + B(p, q)$$

$$\equiv A + B$$

$$(2.160)$$

where $\mathbf{v} = \partial\varepsilon(p)/\partial\mathbf{p} = \frac{p_F}{m^*}\mathbf{p}/p$ is quasiparticle velocity at the Fermi surface, m^* is an effective mass, and $B(p, q)$ does not contain any singularities, and up to the terms of the order of q^2/p_F^2 and ω^2/E_F^2 can be assumed independent of q.

Returning to the analysis of the full vertex Γ, let us rewrite Eq. (2.145) in the following form:

$$\Gamma = U + \Gamma GGU = U + \Gamma(A + B)U = U + U(A + B)\Gamma \qquad (2.161)$$

This equation can be obtained from diagrammatic representation for Γ, if we sum diagrams in "inverse" order. Introduce now the scattering amplitude Γ^ω, defined by the equation:

$$\Gamma^\omega = U + UB\Gamma^\omega = U + \Gamma^\omega BU \qquad (2.162)$$

It is easily seen that Γ^ω can be defined as the following limit:

$$\Gamma^\omega = \lim_{\omega \to 0, \frac{q}{\omega} \to 0} \Gamma \qquad (2.163)$$

The order of limits here is very important, first we have to perform $q \to 0$, and only then put $\omega \to 0$ (L.D. Landau, 1958). In this case we have A in (2.161) going to, which leads to Eq. (2.162).

Multiplying (2.161) from the left side by $1 + \Gamma^\omega B$, we get:

$$\Gamma = \Gamma^\omega + \Gamma^\omega A\Gamma = \Gamma^\omega + \Gamma A\Gamma^\omega \qquad (2.164)$$

It can be checked directly:

$$(1 + \underline{\Gamma^\omega B})\Gamma = (1 + \Gamma^\omega B)U + (1 + \Gamma^\omega B)U(A + B)\Gamma$$

$$= \Gamma^\omega + \Gamma^\omega(A + B)\Gamma = \Gamma^\omega + \Gamma^\omega A\Gamma + \underline{\Gamma^\omega B\Gamma}$$

The underlined terms cancel and we obtain (2.164).

The vertex part Γ^ω depends on p^2, $(p')^2$, $\mathbf{pp'}$ and $\varepsilon, \varepsilon'$ (at the moment we do not discuss spins!), but on the Fermi surface we have $|\mathbf{p}| = |\mathbf{p'}| =$

p_F, $\varepsilon = \varepsilon' = 0$, so that Γ^ω depends only on the angle between vectors $\mathbf{p}$ and $\mathbf{p}'$. Vertex part Γ, taken at the Fermi surface, depends also on the transferred momentum. Integrals in Eq. (2.164) are taken at the Fermi surface (due the explicit form of A given in (2.160)), so that we take $|\mathbf{p}| = |\mathbf{p}'| = p_F$ and $\varepsilon = \varepsilon' = 0$ and obtain the closed equation determining Γ at the Fermi surface (L.D. Landau, 1958).

During this derivation we assumed that the block U (irreducible vertex in particle–hole channel) is non-singular as the transferred momentum $q \to 0$. Thus, strictly speaking, our analysis is invalid in the case of Coulomb interaction between Fermions (e.g. for electrons in metals!), but can be applied for Fermi-liquids with short-range interactions (e.g. for the liquid He^3). Necessary generalizations for the Coulomb case will be given below.

Let us write down Eq. (2.164) explicitly for the limit of small q. Consider first the simplified case, when Γ^ω does not depend on the quasiparticle spins. Then, summation over spin indices of internal lines (particle and hole) leads just to an additional factor of 2 (for Fermions with spin 1/2). Using the explicit form of A from (2.160), we obtain:

$$\Gamma(\mathbf{n}, \mathbf{n}', \mathbf{q}) = \Gamma^\omega(\mathbf{n}, \mathbf{n}')$$
$$+ \frac{Z^2 p_F m^*}{\pi^2} \int \Gamma^\omega(\mathbf{n}, \mathbf{n}_1) \frac{\mathbf{q}\mathbf{v}_1}{\omega - \mathbf{q}\mathbf{v}_1 + i\gamma(\omega)} \Gamma(\mathbf{n}_1, \mathbf{n}', \mathbf{q}) \frac{d\Omega_1}{4\pi} \quad (2.165)$$

where $\gamma(\omega) = \gamma \, sign\omega$, $(\gamma \to +0)$, $\mathbf{n}, \mathbf{n}', \mathbf{n}_1$ are unity vectors for directions of $\mathbf{p}, \mathbf{p}'$ and $\mathbf{v}_1$. Integration in (2.165) is performed over the angles of vector $\mathbf{v}_1$.

Consider as an example the oversimplified case, when $\Gamma^\omega(\mathbf{n}, \mathbf{n}')$ is not dependent on the angle between $\mathbf{n}$ and $\mathbf{n}'$. Then Γ also does not depend on this angle and is easily found from Eq. (2.165):

$$\Gamma(\mathbf{q}\omega) = \frac{\Gamma^\omega}{1 - \frac{1}{2}\Phi_0 \int_{-1}^{1} dx \frac{qvx}{\omega - qvx + i\gamma(\omega)}} \quad (2.166)$$

where $\Phi_0 = Z^2 \Gamma^\omega \frac{m^* p_F}{\pi^2}$. Here $\frac{m^* p_F}{\pi^2}$ is just the density of states at the Fermi level.

Integral in (2.166) is calculated as was done above in (2.23) and (2.25), so that we obtain:

$$\Gamma(\mathbf{q}\omega) = \frac{\Gamma^\omega}{1 + \Phi_0 \left\{ 1 - \frac{\omega}{2qv} \ln \left| \frac{\omega + qv}{\omega - qv} \right| + i\pi \frac{|\omega|}{2qv} \theta(qv - |\omega|) \right\}} \quad (2.167)$$

In case of Coulomb interaction, for $q \to 0$, in first approximation for U we may take only a single diagram shown in Fig. 2.17), so that:

$$U = \frac{4\pi e^2}{q^2} \equiv V_q \qquad (2.168)$$

as it diverges for $q \to 0$. Assuming $Z = 1$ and dropping non-singular contribution to GG, we get:

$$\Gamma(\mathbf{q}\omega) = \frac{V_q}{1 + \frac{mp_F}{\pi^2} V_q \left[1 - \frac{\omega}{2qv} \ln \left|\frac{\omega+qv}{\omega-qv}\right| + i\pi \frac{|\omega|}{2qv}\theta(qv - |\omega|)\right]} \qquad (2.169)$$

For $\omega \gg vq$ this reduces to:

$$\Gamma(\mathbf{q}\omega) = \frac{4\pi e^2}{q^2 - \frac{mp_F}{\pi^2} 4\pi e^2 \frac{1}{3}\frac{v^2 q^2}{\omega^2}} = \frac{V_q}{\left(1 - \frac{\omega_p^2}{\omega^2}\right)} \qquad (2.170)$$

i.e. we obtain effective screening with $\epsilon(\omega) = 1 - \frac{\omega_p^2}{\omega^2}$, where $\omega_p^2 = \frac{4\pi ne^2}{m}$ is the square of plasma frequency. For $vq \gg \omega$ we obtain the usual Debye screening:

$$\Gamma(\mathbf{q}\omega = 0) = \frac{4\pi e^2}{q^2 + \kappa_D^2} \qquad (2.171)$$

where $\kappa_D^2 = \frac{4e^2 mp_F}{\pi}$. These expressions just coincide with those obtained above within RPA.

As we already stressed above, our general analysis of the Fermi-liquid approach assumed the absence of singularity of irreducible vertex part U for $q \to 0$, typical for Coulomb case. Thus, the correct account of Coulomb interaction within the general theory of Fermi-liquids requires special attention (V.P. Silin, 1957; P. Nozieres, J.M. Luttinger, 1962) [Nozieres P. (1964); Nozieres P., Pines D. (1966)]. Consider an arbitrary diagram for the vertex part Γ. Let us call a diagram the "proper" one [Nozieres P. (1964)], if it contains no interaction lines with small momentum transfers q. In the opposite case we shall call a diagram "improper". Typical examples are shown in Fig. 2.20. "Proper" diagrams give regular contributions for $q \to 0$. Let us denote as $\tilde{\Gamma}$ the sum of all "proper" diagrams for the vertex part. This sum is obviously regular for $q \to 0$. Then it is clear that an arbitrary contribution to the full vertex has a structure, shown in Fig. 2.21. Thus we may write:

$$\Gamma = \tilde{\Gamma} + \tilde{\Gamma}\mathcal{V}\tilde{\Gamma} \qquad (2.172)$$

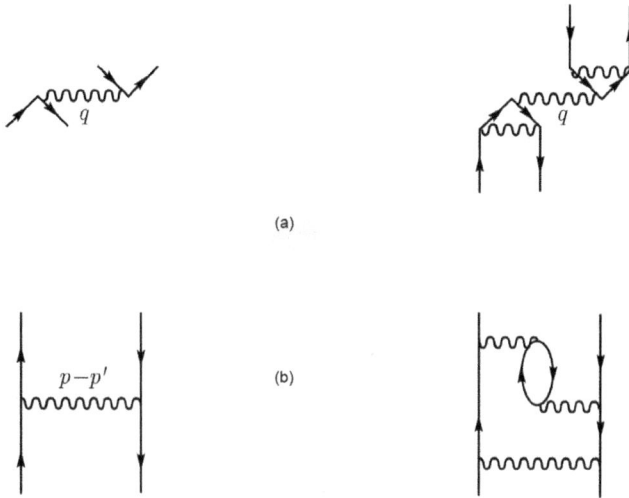

Fig. 2.20 Examples of "improper" (a) and "proper" (b) diagrams for the case of Coulomb interaction.

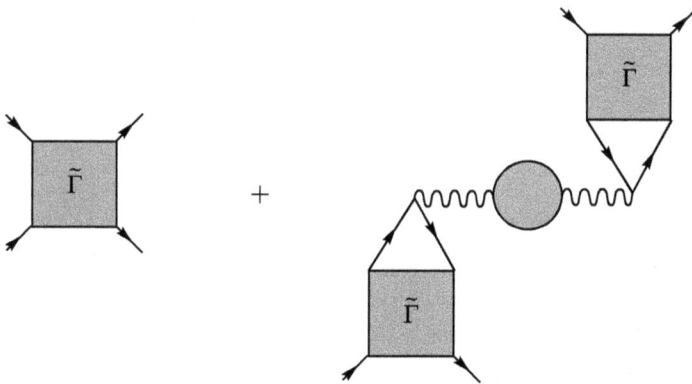

Fig. 2.21 General structure of an arbitrary contribution to the full vertex for the case of Coulomb interaction.

where diagrams for blocks $\mathcal{V}$ (screened interaction!) and $\tilde{T}$ are shown in Fig. 2.22. Analytically (Fig. 2.22(a)):

$$\mathcal{V} = V_q + V_q \tilde{\Pi} V_q + V_q \tilde{\Pi} V_q \tilde{\Pi} V_q + \dots \qquad (2.173)$$

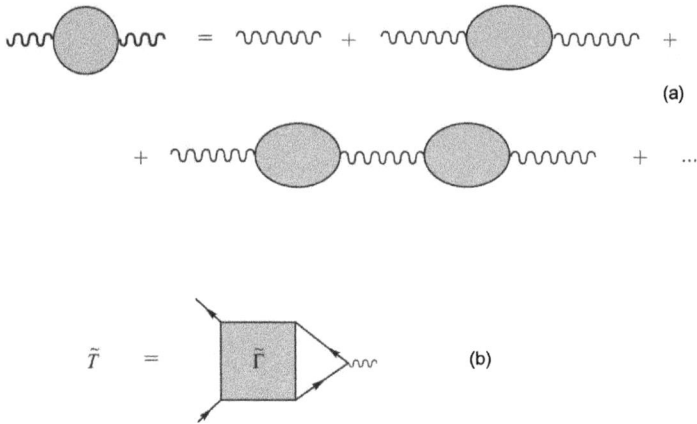

(a)

(b)

Fig. 2.22 Diagrams for the effective (screened) interaction (a) and the definition of the block $\tilde{T}$ (b).

where $\tilde{\Pi}$, as well as $\tilde{T}$ (Fig. 2.22(b)), does not contain "improper" diagrams. It is clear that:

$$\mathcal{V} = \frac{V_q}{1 - V_q \tilde{\Pi}} \qquad (2.174)$$

so that

$$\Gamma = \tilde{\Gamma} + \frac{\tilde{T} V_q \tilde{T}}{1 - V_q \tilde{\Pi}} \qquad (2.175)$$

Blocks $\tilde{\Gamma}$, $\tilde{T}$, $\tilde{\Pi}$ possess well defined limits at $q \to 0$ (of the type of Γ^ω). Thus all the general equation for scattering amplitudes (vertices) of the general theory of Fermi-liquids, derived above, remain, in fact, valid for "proper" vertices (amplitudes), so that in these equations we have only to add "tildas". Physically, this means that we split the full vertex Γ into short-range part $\tilde{\Gamma}$ and the part, describing the self-consistent field (appearing due to long-range forces) Γ_{long}:

$$\Gamma = \tilde{\Gamma} + \Gamma_{long} \qquad (2.176)$$

as it is shown in Fig. 2.23. The value of Γ_{long} is precisely equivalent to the effective self-consistent field, introduced in Landau–Silin theory as the scalar potential to be determined from the solution of Poisson equation [Sadovskii M.V. (2019a); Nozieres P. (1964);

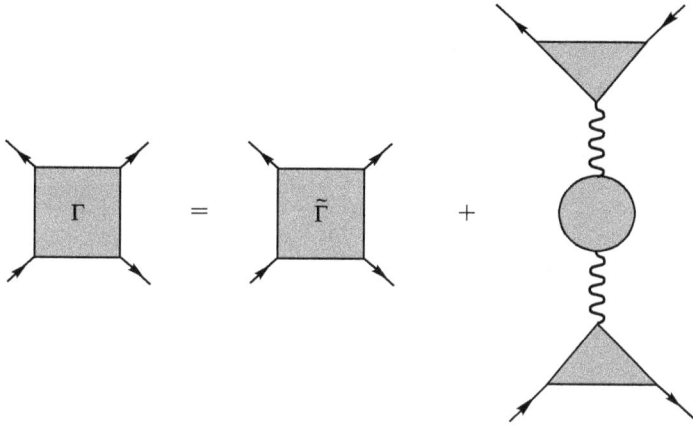

Fig. 2.23 Full interaction vertex in the theory of Fermi-liquids with Coulomb interaction. The second term in the r.h.s. represents an effective self-consistent field (scalar potential).

Nozieres P., Pines D. (1966)]. The "proper" vertex $\tilde{\Gamma}$ describes short-range *correlation* effects, and to determine it in the Fermi-liquid with Coulomb interactions (e.g. electrons in metals) we just write the same phenomenological equations as in the case of short-range interactions (e.g. He^3). These equations will be briefly discussed below. In spatially homogeneous system, the contribution of Γ_{long} at $q = 0$ is just cancelled by the "compensating background" of positive ions (necessary for charge neutrality). However, it becomes quite important in kinetic equation of Landau–Silin theory, which describes collective oscillations in metallic Fermi-liquid [Nozieres P., Pines D. (1966)].

From the previous discussion it is clear that the value of $Z^2\Gamma^\omega$ plays the role of the scattering amplitude of Fermi-liquid quasiparticles. Note that the value of Γ^ω (for $\omega \to 0$) is real (Hermitian in spin indices). Physically Γ^ω represents the scattering amplitude of two particles (in Fermi-liquid) with zero value of scattering angle ($vq \ll E_F$). Imaginary part for the forward scattering amplitude can be expressed via the total scattering cross-section (optical theorem of quantum scattering theory). However, this cross-section goes to zero as momenta of the particles tend to the Fermi momentum, as in this case we have the phase space of the final states tending to zero. Now for Γ^ω we can introduce the standard phenomenological representation (L.D. Landau, 1958):

$$Z^2 \Gamma^\omega \frac{m^* p_F}{\pi^2} \equiv f(\mathbf{p}\sigma; \mathbf{p}'\sigma') = f^s + (\vec{\sigma}\vec{\sigma}') f^a \qquad (2.177)$$

where both f^s and f^a depend only on the angle between $\mathbf{p}$ and $\mathbf{p}'$ (both belonging to the Fermi surface), so that we can introduce the following expansion over Legendre polynomials:

$$f^{s,a}(\theta) = \frac{\pi^2}{m^* p_F} \sum_{l=0}^{\infty} F_l^{s,a} P_l(\cos\theta) \qquad (2.178)$$

where $F^{s,a}$ are dimensionless parameters (Landau constants), describing correlation effects (short-range Fermi-liquid interactions).[12]

Solution of the integral equation (2.165) for Γ can also be sought in the similar form:

$$Z^2 \frac{m^* p_F}{\pi^2} \Gamma = \varphi + (\vec{\sigma}\vec{\sigma}') \psi \qquad (2.179)$$

Then for φ and ψ we obtain the following equations:

$$\varphi(\mathbf{n}, \mathbf{n}', \mathbf{q}) = f^s(\mathbf{n}, \mathbf{n}')$$
$$+ \int f^s(\mathbf{n}, \mathbf{n}_1) \frac{\mathbf{q}\mathbf{v}_1}{\omega - \mathbf{q}\mathbf{v}_1 + i\gamma(\omega)} \varphi(\mathbf{n}_1, \mathbf{n}', \mathbf{q}) \frac{d\Omega_1}{4\pi} \qquad (2.180)$$

$$\psi(\mathbf{n}, \mathbf{n}', \mathbf{q}) = f^a(\mathbf{n}, \mathbf{n}')$$
$$+ \int f^a(\mathbf{n}, \mathbf{n}_1) \frac{\mathbf{q}\mathbf{v}_1}{\omega - \mathbf{q}\mathbf{v}_1 + i\gamma(\omega)} \psi(\mathbf{n}_1, \mathbf{n}', \mathbf{q}) \frac{d\Omega_1}{4\pi} \qquad (2.181)$$

If we are interested in collective excitations of the Fermi-liquid, we must take into account that these are determined by the poles of the *two-particle* Green's function in particle–hole channel, which is determined by the equation (2.140):

$$K = K_0 - GG\Gamma GG \qquad (2.182)$$

Here the term K_0 possesses the pole, corresponding to the sum of energies of two free particles, so that the poles of K, describing collective oscillations can be present only in Γ. Close to that pole we can just neglect inhomogeneous term of (2.145) and write:

$$\Gamma = UGG\Gamma \qquad (2.183)$$

---

[12]Within phenomenological Landau approach these constants are to be determined from the experiments.

Limiting ourselves to excitations with small $q \ll p_F$ and $\omega \ll E_F$, we may use the renormalized equation for Γ:

$$\Gamma = \Gamma^\omega + \Gamma^\omega A \Gamma \qquad (2.184)$$

solution of which can be sought in the form given by (2.179). Then the acoustic type oscillations (zero sound) are possible in our system,[13] described by φ, and spin waves, described by ψ. Consider in more details the case of zero sound. From Eq. (2.180), close to the pole, we obtain:

$$\varphi(\mathbf{n}, \mathbf{n}', \mathbf{q}) = \int f^s(\mathbf{n}, \mathbf{n}_1) \frac{\mathbf{q}\mathbf{v}_1}{\omega - \mathbf{q}\mathbf{v}_1 + i\gamma(\omega)} \varphi(\mathbf{n}_1, \mathbf{n}', \mathbf{q}) \frac{d\Omega_1}{4\pi} \qquad (2.185)$$

Now, close to the pole, describing collective oscillations with the spectrum ω_q, φ can be written as:

$$\varphi(\mathbf{n}, \mathbf{n}') = \frac{\chi(\mathbf{n})\chi(\mathbf{n}')}{\omega^2 - \omega_q^2} 2\omega_q \qquad (2.186)$$

This structure of the solution can be justified on general grounds [Migdal A.B. (1967)], but for us it is sufficient to say, that we just are seeking the solution of this form. Then, for $\chi(\mathbf{n})$ we obtain the following equation:

$$\chi(\mathbf{n}) = \int f^s(\mathbf{n}, \mathbf{n}_1) \frac{\mathbf{q}\mathbf{v}_1}{\omega - \mathbf{q}\mathbf{v}_1 + i\gamma(\omega)} \chi(\mathbf{n}_1) \frac{d\Omega_1}{4\pi} \qquad (2.187)$$

Let us define the function:

$$\rho(\mathbf{n}) = \frac{\mathbf{v}\mathbf{q}}{\omega - \mathbf{v}\mathbf{q} + i\gamma(\omega)} \chi(\mathbf{n}) \qquad (2.188)$$

Then it satisfies the following equation, which is easily obtained from (2.187):

$$(\omega - \mathbf{v}\mathbf{q})\rho(\mathbf{n}) = \mathbf{v}\mathbf{q} \int f^s(\mathbf{n}, \mathbf{n}_1)\rho(\mathbf{n}_1) \frac{d\Omega_1}{4\pi} \qquad (2.189)$$

which coincides with *kinetic equation* of phenomenological Landau theory [Nozieres P., Pines D. (1966)], with $\rho(\mathbf{n})$ being the non-equilibrium part of the distribution function of quasiparticles.

Let us explain this point in more details. Some small change of distribution function of the Fermi-gas (quasiparticles of the Fermi-liquid,

---

[13]To simplify the problem, we are dealing here only with the Fermi-liquid with short-range interactions. Physically, the zero sound corresponds not to the oscillations of the density in the liquid, but to oscillations of the Fermi surface itself [Nozieres P., Pines D. (1966)].

in first approximation, form just such a gas) for small q and ω satisfy the kinetic (transport) equation of the following form:

$$(\omega - \mathbf{vq})\delta f_{\mathbf{q}}(\mathbf{p}) = -\mathbf{q}\frac{\partial f_0}{\partial \mathbf{p}}V_q(\mathbf{p}) \tag{2.190}$$

where f_0 is the equilibrium (Fermi) distribution, while the self-consistent field (potential) $V_q(\mathbf{p})$ is connected with the change of distribution function as:[14]

$$V_q(\mathbf{p}) = 2\int U(\mathbf{p}, \mathbf{p}')\delta f_{\mathbf{q}}(\mathbf{p}')\frac{d\mathbf{p}'}{(2\pi)^3} \tag{2.191}$$

where $U(\mathbf{p}, \mathbf{p}')$ is interaction amplitude of the particles in momentum representation. Using now:

$$\frac{\partial f_0}{\partial \mathbf{p}} = -\frac{\mathbf{p}}{|\mathbf{p}|}\delta(|\mathbf{p}| - p_F) \tag{2.192}$$

and rewriting the non-equilibrium part of distribution function as:

$$\delta f_{\mathbf{q}}(\mathbf{p}) = \delta(|\mathbf{p}| - p_F)\rho(\mathbf{n}) \tag{2.193}$$

we obtain:

$$(\omega - \mathbf{vq})\rho(\mathbf{n}) = \mathbf{vq}\frac{m^*p_F}{\pi^2}\int U(\mathbf{p}, \mathbf{p}_1)\rho(\mathbf{n}_1)\frac{d\Omega_1}{4\pi} \tag{2.194}$$

with $|\mathbf{p}_1| = |\mathbf{p}| = p_F$. Comparison with (2.189) yields:

$$U(\mathbf{p}, \mathbf{p}_1) = \frac{\pi^2}{m^*p_F}f^s(\mathbf{n}, \mathbf{n}_1) = Z^2\Gamma_s^\omega(\mathbf{p}, \mathbf{p}_1) \tag{2.195}$$

where Γ_s^ω denotes the spinless part of the amplitude.

Consider the solution of (2.189) for the simplest case, when $f^s(\mathbf{n}, \mathbf{n}') = F_0$, i.e. is represented by a single constant. Then (2.189) reduces to:

$$\rho(\mathbf{n}) = \frac{\mathbf{vq}}{\omega - \mathbf{vq}}F_0\int \rho(\mathbf{n}_1)\frac{d\Omega_1}{4\pi} \tag{2.196}$$

Performing angular integration (as we have already done before), we obtain the following equation:

$$-\frac{1}{F_0} = 1 - \frac{\omega}{2vq}\ln\left|\frac{\omega + vq}{\omega - vq}\right| + i\pi\frac{|\omega|}{2vq}\theta(vq - |\omega|) \tag{2.197}$$

---

[14] In Landau–Silin theory we have to add to $V_q(\mathbf{p})$ the contribution from self-consistent scalar potential, defining the electric field, and determined by appropriate Poisson equation.

Solution of this equation immediately gives the dispersion law for the zero sound. Real frequencies (no damping!) is obtained for $|\omega| > vq$. Denoting $\omega_q = svq$ we get:

$$\frac{1}{F_0} = \frac{s}{2} \ln \frac{s+1}{s-1} - 1, \quad \text{where} \quad s > 1 \tag{2.198}$$

It is not difficult to see, that the r.h.s. here is positive, so that the zero sound is possible only for $F_0 > 0$. In limiting cases we have:

$$s_{F_0 \to 0} = 1 + 2e^{-\frac{2}{F_0}} \qquad s_{F_0 \to \infty} = \sqrt{F_0/3} \tag{2.199}$$

Finally we just quote a number of basic relations of the standard theory of Fermi-liquids [Nozieres P., Pines D. (1966)]. Using Halilean invariance it can be shown that the effective mass m^* is determined by a simple relation:

$$\frac{m^*}{m} = 1 + \frac{F_1^s}{3} \tag{2.200}$$

Accordingly, the specific heat of Fermi-liquid (at $T = 0$) is given by:

$$c = \frac{m^*}{m} c_0 \tag{2.201}$$

where c_0 is the specific heat of the Fermi-gas.

Magnetic susceptibility is given by:

$$\chi = \frac{m^*}{m} \frac{1}{1 + F_0^a} \chi_0 \tag{2.202}$$

where χ_0 is the susceptibility of an ideal gas. Similarly, compressibility of the Fermi-liquid is:

$$\kappa = \frac{m^*}{m} \frac{1}{1 + F_0^s} \kappa_0 \tag{2.203}$$

where κ_0 is the compressibility of a gas.

Using these relations we can come to some general conclusions. For example, from Eqs. (2.202) and (2.203) we immediately obtain conditions for stability of the homogeneous Fermi-liquid:

$$1 + F_0^a > 0; \qquad 1 + F_0^s > 0 \tag{2.204}$$

If we have $1 + F_0^s < 0$, then we get the negative compressibility (which may mean that the system is unstable to some "structural" phase transition, transforming the system to some new stable state!). The general stability analysis of the Fermi-liquid state (I.Ya. Pomeranchuk, 1957) gives the following stability conditions (which must be satisfied for all values of l):

$$1 + \frac{1}{2l+1} F_l^{s,a} > 0 \qquad (2.205)$$

For $l = 1$ inequality (2.205) guarantees the positiveness of m^* as defined by (2.200). From (2.202) we can see, that for the system, close to magnetic instability, it is possible to have $1 + F_0^a \ll 1$. The value of this parameter can be determined experimentally from the measurements of susceptibility and specific heat, determining the so-called Wilson ratio:

$$R_W = \frac{\pi^2 \chi T}{3\mu_B^2 c} = \frac{1}{1 + F_0^a} \qquad (2.206)$$

2.10 Non-Fermi-liquid behavior

Fermi-liquid is not the only possible ground state of many-electron (Fermion) system. The system may become superconducting, magnetic (antiferromagnetic) ordering or charge (spin) density waves (CDW(SDW)) are also possible. Some of these states may be dielectrics, resulting from the initial metallic state via metal–insulator transitions. Some of these possibilities will be discussed below. However, there is a general question — if Landau Fermi-liquid is the only possible ground state of a normal metal without any type of long-range order? This problem is actively discussed in recent years, mainly due to the problems with an explanation of the anomalies of electronic properties of the normal state of high-temperature copper oxide superconductors (and also the so-called "heavy Fermion" compounds). Non-Fermi-liquid behavior is realized (as a rule!) in one-dimensional models of interacting Fermions. Some of the examples of such systems (models), such as basic Tomonaga–Luttinger model, will be dealt with in the final part of our lectures. However, high-temperature superconducting copper oxides belong to some border-crossing case of two-dimensional (or, more precisely, quasi-two-dimensional) systems, and the question about the proper ground state is still more or less open. There is a number of "scenarios" of non-Fermi-liquid behavior of such systems.

At the moment we shall briefly discuss only one such scenario — that of the so-called "marginal" Fermi-liquid [Varma C.N., Nussinov Z., Wim van Saarloos (2002)]. In this theory it is *assumed*,[15] that the polarization operator of electronic system $\Pi(\mathbf{q}\omega)$ possess no significant dependence on $\mathbf{q}$, while the frequency dependence of its imaginary part has the following form:

$$Im\Pi(\mathbf{q}\omega) = \begin{cases} -N(E_F)\frac{\omega}{T} & \text{for} \quad \omega \ll T \\ -N(E_F) & \text{for} \quad T \ll \omega \ll \omega_c \end{cases} \qquad (2.207)$$

Here ω_c is some cut-off frequency, and it is assumed that $\omega_c \ll E_F$. Using Kramers–Kronig dispersion relations, we can restore the appropriate form of the real part of Π:

$$Re\Pi(\mathbf{q}\omega) \sim N(E_F)\ln\left(\frac{\omega}{T}\right) \qquad (2.208)$$

Now we can estimate the self-energy of an electron, determined by the diagram shown in Fig. 2.24:

$$\Sigma \sim \lambda\varepsilon \left[\ln\frac{x}{\omega_c} + i\frac{\pi}{2}sign\varepsilon\right] \qquad (2.209)$$

where $x = Max(\varepsilon, T)$, while λ is some dimensionless interaction constant.

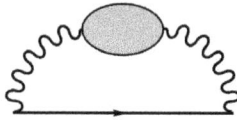

Fig. 2.24   Self-energy of an electron in "marginal" Fermi-liquid.

Then, using (2.50), we immediately obtain:

$$Z = \frac{1}{1 - \frac{\partial Re\Sigma}{\partial\varepsilon}} \sim \frac{1}{1 - \lambda\ln\left(\frac{x}{\omega_c}\right)} \qquad (2.210)$$

Now we see that the residue at the pole of the Green's functions goes to zero at the Fermi surface itself, so that quasiparticles are just not

[15]These assumptions qualitatively correspond to experimentally observed anomalous behavior of copper oxide superconductors in the normal state.

defined there at all! However, everywhere close to the Fermi surface we have more or less "usual" quasiparticle contribution. Important difference with standard Fermi-liquid behavior is that quasiparticle damping, determined by the imaginary part of Σ (2.209), is linear in energy (with respect to the Fermi level): $\gamma \sim \varepsilon$. This means that quasiparticles (close to the Fermi surface) are just "marginally" defined (note that in Landau theory we have obtained $\gamma \sim \varepsilon^2$, leading to well defined quasiparticles).

At present it is not clear what kind of the microscopic mechanism (interaction) can lead to such anomalous behavior, though "marginal" Fermi-liquid gives (phenomenologically) rather satisfactory description of basic anomalies of electronic properties of copper oxides in the normal state and this model is often used to fit experiments [Varma C.N., Nussinov Z., Wim van Saarloos (2002)].

Note, that this problem is linked, in general, with the low dimensionality of the systems under study. We already noted that in one-dimensional (interacting) systems Landau theory just never "works". Many theorists believe that similar situation is typical for two-dimensional systems also. The physical reason for non-Fermi liquid behavior is usually attributed to *strong correlations*, which can not be described by Fermi-liquid phenomenology, as it assumes the qualitative picture of the ground state similar to that of an ideal Fermi-gas. At the same time, a number of detailed studies has shown that Fermi-liquid behavior is mostly conserved in two-dimensional (quasi-two-dimensional) case.

Chapter 3

Electron–Phonon Interaction

3.1 Diagram rules

Phonons are the quanta of lattice oscillations. Phonons can be either acoustical or optical. The difference is that the frequency of acoustical phonons goes to zero as the wave vector $\mathbf{k} \to 0$, while for optical phonons it stays finite. Acoustical phonons are present in any crystal, representing, in fact, Goldstone mode related to broken translation symmetry. Optical phonons appear only in crystals with more than one atom in elementary cell.

There are two standard (simplified!) models to describe phonon spectra — that of Debye and Einstein. In Debye model the phonon spectrum is assumed to be described by $\omega_0(k) = ck$ (c — sound velocity) for all $k < k_D$, where k_D a limiting (cut-off) wave vector of the order of inverse lattice spacing. In Einstein model the phonon frequency is just independent of the wave vector $\omega_0(k) = \Omega_0$ (for all values of k). Debye model gives simplified description of acoustical phonons, while that of Einstein — the same for the optical phonons.

As ions constituting the crystal lattice are charged particles, phonons can interact with electrons. Lattice oscillations induce the deviations of the electric field of ions from the average value dictated by charge neutrality. The potential of this additional field is usually called *deformation potential*. Due to the long-range nature of Coulomb forces electron–phonon interaction can be expected to be strongly non-local. However, as we have seen above, electric field in metals is strongly screened, thus, in most cases, electron–phonon interaction can be assumed local.

Relations of deformation potential to lattice distortion are different for acoustical and optical phonons. For acoustical oscillations nearby lattice ions displacements are almost the same and electric field changes only slightly, proportionally to these (relative) displacements, and the deformation potential $U_{d-ac} \sim div\mathbf{u}$, where $\mathbf{u}(\mathbf{r})$ is atomic displacement. In the case of optical phonons nearby atoms move in opposite directions, so that $U_{d-op} \sim \mathbf{u}(\mathbf{r})$. To describe electron–phonon interaction in a similar way for both types of phonons, an operator of phonon field is usually introduced in the following (Hermitian!) form [Abrikosov A.A., Gorkov L.P., Dzyaloshinskii I.E. (1963)]:

$$\hat{\varphi}(\mathbf{r}t) = i \sum_{\mathbf{k}} \sqrt{\frac{\omega_0(\mathbf{k})}{2V}} \left[\hat{b}_{\mathbf{k}} e^{i\mathbf{kr}-i\omega_0(\mathbf{k})t} - \hat{b}_{\mathbf{k}}^+ e^{-i\mathbf{kr}+i\omega_0(\mathbf{k})t} \right] \qquad (3.1)$$

where $\hat{b}_{\mathbf{k}}^+$, $\hat{b}_{\mathbf{k}}$ are creation and annihilation operators of phonons, V-system volume. Then $\varphi(\mathbf{r}) \sim \nabla\mathbf{u}(\mathbf{r})$ for acoustical phonons and $\varphi(\mathbf{r}) \sim \mathbf{u}(\mathbf{r})$ for the optical phonons. Accordingly, the Hamiltonian of electron–phonon interaction is written as [Abrikosov A.A., Gorkov L.P., Dzyaloshinskii I.E. (1963)]:

$$H_{int} = g \int d\mathbf{r}\hat{\psi}^+(\mathbf{r})\hat{\psi}(\mathbf{r})\hat{\varphi}(\mathbf{r}) \qquad (3.2)$$

where g is the coupling constant. The Hamiltonian density in (3.2) just proportional to the product of electron density and deformation potential.

Diagram rules for electrons and phonons look almost the same as for the case of two-particle interaction [Abrikosov A.A., Gorkov L.P., Dzyaloshinskii I.E. (1963)]. To calculate Green's function at $T = 0$ these rules are formulated as follows:

- Only diagrams of even order give nonzero contributions. Diagram of order $2n$ contains $3n+1$ internal (electron and phonon) lines and $2n$ vertices, with $3n - 1 - (2n - 1) = n$ independent integrations. All lines are attributed with 4-momenta, conserving at the vertices.
- Electron is described by continuous line, denoting free Green's function:

$$G_0(p) = \frac{\delta_{\alpha\beta}}{\varepsilon - \xi(\mathbf{p}) + i\delta sign\xi(\mathbf{p})} \qquad \text{where} \qquad \delta \to +0 \qquad (3.3)$$

where

$$\xi(\mathbf{p}) = \frac{p^2}{2m} - \mu \approx v_F(|\mathbf{p}| - p_F) \qquad (3.4)$$

is the energy spectrum of free electrons, calculated with respect to the Fermi level (chemical potential μ), p_F and v_F are Fermi momentum and velocity.

- *Dashed* line denotes the phonon Green's function:

$$D_0(k) = \frac{\omega_0^2(\mathbf{k})}{\omega^2 - \omega_0^2(\mathbf{k}) + i\delta} \qquad \text{where} \qquad \delta \to +0 \qquad (3.5)$$

- Integration is done over n independent momenta and frequencies (4-momenta).
- The result is multiplied by $(g)^{2n}(2\pi)^{-4n}(i)^n(2s+1)^F(-1)^F$, where F is the number of closed Fermion loops, and s is Fermion spin (for electrons $s = 1/2$, so that, in fact, we always have $2s + 1 = 2$).

For $T > 0$ everything is quite similar:

- Each electronic (continuous) line with momentum $\mathbf{p}$ and Matsubara frequency $\varepsilon_n = (2n + 1)\pi T$ corresponds to:

$$G_0(\varepsilon_n \mathbf{p}) = \frac{\delta_{\alpha\beta}}{i\varepsilon_n - \xi(\mathbf{p})} \qquad (3.6)$$

- Each phonon (dashed) line with momentum $\mathbf{k}$ and frequency $\omega_m = 2\pi m T$ corresponds to:

$$D_0(k) = -\frac{\omega_0^2(\mathbf{k})}{\omega_m^2 + \omega_0^2(\mathbf{k})} \qquad (3.7)$$

- The result is multiplied by $g^{2n}(-1)^n \frac{T^n}{(2\pi)^{3n}}(2s+1)^F(-1)^F$, where F is again the number of Fermion loops and $s = 1/2$ is electronic spin.

The form of phonon Green's function in these rules corresponds to normalization of the operator of phonon field used in (3.1) [Abrikosov A.A., Gorkov L.P., Dzyaloshinskii I.E. (1963)]:[1]

$$\varphi(\mathbf{k}) = \sqrt{\frac{\omega_0(\mathbf{k})}{2}}(b_{\mathbf{k}} + b_{-\mathbf{k}}^+) \qquad (3.8)$$

---

[1]The reader is advised to convince himself that physically (3.1) and (3.8) are just equivalent.

Sometimes in the literature you cam meet with a different normalization [Schrieffer J.R. (1964)]:

$$\varphi(\mathbf{k}) = b_{\mathbf{k}} + b^{+}_{-\mathbf{k}} \tag{3.9}$$

Then the free phonon Green's function takes the following form [Schrieffer J.R. (1964)]:

$$D_0(\mathbf{k}\omega) = \frac{1}{\omega - \omega_0(\mathbf{k}) + i\delta} - \frac{1}{\omega + \omega_0(\mathbf{k}) - i\delta} = \frac{2\omega_0(\mathbf{k})}{\omega^2 - \omega_0^2(\mathbf{k}) + i\delta} \tag{3.10}$$

Accordingly, if we are using different normalizations, we have to take into account some differences in the form of matrix elements of electron–phonon interaction, compensating this difference in normalization, so that physical results are equivalent. This is important to remember, while comparing the results of different authors. If we use (3.9), interaction Hamiltonian (3.2) can be written as [Schrieffer J.R. (1964)] (we take here $V = 1$ for shortness):

$$H_{int} = \sum_{\mathbf{pk}} \bar{g}_k a^{+}_{\mathbf{p+k}} a_{\mathbf{p}} (b_{\mathbf{k}} + b^{+}_{-\mathbf{k}}) \tag{3.11}$$

However, if we use (3.8), then:

$$H_{int} = g \sum_{\mathbf{pk}} \sqrt{\frac{\omega_0(\mathbf{k})}{2}} a^{+}_{\mathbf{p+k}} a_{\mathbf{p}} (b_{\mathbf{k}} + b^{+}_{-\mathbf{k}}) \tag{3.12}$$

which leads to the difference in the definition of electron–phonon coupling constant. In particular, for the appropriate dimensionless constant we mainly use (following the tradition of Russian or rather "Soviet" literature) [Abrikosov A.A., Gorkov L.P., Dzyaloshinskii I.E. (1963)]:

$$\zeta = g^2 \nu_F \tag{3.13}$$

Another common ("Western") definition is:

$$\lambda = \frac{2\bar{g}_k^2 \nu_F}{\omega_0(k)} \tag{3.14}$$

Direct comparison of (3.11) and (3.12) gives:

$$\bar{g}_k = g \sqrt{\frac{\omega_0(k)}{2}} \tag{3.15}$$

Calculation of electron self-energy allows us to determine electron spectrum "renormalization" due to electron–phonon interaction. To determine the phonon spectrum we have to find the poles of phonon Green's function $D(\omega\mathbf{k})$. Similarly to the case of electron Green's function $G(E\mathbf{p})$, for phonon Green's function we may also introduce the self-energy part which, in fact, reduces to the polarization operator

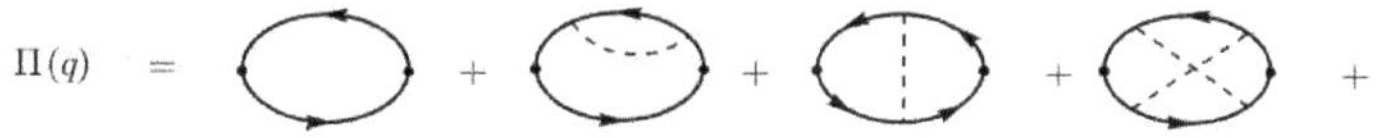

$$\Pi(q) \quad = \quad \bigcirc \quad + \quad \bigcirc \quad + \quad \bigcirc \quad + \quad \bigcirc \quad +$$

Fig. 3.1   Diagrams for phonon self-energy.

$\Pi(\omega\mathbf{k})$, with corrections due to electron–phonon interaction, as shown in Fig. 3.1. Dyson equation for phonon Green's function can be written as:

$$D(\omega\mathbf{k}) = D_0(\omega\mathbf{k}) + D_0(\omega\mathbf{k})g^2\Pi(\omega\mathbf{k})D(\omega\mathbf{k}) \qquad (3.16)$$

with solution:

$$D^{-1}(\omega\mathbf{k}) = D_0^{-1}(\omega\mathbf{k}) - g^2\Pi(\omega\mathbf{k}) \qquad (3.17)$$

Phonon spectrum is defined from:

$$D_0^{-1}(\omega\mathbf{k}) = g^2\Pi(\omega\mathbf{k}) \qquad (3.18)$$

If we introduce the dimensionless coupling constant for electron–phonon interaction as $\zeta = g^2\nu_F$ (where $\nu_F = \frac{mp_F}{2\pi^2\hbar^3}$ is the density of states at the Fermi level for one spin projection), direct estimate gives $\zeta \sim 1$. Thus, it may seem that electron–phonon coupling is always strong enough. However, as we shall see below, there is an additional small parameter in this problem, allowing us to find a simple solution with no assumption of smallness of the coupling constant g. This is so-called adiabaticity parameter $\frac{\omega_D}{E_F} \sim \sqrt{\frac{m}{M}} \ll 1$ (where ω_D is Debye frequency, m-electron mass and M-ion mass). Physically it means that due to a large mass, ions move much slower than electrons. Accordingly, much faster electrons more or less "follow" local ion configuration. As a result, as we shall show below, electron–phonon interaction does not destroy Fermi-liquid behavior.

3.2   Electron self-energy

Consider, following [Levitov L.S., Shitov A.V. (2003)], the simplest contribution to self-energy of an electron, defined by the diagram shown in

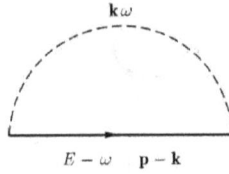

Fig. 3.2 Simplest contribution to electron self-energy due to electron–phonon interaction.

Fig. 3.2. In analytic form we have (for the case of acoustical phonons):

$$\Sigma(E\mathbf{p}) = \frac{ig^2}{(2\pi)^4} \int \frac{d\omega d^3k}{E - \omega - \xi(\mathbf{p}-\mathbf{k}) + i\delta \, sign\xi(\mathbf{p}-\mathbf{k})} \frac{c^2k^2}{\omega^2 - c^2k^2 + i\delta} \tag{3.19}$$

Here we have poles at $\omega_1 = E - \xi(\mathbf{p}-\mathbf{k}) + i\delta \, sign\xi(\mathbf{p}-\mathbf{k})$ and $\omega_{2,3} = \pm(ck - i\delta)$, and we can close integration contour in such a way, that only one of the poles of phonon Green's function is inside. The integral over infinitely far semi-circle is zero and after elementary calculations we obtain:

$$\Sigma(E\mathbf{p}) = \frac{-g^2}{(2\pi)^3} \left\{ \int_{\xi_{\mathbf{p}-\mathbf{k}}<0} \frac{d^3k}{E + ck - \xi(\mathbf{p}-\mathbf{k}) - i\delta} \frac{c^2k^2}{(-2)ck} \right.$$

$$\left. - \int_{\xi_{\mathbf{p}-\mathbf{k}}>0} \frac{d^3k}{E - ck - \xi(\mathbf{p}-\mathbf{k}) + i\delta} \frac{c^2k^2}{2ck} \right\}$$

$$= \frac{g^2c}{16\pi^3} \left\{ \int_{\xi_{\mathbf{p}-\mathbf{k}}>0} \frac{kd^3k}{E - ck - \xi(\mathbf{p}-\mathbf{k}) + i\delta} + \int_{\xi_{\mathbf{p}-\mathbf{k}}<0} \frac{kd^3k}{E + ck - \xi(\mathbf{p}-\mathbf{k}) - i\delta} \right\}$$

$$= \frac{g^2c}{16\pi^3} \left\{ \int_{|\mathbf{p}-\mathbf{k}|>p_F} \frac{kd^3k}{E - ck - v_F(|\mathbf{p}-\mathbf{k}| - p_F) + i\delta} \right.$$

$$\left. + \int_{|\mathbf{p}-\mathbf{k}|<p_F} \frac{kd^3k}{E + ck - v_F(|\mathbf{p}-\mathbf{k}| - p_F) - i\delta} \right\} \tag{3.20}$$

Let us denote as x the cosine of the angle between vectors $\mathbf{k}$ and $\mathbf{p}$, then we have $p_1^2 = |\mathbf{p}-\mathbf{k}|^2 = p^2 + k^2 - 2pkx$ and $d^3k = 2\pi k^2 dkdx$, so that $p_1 dp_1 = -pkdx$. Then (3.20) can be written as:

$$\Sigma(E\mathbf{p}) = -\frac{g^2c}{8\pi^2 p} \left\{ \int_{p_1>p_F} \frac{k^2 dkdp_1 p_1}{E - ck - v_F(p_1 - p_F) + i\delta} \right.$$

$$\left. + \int_{p_1<p_F} \frac{k^2 dkdp_1 p_1}{E + ck - v_F(p_1 - p_F) - i\delta} \right\} \tag{3.21}$$

The main contribution to integrals here comes from the vicinity of the poles, where we can put $p_1 \approx p \approx p_F$, as $\omega_D \ll E_F$. Thus we can neglect the difference between p_1 and p, so that we have:

$$\Sigma(E) = -\frac{g^2 c}{8\pi^2} \left\{ \int\int_{p_1 > p_F} \frac{k^2 dk dp_1}{E - ck - v_F(p_1 - p_F) + i\delta} \right.$$

$$\left. + \int_{p_1 < p_F} \frac{k^2 dk dp_1}{E + ck - v_F(p_1 - p_F) - i\delta} \right\} \quad (3.22)$$

Then for imaginary part of $\Sigma(E)$ we get:

$$Im\Sigma(E) = \frac{g^2 c}{8\pi} \left\{ \int_{p_1 > p_F} \delta(E - ck - v_F(p_1 - p_F)) k^2 dk dp_1 \right.$$

$$\left. - \int_{p_1 < p_F} \delta(E + ck - v_F(p_1 - p_F)) k^2 dk dp_1 \right\} \quad (3.23)$$

First term here (which is nonzero for $E > 0$) gives the lifetime of an electron, while the second one (nonzero for $E < 0$) — the lifetime of a hole. Let us consider now the limiting cases of $E \ll \omega_D$ and $E \gg \omega_D$.

- The case of $E \ll \omega_D$

For $E > 0$ only the first term in (3.23) contributes, and we have to integrate over k in the region of $k < E/c$, or p_1 determined from the argument of the δ-function, will become smaller than p_F. Thus we have:

$$Im\Sigma(E) = \frac{g^2 c}{8\pi} \int_{ck<E} \frac{1}{v_F} k^2 dk = \frac{g^2 c E^3}{24\pi v_F c^3} \quad (3.24)$$

Introducing dimensionless coupling constant of electron–phonon interaction as $\zeta = g^2 \nu_F$, we get:

$$Im\Sigma(E) = \frac{\zeta \pi E^3}{12 p_F^2 c^2} \quad (3.25)$$

For $E < 0$ only the second term in (3.23) contributes. Calculating the integral we again obtain (3.25), due to particle–hole symmetry (valid for $E \ll E_F$) and the imaginary part of $Im\Sigma(E)$ is an odd function of E.

Now we can see that for $E \to 0$ we have $Im\Sigma(E) \ll E$, so that electron–phonon interaction *does not destroy* Fermi-liquid behavior, as due to E^3-dependence of damping, phonon contribution for $E \to 0$

becomes negligible in comparison to electron–electron scattering contribution to damping discussed above, which is $\sim E^2$. At the same time, it is clear that this statement is valid only for $E \to 0$ ($T \to 0$).

- The case of $E \gg \omega_D$

In this case, integration over p_1 in (3.23) does not put any limitations on k-integration, which is now performed up to $k = k_D$. Calculation the integral of the type of (3.24), we obtain:

$$Im\Sigma(E) = \frac{g^2}{8\pi v_F} c \frac{k_D^3}{3} signE = \frac{g^2 k_D^3 mc}{24\pi p_F} signE \qquad (3.26)$$

Again, expressing the damping via dimensionless constant ζ, we get:

$$Im\Sigma(E) = \frac{\zeta \pi k_D^3 c}{12 p_F^2} signE \qquad (3.27)$$

It is easily seen that in this limit $Im\Sigma \sim \zeta \omega_D$. Thus, even for $\zeta \sim 1$ the phonon renormalization is small due to $\omega_D \ll E_F$.

Let us now consider $Re\Sigma(E)$. From (3.22) we have:

$$Re\Sigma(E) = -\frac{g^2 c}{8\pi^2} \left\{ \int_{p_1 > p_F} \frac{k^2 dk dp_1}{E - ck - v_F(p_1 - p_F)} \right.$$

$$\left. + \int_{p_1 < p_F} \frac{k^2 dk dp_1}{E + ck - v_F(p_1 - p_F)} \right\}$$

$$= -\frac{g^2 c}{8\pi^2} \int_{k < k_D} dk k^2 I_1(k) \qquad (3.28)$$

where

$$I_1(k) = \int_{p_1 > p_F} \frac{dp_1}{E - ck - v_F(p_1 - p_F)} + \int_{p_1 < p_F} \frac{dp_1}{E + ck - v_F(p_1 - p_F)} \qquad (3.29)$$

Formally, the first integral here diverges, but this divergence is unphysical, as for large differences between p_1 and p_F we have to take into account the deviations from the linearized form electron spectrum we are using (and also the finiteness of the bandwidth). Thus we may just cut-off integration at $p_1 = p^* \sim p_F$. Exact value of this cut-off parameter is unimportant, as does not influence the form of the spectrum, but only renormalizes the chemical potential (contributing only to $Re\Sigma(0)$). Thus we obtain:

$$I_1(k) = \frac{1}{v_F} \ln \left| \frac{E + ck}{E + ck + v_F p_F} \right| + \frac{1}{v_F} \ln \left| \frac{E - ck - v_F(p^* - p_F)}{E - ck} \right| \tag{3.30}$$

Subtracting this constant renormalization of the chemical potential $\delta\mu = \Sigma(0)$, we obtain:

$$Re(\Sigma(E) - \Sigma(0)) = \frac{g^2 c}{8\pi^2} \int dk k^2 \frac{m}{p_F} \ln \left| \frac{E - ck}{E + ck} \right| \tag{3.31}$$

Characteristic property of an electron self-energy due to electron–phonon interaction is its independence of momentum p. This is due to the "slowness" of phonons, compared to electrons, which leads to the local nature of the processes of phonon emission and absorption by electrons. Let us again analyze limiting cases of $E \ll \omega_D$ and $E \gg \omega_D$.

- The case of $E \ll \omega_D$

In this case we may expand logarithm in (3.31), as $E \ll ck$. Then we have:

$$Re(\Sigma(E) - \Sigma(0)) = -\frac{2mg^2 E}{8\pi^2 p_F} \int_0^{k_D} dkk = -\frac{mg^2 k_D^2}{8\pi^2 p_F} E = -\frac{\zeta}{4} \frac{k_D^2}{p_F^2} E \equiv -\lambda E \tag{3.32}$$

where we have introduced renormalization constant $\lambda = \frac{\zeta k_D^2}{4 p_F^2} \sim \zeta$.

- The case of $E \gg \omega_D$

Now we have $E \gg ck$, so that again, after expanding logarithm in (3.31) we get:

$$Re(\Sigma(E) - \Sigma(0)) = -\frac{mg^2 c^2}{4\pi^2 p_F} \int_0^{k_D} \frac{k^3 dk}{E} = -\frac{mg^2 c^2 k_D^4}{16\pi^2 p_F E} = -\frac{\zeta c^2 k_D^4}{8 p_F^2 E} \tag{3.33}$$

so that at $E \sim \omega_D$ the growth of $Re\Sigma(E)$ with energy changes to decline.

Quasiparticle spectrum for the region of $E \ll \omega_D$ is determined from the equation:

$$E - \xi(p) = Re(\Sigma(E) - \Sigma(0)) \tag{3.34}$$

where $\xi(p) = \frac{p^2}{2m} - \mu$. Then we immediately obtain:

$$E = \frac{p^2}{2m^*} - E_F \qquad E_F = \frac{p_F^2}{2m^*} \qquad (3.35)$$

where the effective mass is defined as:

$$\frac{m^*}{m} = 1 + \frac{\zeta k_D^2}{4 p_F^2} \equiv 1 + \lambda \qquad (3.36)$$

Thus, λ is sometimes called mass renormalization factor. We see that due to electron–phonon interaction an electron becomes "heavier". Accordingly grows the density of states at the Fermi level ($\sim m^*$) and electronic contribution to specific heat.

Let us consider now behavior of electron self-energy at finite temperatures. It is useful also from technical point of view, as we shall be able to study the general method to perform summation over the Matsubara frequencies. So we have to calculate:

$$\Sigma(\varepsilon \mathbf{p}) = -\frac{g^2 T}{(2\pi)^3} \sum_{\varepsilon_1} \int d^3 p_1 G(\varepsilon_1 \mathbf{p}_1) D(\varepsilon - \varepsilon_1, \mathbf{p} - \mathbf{p}_1) \qquad (3.37)$$

where all frequencies are assumed to be Matsubara's!

The general and convenient method to calculate Matsubara sums can be formulated as follows [Schrieffer J.R. (1964)]. The idea is, of course, to go from summation to integration. To be specific, let us start with summation over odd (Fermion) frequencies $i\varepsilon_n = i(2n+1)\pi T$. This sum can be written in the form of the following contour integral in the complex plane of frequency ε:

$$T \sum_{n=-\infty}^{\infty} F(i\varepsilon_n) = -\frac{1}{2\pi i} \int_C d\varepsilon \frac{F(\varepsilon)}{e^{\beta\varepsilon}+1} = \frac{1}{2\pi i} \int_C d\varepsilon \frac{F(\varepsilon)}{e^{-\beta\varepsilon}+1}$$

$$= \frac{1}{4\pi i} \int_C d\varepsilon F(\varepsilon) th\frac{\varepsilon}{2T} \qquad (3.38)$$

where the contour of integration C encircles the imaginary axis, as shown in Fig. 3.3, assuming there is no singularities of $F(\varepsilon)$ inside this contour. Validity of (3.38) follows from Cauchy theorem, as $e^{\beta\varepsilon}+1$ and $e^{-\beta\varepsilon}+1$ (where $\beta = \frac{1}{T}$, as usual) possess simple zeroes at $\varepsilon = i\varepsilon_n$, leading to the poles of the integrand in (3.38) at the discrete set of points on the imaginary axis. Similar poles appear if we use $th\frac{\varepsilon}{2T}$ in the last term of (3.38).

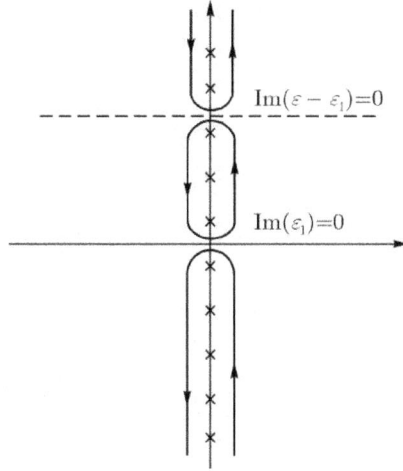

Fig. 3.3   Integration contour used for summation over Matsubara frequencies.

To perform summation over even (Boson) frequencies $i\omega_m = i2\pi T m$ we can use a similar identity:

$$T \sum_{m=-\infty}^{\infty} F(i\omega_m) = \frac{1}{2\pi i} \int_C d\omega \frac{F(\omega)}{e^{\beta\omega} - 1} = -\frac{1}{2\pi i} \int_C d\omega \frac{F(\omega)}{e^{-\beta\omega} - 1}$$

$$= \frac{1}{4\pi i} \int_C d\omega F(\omega) cth \frac{\omega}{2T} \tag{3.39}$$

where the poles of the integrand are at points $i\omega_m = i2\pi T m$.

As the next stage we can, usually, "stretch" integration contour C to infinity. During this operation we have only to calculate contributions from singularities of $F(\varepsilon)$ or $F(\omega)$, which are encircled by the "stretched" contour C. In most cases, the remaining integral over the circle of infinite radius is just zero, due to the fast decrease of $F(\varepsilon)$ and $F(\omega)$ at infinity.

Let us illustrate this method by explicit calculation of (3.37). We have to calculate the following sum over frequencies:

$$S = T \sum_{\varepsilon_1} G(\varepsilon_1 \mathbf{p}_1) D(\varepsilon - \varepsilon_1, \mathbf{p} - \mathbf{p}_1) \tag{3.40}$$

where summation is over Fermion frequencies $i\varepsilon_n = i(2n+1)\pi T$. Thus, we have to use (3.38). Consider, for definiteness, the case of $\varepsilon > 0$, i.e.

belonging to the upper half-plane.[2] Consider the function:

$$f(z) = G(z, \mathbf{p}_1)D(\varepsilon - z, \mathbf{p} - \mathbf{p}_1)th\frac{z}{2T} \qquad (3.41)$$

which has poles at $z = i\varepsilon_n = i(2n+1)\pi T$ and calculate the integral:

$$I = \int_C dz f(z) \qquad (3.42)$$

over the contour C, shown in Fig. 3.4, which encircles the straight lines where $Im(\varepsilon - \varepsilon_1) = 0$ and $Im\varepsilon_1 = 0$, corresponding to the *cuts* of exact Green's functions in (3.40).[3] In the rest of the complex plane of frequency, except these cuts, the function $f(z)$ is analytic. Now the integral in (3.42) can be calculated directly. The residue of $f(z)$ at the pole at $z_n = i(2n+1)\pi T$ is equal to:

$$Res_{z=z_n} f(z) = 2TG(z_n, \mathbf{p}_1)D(\varepsilon - z_n, \mathbf{p} - \mathbf{p}_1) \qquad (3.43)$$

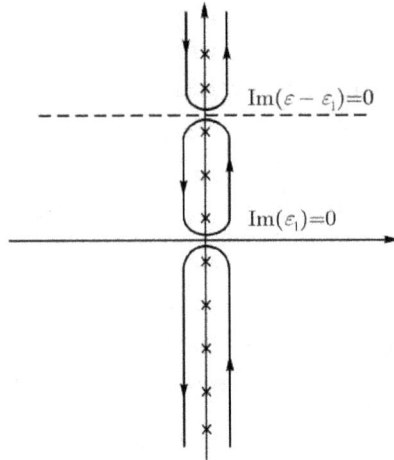

Fig. 3.4   Integration contour used to sum over Matsubara frequencies in electron self-energy.

so that integral in (3.42) reduces to $I = 4\pi i S$ giving us the required sum (3.40). On the other hand, we can consider the "stretched" integration

---

[2] Remember that finally, in most cases, we want to make an analytic continuation $i\varepsilon_n \to \varepsilon + i\delta$!

[3] Analytic continuation from upper and lower half-planes gives *different* Green's functions G^R and G^A [Abrikosov A.A., Gorkov L.P., Dzyaloshinskii I.E. (1963)].

contour, shown in Fig. 3.5. Now our integral reduces to the integrals over the straight lines, shown in Fig. 3.5, and going along the cuts, so that the contribution from the different "sides" of each cut is determined by appropriate discontinuities:

$$I = \int_{-\infty}^{\infty} d\varepsilon_1 \left\{ (G^R(\varepsilon_1 \mathbf{p}_1) - G^A(\varepsilon_1 \mathbf{p}_1)) D^A(\varepsilon - \varepsilon_1, \mathbf{p} - \mathbf{p}_1) th\frac{\varepsilon_1}{2T} \right.$$

$$\left. - G^R(-\varepsilon_1 + \varepsilon, \mathbf{p}_1)(D^R(\varepsilon_1, \mathbf{p} - \mathbf{p}_1) - D^A(\varepsilon_1, \mathbf{p} - \mathbf{p}_1)) th\frac{\varepsilon - \varepsilon_1}{2T} \right\}$$

$$(3.44)$$

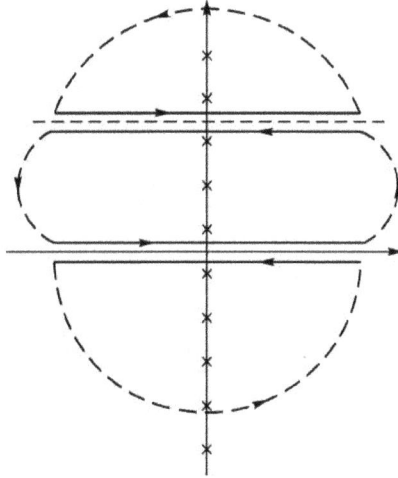

Fig. 3.5 "Stretched" integration contour used for summation over Matsubara frequencies in electron self-energy.

Taking into account $\varepsilon = i(2n+1)\pi T$, we may write $th\frac{\varepsilon - \varepsilon_1}{2T} = -cth\frac{\varepsilon_1}{2T}$. Also, we can use:

$$G^R(\varepsilon \mathbf{p}) - G^A(\varepsilon \mathbf{p}) = 2i ImG^R(\varepsilon \mathbf{p}) \qquad (3.45)$$

and dispersion relation [Abrikosov A.A., Gorkov L.P., Dzyaloshinskii I.E. (1963)]:

$$G^{R(A)}(\varepsilon \mathbf{p}) = \frac{1}{\pi} \int_{-\infty}^{\infty} \frac{ImG^{R(A)}(\omega \mathbf{p})}{\omega - \varepsilon \mp i\delta} d\omega \qquad (3.46)$$

Substituting these relations into (3.44), we obtain:

$$\Sigma(\varepsilon\mathbf{p}) = \frac{g^2}{(2\pi)^4\pi} \int d\varepsilon_1 d\omega d^3 p_1 \left\{ \frac{ImG^R(\varepsilon_1\mathbf{p}_1)ImD^R(\omega\mathbf{p} - \mathbf{p}_1)}{\omega - \varepsilon + \varepsilon_1 - i\delta} th\frac{\varepsilon_1}{2T} \right.$$

$$\left. + \frac{ImG^R(\omega\mathbf{p}_1)ImD^R(\varepsilon_1\mathbf{p} - \mathbf{p}_1)}{\omega - \varepsilon + \varepsilon_1 - i\delta} cth\frac{\varepsilon_1}{2T} \right\} \tag{3.47}$$

Exchanging integration variables ε_1 and ω in the second term, we finally obtain:

$$\Sigma(\varepsilon\mathbf{p}) = \frac{g^2}{(2\pi)^4\pi} \int d\varepsilon_1 d\omega d^3 p_1 \frac{ImG^R(\varepsilon_1\mathbf{p}_1)ImD^R(\omega\mathbf{p} - \mathbf{p}_1)}{\omega - \varepsilon + \varepsilon_1 - i\delta} \left(th\frac{\varepsilon_1}{2T} + cth\frac{\omega}{2T} \right) \tag{3.48}$$

Here we have only integration over real ε_1 and ω. After rather awkward calculations which we drop, it can be shown [Abrikosov A.A., Gorkov L.P., Dzyaloshinskii I.E. (1963)], that for $\varepsilon \ll T \ll \omega_D$ (after analytic continuation $i\varepsilon_n \to \varepsilon + i\delta$) Eq. (3.48) gives:

$$Im\Sigma^R(\varepsilon) \sim \zeta \frac{T^3}{c^2 p_F^2} \tag{3.49}$$

so that, in fact, electron damping due to electron–phonon interaction for $\varepsilon \ll \omega_D$ and $T \ll \omega_D$ can be written in unified form as (remember (3.25)):

$$Im\Sigma^R(\varepsilon) \sim \zeta \frac{Max[T^3, \varepsilon^3]}{c^2 p_F^2} \tag{3.50}$$

For $\varepsilon \gg \omega_D$ it follows from (3.48) that:

$$Im\Sigma^R(\varepsilon) = const \sim \omega_D \tag{3.51}$$

From these expressions it is clear that the damping of quasiparticles (electrons) becomes comparable to their energy for $\varepsilon \sim \omega_D$. At the same time it is clear that with the further growth of energy, damping again becomes smaller than the quasiparticle energy. Thus, we have two regions, where the notion of quasiparticles is meaningful: $|\varepsilon| \ll \omega_D$ and $|\varepsilon| \gg \omega_D$. In both regions the energy of electrons can be written as $v_F(p - p_F)$, but velocities v_F (effective masses) are different.

3.3 Migdal theorem

Up to now we limited ourselves to the simplest contribution to electron self-energy, shown in Fig. 3.2. It may seem that we have to add also numerous diagrams with higher-order vertex corrections. But in fact we do not need these (!), as in the case of electron–phonon interaction all these corrections are small over the adiabaticity parameter $\frac{\omega_D}{E_F} \sim \sqrt{\frac{m}{M}} \ll 1$. This statement is usually referred to as Migdal theorem (A.B. Migdal, 1957). Let us show the validity of this claim, making a simple estimate of the vertex correction, shown by the diagram of Fig. 3.6. Let us write down an analytic expression, corresponding to this diagram:

$$\Gamma^{(1)} = -g^3 \int G(\mathbf{p}_1 \varepsilon_1) G(\mathbf{p}_1 + \mathbf{k}, \varepsilon_1 + \omega) D(\varepsilon - \varepsilon_1, \mathbf{p} - \mathbf{p}_1) \frac{d^3 p_1 d\varepsilon_1}{(2\pi)^4} \tag{3.52}$$

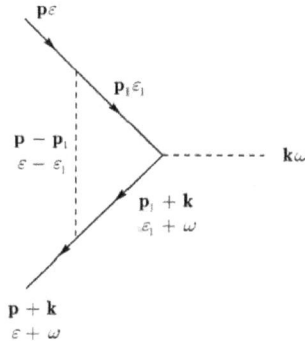

Fig. 3.6 Simplest vertex correction due to electron–phonon interaction.

Now make a crude estimate of this expression. Consider first the integral over ε_1. Assuming that the characteristic momentum transfer due to phonon exchange is of the order of $k_D \sim p_F$, and taking into account that $D(\varepsilon - \varepsilon_1)$ decreases quadratically for $|\varepsilon - \varepsilon_1| \gg \omega_D$, we understand that the main contribution to the integral comes from the region of $|\varepsilon - \varepsilon_1| \sim \omega_D$. Then the integral over ε_1 is of order of ω_D, and we can write:

$$\Gamma^{(1)} \sim g^3 \omega_D \int \frac{d^3 p_1}{(\varepsilon_1 - \xi(\mathbf{p}_1) + i\delta sign\xi(\mathbf{p}_1))(\varepsilon_1 + \omega - \xi(\mathbf{p}_1 + \mathbf{k}) + i\delta sign\xi(\mathbf{p}_1 + \mathbf{k}))} \tag{3.53}$$

Consider now the remaining integral over p_1. Characteristic momentum transfer here is also of the order $k_D \sim p_F$. Thus we may estimate all denominators to be of the order of $\sim E_F$, and $\int d^3 p_1 \sim p_F^3$. Then we have:

$$\Gamma^{(1)} \sim g^3 \omega_D \frac{p_F^2}{v_F} \frac{E_F}{E_F^2} \sim g^3 \frac{p_F^2}{v_F E_F} \omega_D \tag{3.54}$$

and the relative vertex correction is:

$$\frac{\Gamma^{(1)}}{g} \sim g^2 \frac{p_F^2}{v_F E_F} \omega_D \sim \zeta \frac{\omega_D}{E_F} \sim \zeta \sqrt{\frac{m}{M}} \tag{3.55}$$

where we have used $\frac{\omega_D}{E_F} \sim \sqrt{\frac{m}{M}}$. Electrons are much lighter than ions, so this correction is practically negligible! Of course, our analysis is too crude, e.g. it is invalid if $\omega \sim v_F k$ and $\omega \ll \omega_D$, when the poles of Green's functions in (3.53) are close to each other and more refined considerations are necessary. However, in most cases, the contribution from this region is also small due to $c \ll v_F$.

For better understanding of the situation, it is instructive to make estimates of the vertex correction in the "mixed" momentum–time representation. This will allow us to show the importance of different time-scales. First, let us introduce the appropriate free phonon and electron Green's functions:

$$D(\mathbf{k}t) = \int \frac{d\omega}{2\pi} D(\omega \mathbf{k}) e^{-i\omega t} = -\frac{ick}{2} e^{-ick|t|} \tag{3.56}$$

$$G(\mathbf{p}t) = -i e^{-i\xi(p)t} \begin{cases} \theta(\xi(p)) & \text{for} \quad t > 0 \\ -\theta(-\xi(p)) & \text{for} \quad t < 0 \end{cases} \tag{3.57}$$

Note that $D(\mathbf{k}t)$ is much more slowly changing function of t, than $G(\mathbf{p}t)$.

Now write the vertex correction shown in Fig. 3.7 in analytic form:

$$\Gamma^{(1)} = -g^3 \int \frac{d^3 p_1}{(2\pi)^3} \int dt G(\mathbf{p}_1, t - t_1) G(\mathbf{p}_1 + \mathbf{k}, t_2 - t) D(\mathbf{p} - \mathbf{p}_1, t_1 - t_2) \tag{3.58}$$

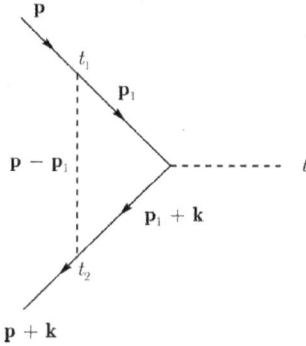

Fig. 3.7 Simplest correction to electron–phonon vertex in momentum–time representation.

For $p_1 \sim p_F$ characteristic time scale for the change of electron Green's function is $\sim E_F^{-1}$. Thus in (3.58) we may put $|t_1 - t| \sim |t_2 - t| \sim |t_1 - t_2| \sim E_F^{-1}$. On such a time scale, phonon Green's function practically does not change at all and we may estimate its value putting $t_1 \approx t_2$, so that it is simply proportional to $c|\mathbf{p} - \mathbf{p}_1| \sim \omega_D$. These estimates immediately lead to the appearance of the small (adiabaticity) parameter $\frac{\omega_D}{E_F}$. In other words, electron quickly (during the time of the order of $\sim E_F^{-1}$) absorbs phonon, and "following" the phonon induced lattice deformation. During this short time interval, electrons just are not able to induce any strong changes in the local configuration of ions — this requires the time of the order of ω_D^{-1}. Electrons are moving adiabatically in slowly changing field of heavy ions.

Migdal theorem is very important, as it allows us to neglect numerous diagrams, without assumption of smallness of electron–phonon coupling.

3.4 Eliashberg–McMillan approximation

Migdal's theorem allows neglecting vertex corrections in all calculations related to electron–phonon interaction in typical metals. The real small parameter of perturbation theory is $\lambda \frac{\Omega_0}{E_F} \ll 1$, where λ — is dimensionless coupling constant of electron–phonon interaction, $\Omega_0 \sim \omega_D$ — is characteristic phonon frequency, and E_F — is Fermi energy of electrons. In particular, this leads to a common view, that vertex corrections can be neglected in this theory even for the case of $\lambda > 1$, due to the validity of inequality $\frac{\Omega_0}{E_F} \ll 1$ in typical metals. In fact, this means that diagram of Fig. 3.1 is sufficient to describe even the case of rather strong electron–phonon coupling. This conclusion is of major importance for the derivation of Eliashberg–McMillan theory of strong coupling superconductors, which will be discussed below in Chapter 5. Here we consider the main expressions used in this theory for the case of normal metal, where all analysis is much simpler.

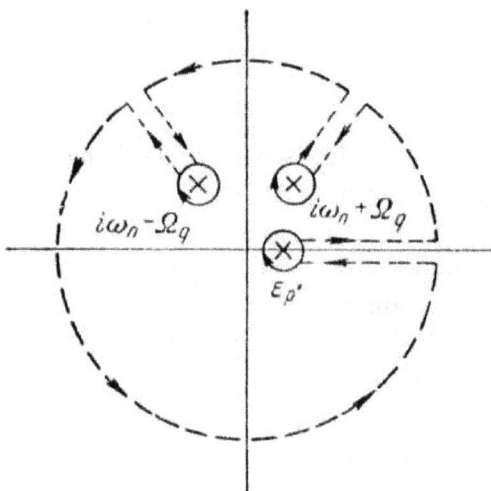

Fig. 3.8 "Stretched" integration contour used to calculate sums over Matsubara frequencies in (3.60).

Consider once again the second order (over electron–phonon interaction) diagram shown in Fig. 3.1. Calculations will be performed in Matsubara formalism ($T \neq 0$), using notations (3.9)–(3.15) [Schrieffer J.R. (1964)]. Analytic expression, corresponding to this diagram, is written as (cf. (3.37)):

$$\Sigma(i\omega_n, \mathbf{p}) = -T \sum_{n=-\infty}^{\infty} \sum_{\mathbf{p}'} |\bar{g}_{\mathbf{p}\mathbf{p}'}|^2 G_0(i\omega_{n'}, \mathbf{p}') D_0(i\omega_n - i\omega_{n'}, \mathbf{p} - \mathbf{p}')$$

$$(3.59)$$

Here $\bar{g}_{\mathbf{p},\mathbf{p'}}$ is Fröhlih coupling constant of electron–phonon interaction (3.15) with explicit, in general, dependence on momenta, $\varepsilon_{\mathbf{p}}$ is electronic spectrum reckoned from the Fermi level, while phonon spectrum in Green's function (3.10) in the following is denoted as $\Omega_{\mathbf{q}}$ ($\mathbf{q} = \mathbf{p} - \mathbf{p'}$). Index 0 of Green's functions here stresses, that we consider here Green's functions of free particles.

Summation over Matsubara frequencies can again be performed with the help of (3.38):

$$-T \sum_{n=-\infty}^{\infty} G_0(i\omega_{n'}, \mathbf{p'})D_0(i\omega_n - i\omega_{n'}, \mathbf{p} - \mathbf{p'})$$

$$= \frac{1}{2\pi i} \int_C G_0(i\omega, \mathbf{p'})D_0(i\omega_n - i\omega, \mathbf{p} - \mathbf{p'})f(\omega)d\omega \qquad (3.60)$$

where $f(\omega) = \frac{1}{e^{\frac{\omega}{T}}+1}$ is Fermi function, while the "stretched" integration contour C here should be taken as shown in Fig. 3.8, encircling the poles of the integrand at $\omega = \varepsilon_p$ and $\omega = i\omega_n \pm \Omega_q$, which contribution determines the value of the integral. Then, applying Cauchy theorem we immediately obtain the result of the summation in (3.60) in the following form:

$$\frac{-2\Omega_q f(\varepsilon_{p'})}{(\varepsilon_{p'} - i\omega_n)^2 - \Omega_q^2} - \frac{f(i\omega_n + \Omega_q)}{i\omega_n + \Omega_q - \varepsilon_{p'}} + \frac{f(i\omega_n - \Omega_q)}{i\omega_n - \Omega_q - \varepsilon_{p'}} \qquad (3.61)$$

Taking into account the equality:

$$f(i\omega_n \pm \Omega_q) = \frac{1}{e^{i\frac{\omega_n}{T}}e^{\pm\frac{\Omega_q}{T}}+1} = \frac{1}{1 - e^{\pm\frac{\Omega_q}{T}}} \qquad (3.62)$$

and substituting everything into (3.59) we obtain:

$$\Sigma(i\omega_n, \mathbf{p}) = \sum_{\mathbf{p'}} |\bar{g}_{\mathbf{pp'}}|^2 \left\{ \frac{f_{\mathbf{p'}} + n_{\mathbf{q}}}{z - \varepsilon_{p'} + \Omega_{\mathbf{q}}} + \frac{1 - f_{\mathbf{p'}} + n_{\mathbf{q}}}{i\omega_n - \varepsilon_{p'} - \Omega_{\mathbf{q}}} \right\} \qquad (3.63)$$

where $n_q = \frac{1}{e^{\frac{\Omega_q}{T}}+1}$ Bose (Planck) distribution for phonons. For temperatures $T \to 0$, when Fermi-distribution becomes a step-function, while Planckian function for phonons becomes zero, the first term in figure brackets is finite only for $\varepsilon_{p'} < 0$ and the second one — for $\varepsilon_{p'} > 0$. Correspondingly, in the limit of $T = 0$, after the substitution

$i\omega_n \to \varepsilon + i\delta sign\varepsilon_{p'}$, the contribution of the diagram of Fig. 3.1 can be written in the standard form of zero-temperature technique:

$$\Sigma(\varepsilon, \mathbf{p}) = \sum_{\mathbf{p'}} |\bar{g}_{\mathbf{pp'}}|^2 \left\{ \frac{f_{\mathbf{p'}}}{\varepsilon - \varepsilon_{\mathbf{p'}} + \Omega_{\mathbf{p-p'}} - i\delta} + \frac{1 - f_{\mathbf{p'}}}{\varepsilon - \varepsilon_{\mathbf{p'}} - \Omega_{\mathbf{p-p'}} + i\delta} \right\}$$

(3.64)

In particular, for the imaginary part of self-energy at positive frequencies we obtain:

$$Im\Sigma(\varepsilon > 0, \mathbf{p}) = -\pi \sum_{\mathbf{p'}} |\bar{g}_{\mathbf{pp'}}|^2 (1 - f_{\mathbf{p'}}) \delta(\varepsilon - \varepsilon_{\mathbf{p'}} - \Omega_{\mathbf{p-p'}}) \quad (3.65)$$

Eq. (3.64) can be identically rewritten as:

$$\Sigma(\varepsilon, \mathbf{p}) = \int d\omega \sum_{\mathbf{p'}} |\bar{g}_{\mathbf{pp'}}|^2 \delta(\omega - \Omega_{\mathbf{p-p'}})$$

$$\times \left\{ \frac{f_{\mathbf{p'}}}{\varepsilon - \varepsilon_{\mathbf{p'}} + \omega - i\delta} + \frac{1 - f_{\mathbf{p'}}}{\varepsilon - \varepsilon_{\mathbf{p'}} - \omega + i\delta} \right\} \quad (3.66)$$

Electron scattering by phonons actually takes place in a narrow energy layer close to the Fermi level of the width of the order of double Debye frequency $2\Omega_D$, and in typical metals we always have $\Omega_D \ll E_F$. In this situation we can with high accuracy assume, that both initial and final momenta $\mathbf{p}$ and $\mathbf{p'}$ of the scattered electron are at the Fermi surface. The basic idea of Eliashberg–McMillan approach is that we can get rid of explicit dependence on momenta by averaging the matrix element of electron–phonon interaction over surfaces of constant energy, corresponding to initial and final $\mathbf{p}$ and $\mathbf{p'}$, which in practice reduces to averaging over corresponding Fermi surfaces, determined by the equations $\varepsilon(\mathbf{p}) = 0$ and $\varepsilon(\mathbf{p'}) = 0$. This is achieved by the following substitution ($N(0)$ is the density of states at the Fermi level):

$$|\bar{g}_{\mathbf{pp'}}|^2 \delta(\omega - \Omega_{\mathbf{p-p'}}) \Longrightarrow$$

$$\frac{1}{N(0)} \sum_{\mathbf{p}} \frac{1}{N(0)} \sum_{\mathbf{p'}} |\bar{g}_{\mathbf{pp'}}|^2 \delta(\omega - \Omega_{\mathbf{p-p'}}) \delta(\varepsilon_{\mathbf{p}}) \delta(\varepsilon_{\mathbf{p'}})$$

$$\equiv \frac{1}{N(0)} \alpha^2(\omega) F(\omega) \quad (3.67)$$

where we have introduced the standard *definition* of Eliashberg function $\alpha^2(\omega)$, which reflects the strength of electron–phonon interaction, while

$F(\omega) = \sum_{\mathbf{q}} \delta(\omega - \Omega_{\mathbf{q}})$ is the phonon density of states. In principle, these functions can be directly determined from some experiments [White R.M., Geballe T.H. (1979)].

After the substitution like (3.67) the explicit dependence of self-energy on momenta just disappears and in the following we, in fact, are dealing with self-energy averaged over the real Fermi surface: $\Sigma(\varepsilon) \equiv \frac{1}{N(0)} \sum_{\mathbf{p}} \delta(\varepsilon_{\mathbf{p}}) \Sigma(\varepsilon, \mathbf{p})$, which is now written as:

$$\Sigma(\varepsilon) = \int d\varepsilon' \int d\omega \alpha^2(\omega) F(\omega) \left\{ \frac{f(\varepsilon')}{\varepsilon - \varepsilon' + \omega - i\delta} + \frac{1 - f(\varepsilon')}{\varepsilon - \varepsilon' - \omega + i\delta} \right\}$$
(3.68)

In the case of self-energy depending only on frequency (and not on momentum) we can use the previously derived relations (2.50) and (2.56) for the residue in the pole of the Green's function and electron mass renormalization:

$$Z^{-1} = 1 - \left. \frac{\partial \Sigma(\varepsilon)}{\partial \varepsilon} \right|_{\varepsilon=0}$$
(3.69)

$$m^\star = \frac{m}{Z} = m \left(1 - \left. \frac{\partial \Sigma(\varepsilon)}{\partial \varepsilon} \right|_{\varepsilon=0} \right)$$
(3.70)

and directly obtain from Eq. (3.68) (all integrals here are in infinite limits):

$$-\left. \frac{\partial \Sigma(\varepsilon)}{\partial \varepsilon} \right|_{\varepsilon=0} = \int d\varepsilon' \int d\omega \alpha^2(\omega) F(\omega) \left\{ \frac{f(\varepsilon')}{(\omega - \varepsilon' - i\delta)^2} + \frac{1 - f(\varepsilon')}{(\omega + \varepsilon' + i\delta)^2} \right\}$$

$$= 2 \int_0^\infty \frac{d\omega}{\omega} \alpha^2(\omega) F(\omega)$$
(3.71)

Introducing now the dimensionless electron–phonon coupling constant of Eliashberg–McMillan theory as:

$$\lambda = 2 \int_0^\infty \frac{d\omega}{\omega} \alpha^2(\omega) F(\omega)$$
(3.72)

we immediately obtain the standard expression for electron mass renormalization due to interactions with phonons:

$$m^\star = m(1 + \lambda)$$
(3.73)

The function $\alpha^2(\omega)F(\omega)$ in the expression for Eliashberg coupling constant of electron–phonon interaction (3.72) should be calculated according to (3.67) or determined from experiments.

Using (3.67) we can rewrite (3.72) in the following form:

$$\lambda = \frac{2}{N(0)} \int_0^\infty \frac{d\omega}{\omega} \sum_{\mathbf{p}} \sum_{\mathbf{p'}} |\bar{g}_{\mathbf{pp'}}|^2 \times \delta(\omega - \Omega_{\mathbf{p-p'}}) \delta(\varepsilon_{\mathbf{p}}) \delta(\varepsilon_{\mathbf{p'}}) \quad (3.74)$$

which gives the general recipe to calculate electron–phonon coupling constant λ, which determines the Cooper pairing in Eliashberg–McMillan theory (also for real metals).

Obviously, we can act similarly to (3.66), (3.67) in expression (3.64), which is valid for finite temperatures, so that instead of (3.68) we get:

$$\Sigma(i\omega_n) = \int d\varepsilon' \int_0^\infty d\omega \alpha^2(\omega) F(\omega) \left\{ \frac{f(\varepsilon') + n(\omega)}{i\omega_n - \varepsilon' + \omega} + \frac{1 - f(\varepsilon') + n(\omega)}{i\omega_n - \varepsilon' - \omega} \right\}$$
$$(3.75)$$

The above expressions in fact determine the structure of Eliashberg–McMillan theory for superconductors with strong electron–phonon coupling (cf. below in Chapter 5).

3.5 Self-energy and spectrum of phonons

Return now to the analysis of Dyson equations for the phonon Green's function (3.16), (3.17) and (3.18), which determine the phonon spectrum renormalization due to electron–phonon interaction in metals. Using the simplest approximation for the polarization operator of electron gas, we can write:

$$g^2 \Pi_0(\omega \mathbf{k}) = -\frac{2ig^2}{(2\pi)^4} \int \frac{dE d^3p}{(E - \xi(\mathbf{p}) + i\delta sign\xi(\mathbf{p}))(E + \omega - \xi(\mathbf{p+k}) + i\delta Sign\xi(\mathbf{p+k}))}$$
$$(3.76)$$

Above we have already calculated this polarization operator, obtaining Eq. (2.25), so that we have:

$$g^2 \Pi_0(\omega \mathbf{k}) = -\frac{g^2 m p_F}{\pi^2} \left\{ 1 - \frac{\omega}{2v_F k} \ln \left| \frac{\omega + v_F k}{\omega - v_F k} \right| + \frac{i\pi |\omega|}{2 v_F k} \theta \left(1 - \frac{|\omega|}{v_F k} \right) \right\}$$
$$(3.77)$$

According to Eq. (3.17), the phonon Green's function in the system with electron–phonon interaction is determined by Dyson equation of the form:

$$D^{-1}(\omega\mathbf{k}) = D_0^{-1}(\omega\mathbf{k}) - g^2\Pi(\omega\mathbf{k}) \qquad (3.78)$$

Then, the phonon spectrum is determined by the equation $D^{-1}(\omega\mathbf{k}) = 0$. As sound velocity is much smaller than Fermi velocity of electrons, we may safely assume that $\omega \ll v_F k$. Then, polarization operator, determining the phonon self-energy, can be taken in static approximation ($\omega = 0$) and we can write:

$$g^2\Pi_0 \approx -\frac{g^2 m p_F}{\pi^2} = -2\zeta \qquad (3.79)$$

Then (3.78) reduces to:

$$D^{-1}(\omega\mathbf{k}) = D_0^{-1}(\omega\mathbf{k}) - g^2\Pi = \frac{\omega^2 - c_0^2 k^2}{c_0^2 k^2} + 2\zeta \qquad (3.80)$$

where c_0 is "bare" sound velocity, while the renormalized phonon spectrum is written as $\omega = ck$, where the sound velocity is defined as:

$$c^2 = c_0^2(1 - 2\zeta) \qquad (3.81)$$

We see that electron–phonon interaction leads to the "softening" of the lattice (decrease of phonon frequency).

It may seem that Eq. (3.81) leads to the instability of the lattice ($\omega^2 < 0$!) for $\zeta > 1/2$. However, this instability is, in fact, unphysical. More elaborated analysis [Ginzburg V.L., Kirzhnits D.A. (1982)] shows that we have to introduce physical (renormalized) electron–phonon coupling constant λ, which can be expressed via ζ by the following relation:

$$\lambda = \zeta \frac{\omega_0^2}{\omega^2} = \frac{\zeta}{1 - 2\zeta} \qquad (3.82)$$

Then it is clear that $\lambda \approx \zeta$ only for $\zeta \ll 1$, while with the growth of ζ the coupling constant λ just grows continuously, diverging only at "instability" point itself. Thus, the condition of $\zeta < 1/2$, in fact, does not lead to any limitation of the value of λ. Note that the condition of $\zeta < 1/2$, in some sense, is also unphysical, as in the framework of the standard Fröhlih model of electron–phonon interaction we have no rigorous way to define the "bare" coupling constant ζ, and experimentally "observable" is only renormalized coupling λ. Using (3.82) we can write inverse relation:

$$\zeta = \frac{\lambda}{1 + 2\lambda} \qquad (3.83)$$

so that for any $\lambda > 0$ we, in fact, have $\zeta < 1/2$.

Physical meaning of "bare" parameters of the Fröhlih model, such as frequency $\omega_0(k) = c_0 k$, is not clear at all, while the "real" spectrum of phonons $\omega(\mathbf{k})$ is determined by Dyson equation with the account of electron–phonon coupling and Green's function:

$$D(\omega \mathbf{k}) = \frac{\omega_0^2(k)}{\omega^2 - \omega^2(k) + i\delta} \qquad (3.84)$$

Then the physical constant of electron–phonon interaction can be defined [Ginzburg V.L., Kirzhnits D.A. (1982)] by the following integral expression:

$$\lambda = \zeta \int_0^{2p_F} \frac{dk k}{2p_F^2} \frac{\omega_0^2(k)}{\omega^2(k)} \qquad (3.85)$$

If we neglect the relatively weak dependence of Π_0 on k and use (3.79), Eq. (3.85) immediately gives (3.82). It is believed that this coupling constant λ enters e.g. into the famous expression for transition temperature in BCS theory of superconductivity [Ginzburg V.L., Kirzhnits D.A. (1982)].

To find phonon damping we have to take into account the imaginary part of polarization operator (3.77):

$$g^2 Im\Pi_0(\omega \mathbf{k}) = -\pi \zeta \frac{|\omega|}{k v_F} \qquad (3.86)$$

Substituting this into Dyson equation for the phonon Green's function and seeking the solution for the spectrum as $\omega = ck + i\gamma$, we find:

$$\gamma = \frac{\pi}{2} \zeta \frac{c^2}{v_F} k = \frac{\pi}{2} \zeta \frac{c}{v_F} \omega \qquad (3.87)$$

Though damping is proportional to frequency, it is in fact small in comparison with $Re\omega$ due to smallness of $c/v_F \sim \sqrt{m/M}$.

In usual liquids and gases sound damping is of the order of:

$$\gamma \sim \frac{\eta \omega^2}{\rho c^3} \qquad (3.88)$$

where η is the viscosity of the medium and ρ its density. Thus we may say that in electron–phonon system the effective viscosity of electron gas grows with the decrease of frequency: $\eta(\omega) \sim \omega^{-1}$. Physically the effective viscosity here is due to high density of electron–hole excitations with energy $\omega < ck$, which are excited by phonons.

In the previous chapter we have noted, that at $q = 2p_F$ polarization operator $\Pi_0(q0)$ has the logarithmic singularity in its derivative $\frac{\partial \Pi_0(q0)}{\partial q}\big|_{q=2p_F}$. This singularity becomes more strong in two-dimensional system $(d = 2)$ and, especially strong, for one-dimensional case $(d = 1)$,

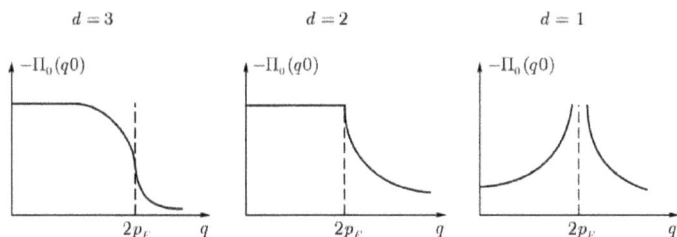

Fig. 3.9 Qualitative behavior of static polarization operator (as a function of q) for the free electron gas in different space dimensionalities.

when we have logarithmic singularity in polarization potential $\Pi_0(q0)$ itself (A.M. Afanas'ev, Yu.M. Kagan, 1962):

$$\Pi_0(q0) \sim \ln|q - 2p_F| \qquad (3.89)$$

Qualitative behavior of $\Pi_0(q0)$ fro different dimensionalities is shown in Fig. 3.9. The presence of these singularities leads to important anomalies of physical properties. The essence of the previous discussion was that the phonon Green's function, with the account of electron–phonon interaction, is given by:

$$D(\omega\mathbf{q}) = \frac{1}{D_0^{-1}(\omega\mathbf{q}) - g^2\Pi_0(\omega\mathbf{q})} = \frac{\omega_0^2(q)}{\omega^2 - \omega_0^2(q) - g^2\omega_0^2(q)\Pi_0(\omega q)} \qquad (3.90)$$

so that the phonon spectrum is:

$$\omega^2(q) = \omega_0^2(q)[1 + g^2\Pi_0(\omega q)] \qquad (3.91)$$

Then it is clear that due to $\Pi_0(q0) \to -\infty$ at $q = 2p_F$ (for $d = 1$) the frequency of a phonon with $q = 2p_F$ becomes *imaginary* ($\omega^2 < 0$), for any (even infinitesimally small) value of the coupling constant g. This signifies an *instability* of the system, leading to the appearance of spontaneous static deformation of the lattice (superstructure) with the wave vector $Q = 2p_F$ (i.e. with period $L = \frac{2\pi}{Q}$). This is so-called *Peierls* instability, which will be discussed in detail in the last chapter.[4] Even for $d = 3$, when we have singularity only in the derivative

---

[4]For $d = 1$ such instability leads to the appearance of the gap in an energy spectrum of electrons at $\pm p_F$, i.e. to the metal-insulator transition.

of polarization operator, there appears an anomaly in the phonon spectrum at $q = 2p_F$ (W. Kohn, 1959) (so-called Kohn anomaly, for $d = 1$ it is sometimes called the "giant" Kohn anomaly). These anomalies are directly observed in phonon spectra of metals in experiments with inelastic neutron scattering.

Up to now we have dealt only with isotropic electron spectrum of the type of $\varepsilon(p) = \frac{p^2}{2m^*}$. In real materials this spectrum may be anisotropic, and Fermi surfaces are not spheres ($d = 3$) or circles ($d = 2$). In general, the topology of the Fermi surface can be rather complicated. Especially interesting is the case, when *flat* parts (sometimes called "patches") appear on the Fermi surface. For example, for $d = 2$ and simple square lattice, the tight binding electron spectrum (with the account of only nearest neighbor transfers) takes the form:

$$\varepsilon(\mathbf{p}) - \mu = -2t(\cos p_x a + \cos p_y a) - \mu \qquad (3.92)$$

where t is transfer integral between nearest neighbors. Curves of constant energy inside the Brillouin zone, corresponding to this spectrum for different values of chemical potential μ (electron concentration), are shown in Fig. 3.10. In particular case of $\mu = 0$ (half-filled band, one electron per lattice site) we have the Fermi surface in the form of the plane

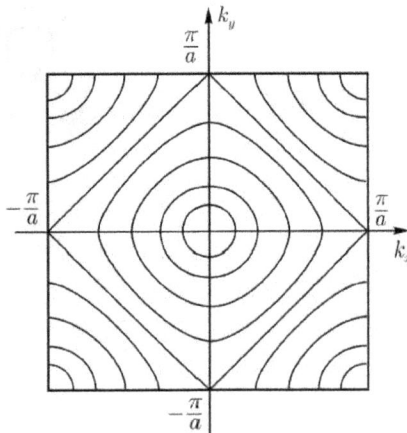

Fig. 3.10 Curves of constant energy in the Brillouin zone of the square lattice, corresponding to the simple tight-binding spectrum with only nearest neighbors transfers.

square. Direct calculation show, that in this case $\Pi_0(\mathbf{q}0)$ for $\mathbf{q} \to \left(\frac{\pi}{a}, \frac{\pi}{a}\right)$ possess a singularity of "one-dimensional" type: $\Pi_0(\mathbf{q}0) \sim \ln|\mathbf{q} - \mathbf{Q}|$, where $\mathbf{Q} = \left(\frac{\pi}{a}, \frac{\pi}{a}\right)$, which naturally leads to the "giant" Kohn anomaly of the phonon spectrum and structural transition of the Peierls type (period doubling).

In general case, a special property of the Fermi surface is needed for the appearance of such "giant" anomalies, which is called "nesting". Nesting property of the Fermi surface means that certain parts of the Fermi surface are congruent (completely coincide with each other) after the translation by some specific vector $\mathbf{Q}$ in momentum space (vector of nesting). For the square Fermi surface of the tight-binding spectrum at half-filling $\mathbf{Q} = \left(\frac{\pi}{a}, \frac{\pi}{a}\right)$, but other, more general, situations are also possible. Mathematically it is expressed by the following property of electronic spectrum:

$$\varepsilon(\mathbf{p} + \mathbf{Q}) - \mu = -\varepsilon(\mathbf{p}) + \mu \qquad (3.93)$$

which is usually called the nesting condition. We can see that the spectrum given by Eq. (3.92) satisfies this condition for $\mu = 0$ (half-filled band) and $\mathbf{Q} = \left(\frac{\pi}{a}, \frac{\pi}{a}\right)$. Similarly, this condition is satisfied for the tight-binding spectrum for the simple cubic lattice ($d = 3$), analogous to (3.92):

$$\varepsilon(\mathbf{p}) - \mu = -2t(\cos p_x a + \cos p_y a + \cos p_z a) - \mu \qquad (3.94)$$

for $\mu = 0$ and $\mathbf{Q} = \left(\frac{\pi}{a}, \frac{\pi}{a}, \frac{\pi}{a}\right)$. Fermi surface in this case possess nesting property, though there are no "flat" parts.

In all cases with nesting, calculation of polarization operator shows the divergence at $\mathbf{q} = \mathbf{Q}$, leading to the appearance of the giant Kohn anomaly in phonon spectrum and lattice instability (structural phase transition, leading to static superstructure with wave vector $\mathbf{Q}$).

3.6 Plasma model

Let us consider now the simplest "plasma" model of a metal, where both phonons and electron phonon interactions appear self-consistently [Schrieffer J.R. (1964); Ginzburg V.L., Kirzhnits D.A. (1982)]. Start with plasma consisting of electrons and ions, interacting via (non-screened) Coulomb forces. In first approximation, collective oscillations in this system are just independent plasma oscillations of electrons and ions. We shall show how the account of screening allows to introduce the "usual" phonons and obtain the coherent description of electron–phonon interaction.

Let us write the Hamiltonian of electron–ion plasma as:

$$H = \sum_{\mathbf{k}} E_{\mathbf{k}} a_{\mathbf{k}}^{+} a_{\mathbf{k}} + \sum_{\mathbf{q}\lambda} \Omega_{\mathbf{q}\lambda} \left(b_{\mathbf{q}\lambda}^{+} b_{\mathbf{q}\lambda} + \frac{1}{2} \right)$$

$$+ \sum_{\mathbf{k}\mathbf{k}'\lambda} g_{\mathbf{k}\mathbf{k}'\lambda} a_{\mathbf{k}}^{+} a_{\mathbf{k}'} \left(b_{\mathbf{k}-\mathbf{k}'\lambda} + b_{\mathbf{k}'-\mathbf{k}\lambda}^{+} \right)$$

$$+ \frac{1}{2} \sum_{\mathbf{pkq}} V_{\mathbf{q}} a_{\mathbf{p}+\mathbf{q}}^{+} a_{\mathbf{k}-\mathbf{q}}^{+} a_{\mathbf{k}} a_{\mathbf{p}} \tag{3.95}$$

where $V_{\mathbf{q}} = \frac{4\pi e^2}{q^2}$ and $E_{\mathbf{k}}$ is the energy of (Bloch) electron, define by the solution of Schroedinger equation:

$$\left\{ \frac{k^2}{2m} + \sum_{n} V_{ei}(\mathbf{r} - \mathbf{R}_n) + U_H(\mathbf{r}) \right\} \psi_{\mathbf{k}}(\mathbf{r}) = E_{\mathbf{k}} \psi_{\mathbf{k}}(\mathbf{r}) \tag{3.96}$$

where $V_{ei}(\mathbf{r} - \mathbf{R}_n)$ is the potential of electron–ion interaction, $U_H(\mathbf{r})$ — Hartree contribution from electron–electron interaction, $\Omega_{\mathbf{q}\lambda}$ — "bare" frequencies of ion plasma oscillations.

In the simplest possible *jellium* model we assume ions to form a homogeneous structureless medium, so that:

$$\Omega_{q\lambda}^2 = \frac{4\pi n (Ze)^2}{M} \tag{3.97}$$

where n is ion density, Z-ion charge, M-ion mass. In jellium model this is the only (longitudinal) mode of ion oscillations.[5]

The "bare" electron–phonon coupling $g_{\mathbf{k}\mathbf{k}'\lambda}$ is defined as:

$$g_{\mathbf{k}\mathbf{k}'\lambda} = - \left(\frac{n}{M\Omega_{\mathbf{k}\lambda}^2} \right)^{1/2} < \mathbf{k}' |\nabla_i V_{ei}| \mathbf{k} > \mathbf{e}_{\mathbf{q}\lambda}, \qquad (\mathbf{q} = \mathbf{k} - \mathbf{k}') \tag{3.99}$$

---

[5] In crystals there exist three branches of ion oscillations, which we number as $\lambda = 1, 2, 3$. Two branches are transverse, while the "bare" longitudinal branch represents the optical plasma oscillations. There is a general sum rule:

$$\sum_{\lambda} \Omega_{q\lambda}^2 = \frac{4\pi n (Ze)^2}{M} \tag{3.98}$$

where $\mathbf{e}_{\mathbf{q}\lambda}$ is polarization vector of "bare" phonons. It can be easily seen that in the simplest jellium type model the "bare" electron–phonon coupling is $g_{\mathbf{kk'}\lambda}^2$ possess Coulomb type singularity:

$$g_{\mathbf{kk'}\lambda}^2 \sim \frac{1}{(\mathbf{k} - \mathbf{k'})^2} \tag{3.100}$$

Now we have to make renormalizations, accounting for screening and regularizing such singularities. For Coulomb interaction between electrons we can just use the RPA expression:

$$\mathcal{V}(q\omega) = \frac{4\pi e^2}{q^2 \epsilon_e(q\omega)} \tag{3.101}$$

where

$$\epsilon_e(q\omega) = 1 - \frac{4\pi e^2}{q^2} \Pi_0(q\omega) \tag{3.102}$$

is the dielectric function of free electrons, corresponding to diagrams shown in Fig. 2.4(b). In a similar way, as shown by diagrams of Fig. 3.11, we may describe the screening of electron–phonon vertex:

$$\tilde{g}(q, \lambda) = g + gV_q\Pi_0 + gV_q\Pi_0 V_q\Pi_0 + \ldots = \frac{g(q, \lambda)}{\epsilon_e(q\omega)} \tag{3.103}$$

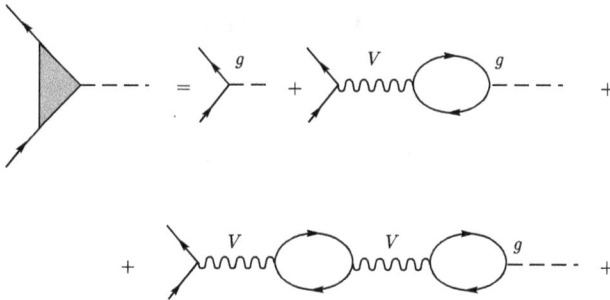

Fig. 3.11   Screening of electron–phonon vertex.

To define the "physical" phonon spectrum we can write Dyson equation, shown in Fig. 3.12:

$$D^{-1}(q\lambda, \omega) = D_0^{-1}(q\lambda, \omega) - g^2\Pi_0(q\omega) - g^2\Pi_0(q\omega)V_q\Pi_0(q\omega) - \ldots$$
$$= D_0^{-1}(q\lambda, \omega) - \frac{g^2(q, \lambda)}{V_q}\left(\frac{1}{\epsilon_e(q\omega)} - 1\right) \tag{3.104}$$

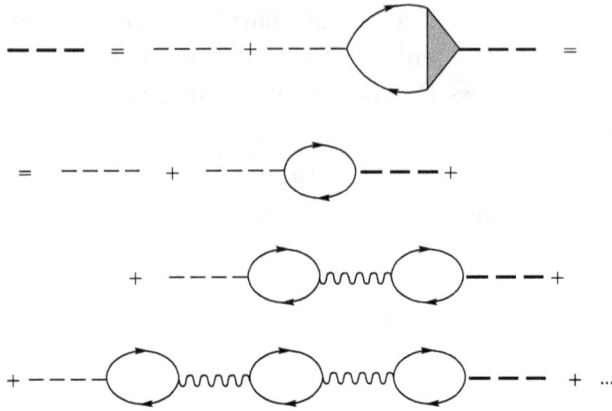

Fig. 3.12 Dyson equation for phonon Green's function in generalized jellium model.

where

$$D_0(q\lambda, \omega) = \frac{\Omega_{q\lambda}^2}{\omega^2 - \Omega_{q\lambda}^2 + i\delta} \qquad (3.105)$$

Then, from (3.102), (3.104) and (3.105) we immediately obtain:

$$D(q\lambda, \omega) = \frac{\Omega_{q\lambda}^2}{\omega^2 - \frac{g^2(q,\lambda)\Omega_{q\lambda}^2}{V_q \epsilon_e(q0)} - \Omega_{q\lambda}^2 \left[1 - \frac{g^2(q,\lambda)}{V_q}\right] + i\delta} \qquad (3.106)$$

Here we neglected frequency dependence of $\epsilon_e(q\omega)$, as this is unimportant for small ω of the order of phonon frequencies.

For a simple jellium model from (3.97) and (3.99) we can easily obtain the following identity :

$$\frac{g^2(q, \lambda)}{V_q} = 1 \qquad (3.107)$$

In this case (3.106) reduces to:

$$D(q\lambda, \omega) = \frac{\Omega_{q\lambda}^2}{\omega^2 - \frac{\Omega_{q\lambda}^2}{\epsilon_e(q0)} + i\delta} \qquad (3.108)$$

The poles of this expression define the frequencies of renormalized ("physical") phonons (D. Bohm, T. Staver, 1950):

$$\omega^2(q\lambda) = \frac{\Omega_{q\lambda}^2}{\epsilon_e(q0)} \approx \frac{\Omega_{q\lambda}^2}{1 + \frac{\kappa_D^2}{q^2}} = \frac{mZ}{3M} v_F^2 q^2 \qquad (3.109)$$

where we have used (2.29) and (2.30), with electron density equal to Zn (charge neutrality!). Now we see that renormalized phonons in jellium model acquire the acoustical dispersion with sound velocity $c = \left(\frac{mZ}{3M}\right)^{1/2} v_F$. This result can also be obtained in a more general case, when the potential $V_{ei}(q)$ differs from purely Coulomb form, as for small q the charge neutrality condition still requires $V_{ei}(q)$ to be equal to Ze^2/q^2.

In this model we can also determine the full (effective) interelectron interaction, which is necessary e.g. for calculations of superconducting properties of metals. This interaction can be described by diagrams shown in Fig. 3.13 and is given by:

$$V_{eff}(q\omega) = \frac{4\pi e^2}{q^2 \epsilon_e(q\omega)} + \frac{g^2(q,\lambda)}{\epsilon_e(q0)} \frac{\Omega_{q\lambda}^2}{\omega - \omega^2(q\lambda)} \qquad (3.110)$$

where $\omega^2(q\lambda)$ is the spectrum of renormalized phonons, following from (3.104):

$$\omega^2(q,\lambda) = \Omega_{q\lambda}^2 \left\{ 1 - \frac{g^2(q,\lambda)}{V_q} \left(1 - \frac{1}{\epsilon_e(q0)} \right) \right\} \qquad (3.111)$$

Fig. 3.13   Effective interaction between electrons in metals.

In jellium model $V_{eff}(q\omega)$ reduces to:

$$V_{eff}(q\omega) = \frac{4\pi e^2}{q^2 \epsilon_{eff}(q\omega)} \qquad (3.112)$$

where $\epsilon_{eff}(q\omega)$ is *full* dielectric function:

$$\epsilon_{eff}(q\omega) = \epsilon_e(q\omega) - \frac{\Omega_{q\lambda}^2}{\omega^2} \qquad (3.113)$$

which includes both electron and ion contributions. In more general (than jellium) case, interelectron interaction also reduces to (3.112), but with $\epsilon_{eff}(q\omega)$ given by:

$$\epsilon_{eff}(q\omega) = \epsilon_e(q\omega) - \frac{g^2(q,\lambda)}{V_q} \frac{\Omega_{q\lambda}^2}{\omega^2 - \Omega_{q\lambda}^2 \left[1 - \frac{g^2(q,\lambda)}{V_q} \right]} \qquad (3.114)$$

Stability of the lattice requires $\omega^2(q\lambda) > 0$, so that from (3.111) we get:

$$1 + \frac{g^2(q\lambda)}{V_q} \frac{1 - \epsilon_e(q0)}{\epsilon_e(q0)} > 0 \qquad (3.115)$$

These expressions allow to determine conditions, when this effective interaction may become *attractive*, in particular for $q \sim 2p_F$, which is necessary for the appearance of superconductivity [Ginzburg V.L., Kirzhnits D.A. (1982)]. Of course, in *real* metals we need something more, e.g. we have to overcome somehow limitations due to our use of RPA.

Eq. (3.111) determining the phonon spectrum can be written in more general form:

$$\omega^2(q\lambda) = \Omega_{q\lambda}^2 \left\{ 1 + g^2(q, \lambda)\chi(q, \omega(q\lambda)) \right\} \qquad (3.116)$$

where we have introduced the generalized susceptibility of electronic subsystem, expressed via appropriate dielectric function as:

$$\chi_e(q\omega) = \frac{1}{V_q} \left(\frac{1}{\epsilon_e(q\omega)} - 1 \right) \qquad (3.117)$$

Here we also take into account the ω-dependence, neglected above in the adiabatic approximation. Calculations of *non-adiabatic* corrections should be done solving Eq. (3.116). It is not difficult to convince oneself, that due to small velocities of ions (compared to Fermi velocity of electrons), the account of frequency dependence of $\chi_e(q\omega)$ will lead to small change of phonon frequencies of the order of $\sim \sqrt{\frac{m}{M}}$.

It is clear now, that the "softening" of the frequencies of real phonons as well as lattice instability can be expressed via the changes of effective inter-ion interaction, which is in turn connected with the change of static dielectric function of electrons. This situation is typical for quasi-one-dimensional conductors (and also in the case of nesting for $d = 2, 3$), when, as noted above, both polarization operator and $\epsilon_e(q\omega)$ at $T = 0$ possess logarithmic singularity and diverge at $q = 2p_F$. In this case, both phonon frequency and Fourier-component of inter-ion interaction at $q = 2p_F$ may become zero.

3.7   Phonons and fluctuations

Let us consider, following [Levitov L.S., Shitov A.V. (2003)], correlation function of atomic displacements:

$$C_T(r) = \sum_{\alpha\beta} < u_\alpha(r)u_\beta(0) > \qquad (3.118)$$

and study its asymptotic behavior at large distances $r \to \infty$. Correlation function (3.118) can be expressed via the Matsubara Green's function of phonons. We only have to take into account that the standard phonon Green's function [Abrikosov A.A., Gorkov L.P., Dzyaloshinskii I.E. (1963)] defines, in fact, the correlator of *gradients* of atomic displacements, which allows convenient introduction of electron–phonon interaction Hamiltonian. Green's function of atomic displacements can be obtained by dividing the standard phonon Green's function by $\rho \omega_k^2$ (where ρ is the density of continuous medium of ions) and changing the sign.[6] Then we obtain:

$$C_T(r) = \frac{T}{\rho} \sum_m \int \frac{d^d k}{(2\pi)^d} \frac{e^{i\mathbf{kr}}}{\omega_m^2 + \omega_k^2} \qquad (3.119)$$

where ω_k is the phonon spectrum, which we assume here to be acoustic. Summation over (even) Matsubara frequencies in (3.119) can be done using the identity:

$$\sum_{m=-\infty}^{\infty} \frac{1}{m^2 + a^2} = \frac{\pi}{a} cth\pi a \qquad (3.120)$$

Then we get:

$$C_T(r) = \frac{1}{2\rho} \int \frac{d^d k}{(2\pi)^d} \frac{1}{\omega_k} cth\frac{\omega_k}{2T} e^{i\mathbf{kr}} \qquad (3.121)$$

From these expression we can separate contributions of thermal and quantum (zero-point, $T = 0$) fluctuations (displacements) using the formula:

$$\frac{1}{2} cth\frac{\omega}{2T} = \frac{1}{2} + n_B(\omega) \qquad (3.122)$$

where $n_B(\omega) = \frac{1}{e^{\frac{\omega}{T}} - 1}$ is Bose distribution. Obviously, we have $n_B \to 0$ for $T \to 0$, so that the appropriate contribution defines thermal fluctuations.[7]

Thus, we can write two contributions to our correlator:

$$C_0(r) = C_{T=0}(r) = \frac{1}{2\rho} \int \frac{d^d k}{(2\pi)^d} \frac{e^{i\mathbf{kr}}}{\omega_k} \qquad (3.123)$$

---

[6]It is immediately seen if we compare Eqs. (7.9), (7.10) and (7.13) of [Abrikosov A.A., Gorkov L.P., Dzyaloshinskii I.E. (1963)].

[7]Note, that all these expressions can be derived without the use of Matsubara functions, just performing Gibbs averaging of operators of atomic displacements.

$$\Delta C(r,T) = \frac{1}{\rho} \int \frac{d^d k}{(2\pi)^d} \frac{n_B(\omega_k)}{\omega_k} e^{i\mathbf{kr}} \tag{3.124}$$

We are interested in the behavior of these functions for $r \to \infty$. Leading contributions to this asymptotics of $\Delta C(r,T)$ came only from very small $k \sim \frac{1}{r}$, corresponding to $\omega_k \ll T$. Thus we can approximate Bose distribution here as $n_B(\omega_k) \approx \frac{T}{\omega_k}$ and write:

$$\Delta C(r,T) \approx \frac{T}{\rho} \int \frac{d^d k}{(2\pi)^d} \frac{e^{i\mathbf{kr}}}{\omega_k^2} \tag{3.125}$$

It is clear that this expression directly follows from equipartition law of classical statistics [Sadovskii M.V. (2019a)].

Calculate now these correlators for different spatial dimensionalities. Consider first quantum correlations described by $C_0(r)$.

For $d = 3$ we have:

$$C_0^{(3)}(r) = \frac{4\pi}{(2\pi)^3 \rho} \int_0^\infty \frac{dk k^2}{ck} \frac{\sin kr}{kr} \sim \frac{1}{4\pi^2 \rho c r^2} \tag{3.126}$$

where we have to cut-off divergence at the upper limit at $k \sim \frac{1}{r}$, as for larger values of k integrand oscillations just compensate each other.

For $d = 2$:

$$C_0^{(2)}(r) = \frac{1}{4\pi\rho c} \int_0^{2\pi} \frac{d\theta}{2\pi} \int_0^\infty dk e^{ikr\cos\theta} = \frac{1}{4\pi\rho c} \int_0^\infty dk J_0(kr) = \frac{1}{4\pi\rho c r} \tag{3.127}$$

where $J_0(r)$ is appropriate Bessel function.

Finally, for $d = 1$ we obtain:

$$C_0^{(1)}(r) = \frac{1}{4\pi\rho c} \int_{-\infty}^\infty \frac{dk}{|k|} e^{ikr} = \frac{1}{2\pi\rho c} \int_0^\infty \frac{dk}{k} \cos kr = \frac{1}{2\pi\rho c} \ln \frac{L}{r} \tag{3.128}$$

In the last integration (similarly to (3.126)) we again have to cut-off upper limit logarithmic divergence at $k \sim \frac{1}{r}$, and at $k \sim \frac{1}{L}$ for lower limit (L is the size of the system).

Consider now thermal fluctuations — $\Delta C(r,T)$.

For $d = 3$ we have:

$$\Delta C_T^{(3)}(r) = \frac{T}{2\pi^2 \rho c^2 r} \int_0^\infty \frac{dk}{k} \sin kr = \frac{T}{4\pi\rho c^2 r} \tag{3.129}$$

For $d = 2$:

$$\Delta C_T^{(2)}(r) = \frac{T}{(2\pi)^2 \rho c^2} \int_0^{2\pi} d\theta \int_0^\infty \frac{dk}{k} e^{ikr\cos\theta} \tag{3.130}$$

This integral also diverges and we have to introduce a cut-off, similarly to the case of $C_1(r)$. Then we get:

$$\Delta C_T^{(2)}(r) = \frac{T}{2\pi\rho c^2} \ln \frac{L}{r} \tag{3.131}$$

For $d = 1$, in a similar way we obtain:

$$\Delta C_T^{(1)}(r) = \frac{T}{\pi\rho c^2} \int_0^\infty \frac{dk}{k^2} \cos kr = Const L \tag{3.132}$$

Our calculations are summarized in Table 3.1.

Table 3.1 Asymptotic ($r \to \infty$)
behavior of correlation functions.

d	$C_0(r)$	$\Delta C_T(r)$
3	$\sim \frac{1}{r^2}$	$\sim \frac{T}{r}$
2	$\sim \frac{1}{r}$	$\sim T \ln \frac{L}{r}$
1	$\sim \ln \frac{L}{r}$	$\sim TL$

These results allow us to study the problem of possible destruction of the long-range (crystalline here!) order by quantum and thermal fluctuations (atomic displacements). We only have to look at the asymptotic behavior of $C(r)$ for $r \to \infty$. If we have $C(r) \to 0$, long-range order (crystalline lattice) survives, as even rather large initial displacement $u(0)$ of an atom from its average position does not lead to a strong change of $u(r)$, at some far away position. However, if $C(r) \to \infty$, this means that the long-range order is destroyed. This situation is typical for quantum fluctuations for $d = 1$, and for thermal fluctuations for $d = 1, 2$!

Chapter 4

Electrons in Disordered Systems

4.1 Diagram technique for "impurity" scattering

Consider an electron moving in a random potential field, created by N_i scatterers ("impurities"), which are randomly placed in space with some fixed density (concentration) $\rho_i = \frac{N_i}{V}$, where V is the system volume. Total potential (random field!), created by these impurities is given by:

$$V(\mathbf{r}) = \sum_{j=1}^{N_i} v(\mathbf{r} - \mathbf{R}_j) \qquad (4.1)$$

where $v(\mathbf{r} - \mathbf{R}_j)$ is the potential of a single scatterer, situated at the (random!) point $\mathbf{R}_j$. Absolutely random distribution of scatterers corresponds to the following distribution function in coordinate space:

$$\mathcal{P}\{\mathbf{R}_j\} = V^{-N_i} \qquad (4.2)$$

For some given configuration of scatterers, electronic Green's function satisfies the following equation:

$$\left\{ i\hbar \frac{\partial}{\partial t} + \frac{\hbar^2}{2m} \nabla^2 - \sum_{j=1}^{N_i} v(\mathbf{r} - \mathbf{R}_j) \right\} G(\mathbf{rr}'t\{\mathbf{R}_j\}) = \delta(\mathbf{r} - \mathbf{r}')\delta(t)$$

$$(4.3)$$

and is functionally dependent on all $\mathbf{R}_j$. Usually, in the theory of disordered systems it is assumed [Lifshits I.M., Gredeskul S.A., Pastur L.A. (1988)], that (experimentally measurable) physical characteristics of a system are determined as averages over the ensemble of samples with all possible configurations of "impurities" (impurity averaging). Thus, we shall be mainly interested in studying of the averaged Green's function, defined as:

$$G(\mathbf{r} - \mathbf{r}', t) = < G(\mathbf{rr}'t) > = \frac{1}{V^{N_i}} \int \cdots \int \prod_{j=1}^{N_i} d\mathbf{R}_j G(\mathbf{rr}'t\{\mathbf{R}_j\}) \quad (4.4)$$

Assuming the scattering potential to be weak enough, we may develop perturbation theory, writing down the second-quantized Hamiltonian for electron interaction with (random) field (4.1) as:

$$H_{int} = \int d\mathbf{r}\psi^+(\mathbf{r})V(\mathbf{r})\psi(\mathbf{r}) \quad (4.5)$$

This perturbation theory (over "external" field [Abrikosov A.A., Gorkov L.P., Dzyaloshinskii I.E. (1963)]) is very simple, and appropriate expansion for the Green's function (4.3) has the well known form:

$$G(1, 1') = G_0(1, 1') + \int d2 G_0(1, 2)V(2)G_0(2, 1')$$

$$+ \int d2d3 G_0(1, 2)V(2)G_0(2, 3)V(3)G_0(3, 1') + \dots \quad (4.6)$$

where $1 = (\mathbf{r}, t)$, $1' = (\mathbf{r}', t')$ etc. Graphically this expansion is shown in Fig. 4.1. But we are interested in the averaged Green's function $< G(\mathbf{rr}'t) >$, defined in (4.4). Then, in the process of averaging of the series given by (4.6) over the distribution function (4.2), we need to calculate the following averages:

$$< V(2) >, \qquad < V(2)V(3) >, \qquad < V(2)V(3)V(4) >, \dots \quad (4.7)$$

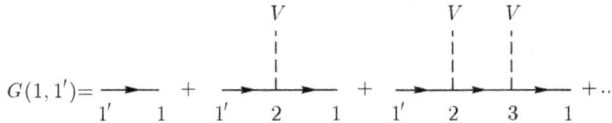

Fig. 4.1 Electron scattering by fixed configuration of scatterers.

For random impurities (the random field (4.1) distributed according to (4.2)) all these averages can be calculated explicitly. First of all we introduce the Fourier representation:

$$V(\mathbf{r}) = \sum_{\mathbf{p}} \sum_{j} v(\mathbf{p})e^{i\mathbf{p}(\mathbf{r}-\mathbf{R}_j)} \quad (4.8)$$

where $v(\mathbf{p})$ is the Fourier transform of the potential of a single scatterer, $v(-\mathbf{p}) = v^*(\mathbf{p})$. For simplicity, we mostly assume this potential to be point-like, so that $v(\mathbf{p}) = v = const$. This limitation is actually not very important.

Using (4.8), we reduce our task of calculating the averages of the type of (4.7) to calculations of:

$$M_s(\mathbf{p}_1, \mathbf{p}_2, ..., \mathbf{p}_s) = < \rho(\mathbf{p}_1)\rho(\mathbf{p}_2)...\rho(\mathbf{p}_s) >$$
$$\equiv \left\langle \sum_{l_1} \sum_{l_2} ... \sum_{l_s} \exp(-i \sum_j \mathbf{p}_j \mathbf{R}_{l_j}) \right\rangle \qquad (4.9)$$

It is convenient to consider slightly different (actually more general and "realistic") version of our model. Let N_i scatterers (impurities) be distributed randomly over N sites of a regular (e.g. simple cubic) lattice. Then, instead of (dimensional) volume density ρ_i, introduced above, we can introduce dimensionless concentration of impurities $\rho = \frac{N_i}{N}$, which may change in the interval between 0 and 1. Then, the averaging of the arbitrary sum over the impurity positions is obviously calculated as:

$$\left\langle \sum_{l_i} ... \right\rangle \to \frac{N_i}{N} \sum_l ... = \rho \int \frac{d\mathbf{R}_l}{a^3} ... = \rho_i \int d\mathbf{R}_l ... \qquad (4.10)$$

where the second sum is already done over all sites of the lattice, and a denotes the lattice spacing. Here we also have taken into account that the dimensional (volume) density of scatterers $\rho_i = \frac{N_i}{V} = \frac{N_i}{Na^3} = \rho a^{-3}$. Transition to the "continuous" model, discussed above, is obtained as the limit of $a \to 0$, so that the fixed value of ρ corresponds to the limit of $\rho_i \to \infty$. At the same time, if we fix ρ_i, the limit of $a \to 0$ gives $\rho = \rho_i a^3 \to 0$. If we put (as is done very often) the system volume $V = 1$, we have $N = a^{-3}$, and the difference in definitions of concentrations just vanish. Thus, in the future discussion we shall use the single notation of ρ.

The following calculations are more or less simple, we need only to separate accurately the special cases, when summation (impurity) indices in (4.9) coincide. Then, direct calculations (for the lattice model) show:

$$M_1(\mathbf{p}) = \left\langle \sum_l \exp(-i\mathbf{p}\mathbf{R}_l) \right\rangle = \rho \int d\mathbf{R} e^{-i\mathbf{p}\mathbf{R}} = (2\pi)^3 \rho \delta(\mathbf{p}) \qquad (4.11)$$

$$M_2(\mathbf{p}_1, \mathbf{p}_2) = \left\langle \sum_l \exp[-i(\mathbf{p}_1 + \mathbf{p}_2)\mathbf{R}_l] + \sum_{l \neq m} \exp[-(\mathbf{p}_1\mathbf{R}_l + \mathbf{p}_2\mathbf{R}_m)] \right\rangle$$

$$= (2\pi)^3 \rho\delta(\mathbf{p}_1 + \mathbf{p}_2) + \rho^2[(2\pi)^3\delta(\mathbf{p}_1)(2\pi)^3\delta(\mathbf{p}_2) - (2\pi)^3\delta(\mathbf{p}_1 + \mathbf{p}_2)]$$

$$= (2\pi)^6 \rho^2 \delta(\mathbf{p}_1)\delta(\mathbf{p}_2) + (2\pi)^3(\rho - \rho^2)\delta(\mathbf{p}_1 + \mathbf{p}_2)$$

$$\equiv <\rho(\mathbf{p}_1)>_c<\rho(\mathbf{p}_2)>_c + <\rho(\mathbf{p}_1)\rho(\mathbf{p}_2))>_c \qquad (4.12)$$

where, by definition, we have introduced *cumulant averages* $<\ ...\ >_c$.[1]
Similarly we get:

$$M_3(\mathbf{p}_1, \mathbf{p}_2, \mathbf{p}_3) = <\rho(\mathbf{p}_1)>_c<\rho(\mathbf{p}_2)>_c<\rho(\mathbf{p}_3)>_c$$

$$+ <\rho(\mathbf{p}_1)>_c<\rho(\mathbf{p}_2)\rho(\mathbf{p}_3)>_c + <\rho(\mathbf{p}_2)>_c<\rho(\mathbf{p}_1)\rho(\mathbf{p}_3)>_c$$

$$+ <\rho(\mathbf{p}_3)>_c<\rho(\mathbf{p}_1)\rho(\mathbf{p}_2)>_c + <\rho(\mathbf{p}_1)\rho(\mathbf{p}_2)\rho(\mathbf{p}_3)>_c \qquad (4.14)$$

Finally, after the averaging of an expansion in (4.6) we obtain the following elements of the new (averaged) perturbation series:

$$v<\rho(\mathbf{p}_1)>_c = (2\pi)^3 \rho v \delta(\mathbf{p}_1) \qquad (a) \qquad\qquad (4.15)$$

$$v^2<\rho(\mathbf{p}_1)\rho(\mathbf{p}_2)>_c = (2\pi)^3(\rho - \rho^2)v^2\delta(\mathbf{p}_1 + \mathbf{p}_2) \quad (b) \qquad (4.16)$$

$$v^3<\rho(\mathbf{p}_1)\rho(\mathbf{p}_2)\rho(\mathbf{p}_3)>_c = (2\pi)^3 v^3(\rho - 3\rho^2 + 2\rho^3)\delta(\mathbf{p}_1 + \mathbf{p}_2 + \mathbf{p}_3) \qquad (c)$$

$$\qquad\qquad\qquad\qquad\qquad\qquad\qquad\qquad\qquad\qquad (4.17)$$

$$v^4<\rho(\mathbf{p}_1)\rho(\mathbf{p}_2)\rho(\mathbf{p}_3)\rho(\mathbf{p}_4)>_c = (2\pi)^3 v^4(\rho - 7\rho^2 + 12\rho^3 - 6\rho^4)$$

$$\times \delta(\mathbf{p}_1 + \mathbf{p}_2 + \mathbf{p}_3 + \mathbf{p}_4) \qquad (d) \qquad\qquad (4.18)$$

which can be represented diagrammatically as shown in Figs. 4.2(a–d). Cumulants of higher orders are even more awkward.

4.2   Single-electron Green's function

The main conclusion of the previous discussion is that diagrammatic expansion for the single-electron Green's function, averaged over random configurations of scatterers (impurities), can be represented by diagrams shown in Fig. 4.3. Sometimes it is said that interaction lines (denoting interaction with impurities) are grouped into "bunches" attached to "crosses" (impurity diagram technique (S.F. Edwards, 1958)).

----

[1] Formal correspondence between the average moments and cumulants is given by:

$$\left\langle \exp\sum_j \alpha_j \rho(\mathbf{p}_j) \right\rangle = \exp\left\langle \exp\left[\sum_j \alpha_j\rho(\mathbf{p}_j)\right] - 1 \right\rangle_c \qquad (4.13)$$

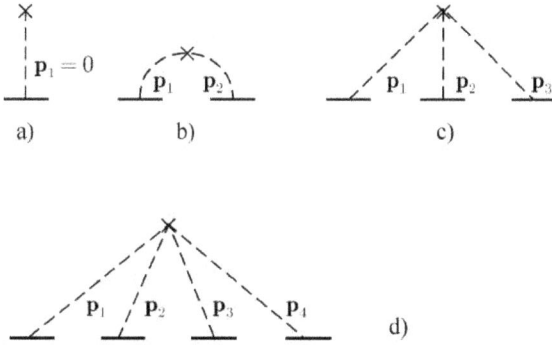

Fig. 4.2 Diagrams representing different cumulants in the averaged perturbation series.

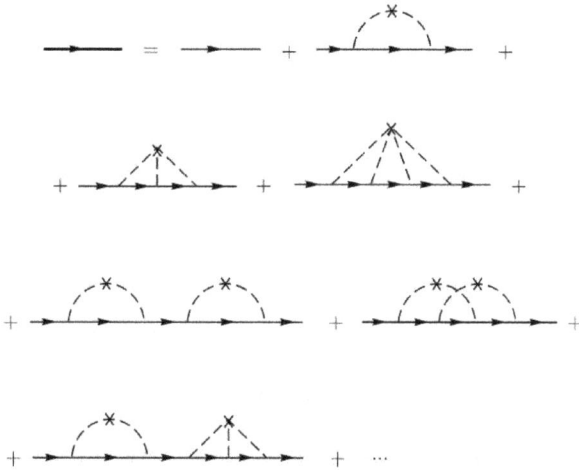

Fig. 4.3 Diagrammatic expansion for the averaged Green's function of an electron in a random field of impurities.

For small concentration of impurities ($\rho \to 0$) (or in the "continuous" model) we may limit ourselves to terms linear in ρ. Then, assuming also the smallness of the potential ($v \to 0$), we may consider only contributions from (4.15), (4.16), or, accordingly, diagrams (cumulants), shown in Figs. 4.2(a,b). Note that the contribution of (4.15) is trivial and reduces to a constant, which only changes the origin of the energy

axis by ρv (or, equivalently, just renormalizes the chemical potential).[2]
The second cumulant (Fig. 4.2(b)) reduces now to $(2\pi)^3 \rho v^2 \delta(\mathbf{p}_1 + \mathbf{p}_2)$.
Then, the expansion for the averaged Green's function reduces to the
sum of diagrams, shown in Fig. 4.4. This case corresponds to the simplest "Wick-like" *factorization* of random field correlators (4.7):

$$< V(1)V(2) > \neq 0$$
$$< V(1) >= 0 \qquad < V(1)V(2)V(3) >= 0$$

– etc., for all odd products,

$$< V(1)V(2)V(3)V(4) >=< V(1)V(2) >< V(3)V(4) >$$
$$+ < V(1)V(4) >< V(2)V(3) >$$

– etc., for all even products. $\qquad\qquad$ (4.19)

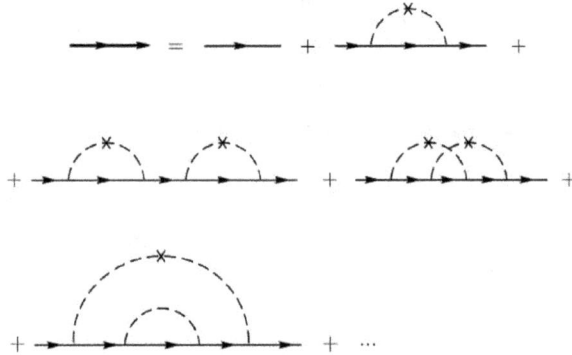

Fig. 4.4 Diagrammatic expansion of the averaged Green's function in the Gaussian
random field.

From mathematical (statistical) point of view this means that we are
dealing with the *Gaussian* random field.[3]

---

[2]Thus, we may just put $< V(2) >= 0$ in (4.7) and calculate energies with respect to
the average level of the random field. In fact, if we limit ourselves to the self-energy
given by diagram of Fig. 4.2(a), we obtain $\Sigma = \rho v(0) = \rho \int d\mathbf{r} v(\mathbf{r})$ and, accordingly, the
Green's function is: $G(\varepsilon_n \mathbf{p}) = \frac{1}{i\varepsilon_n - \xi(p) - \rho v(0)}$, which proves our statement.

[3]It can be shown that the same result follows from the sum of all perturbation series for
the "continuous" model of impurity distribution in the formal limit of $\rho_i \to \infty$, $v^2 \to 0$,
with $\rho_i v^2 \to const$!

For the problem under consideration the "two-point" correlator of the random field in coordinate space has the following form:

$$< V(\mathbf{r}_1)V(\mathbf{r}_2) > = (2\pi)^3 \rho v^2 \int \frac{d^d p_1}{(2\pi)^3} e^{i\mathbf{p}_1 \mathbf{r}_1} \int \frac{d^d p_2}{(2\pi)^3} e^{i\mathbf{p}_2 \mathbf{r}_2} \delta(\mathbf{p}_1 + \mathbf{p}_2)$$

$$= \rho v^2 \int \frac{d^d p_1}{(2\pi)^3} e^{i\mathbf{p}_1 (\mathbf{r}_1 - \mathbf{r}_2)} = \rho v^2 \delta(\mathbf{r}_1 - \mathbf{r}_2) \qquad (4.20)$$

Thus, it is usually said, that here we are dealing with the problem of an electron moving in the Gaussian random field with "white-noise" correlation.

Expansion shown in Fig. 4.3 can be written in the form of Dyson equation:

$$G(1, 1') = G_0(1, 1') + \int d2 d3 G_0(1, 2) \Sigma(2, 3) G(3, 1') \qquad (4.21)$$

or in momentum space (Matsubara technique):[4]

$$G(\mathbf{p}\varepsilon_n) = G_0(\mathbf{p}\varepsilon_n) + G_0(\mathbf{p}\varepsilon_n) \Sigma(\mathbf{p}\varepsilon_n) G(\mathbf{p}\varepsilon_n) \qquad (4.22)$$

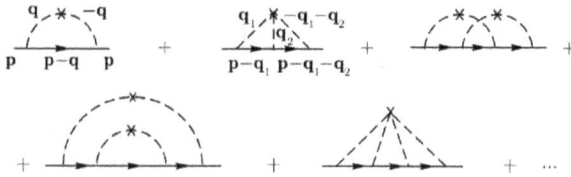

Fig. 4.5 Diagrams for electron self-energy in a random field.

where the self-energy part $\Sigma(1, 2)$ is given by diagrams, shown in Fig. 4.5. Consider the contribution of the first diagram of Fig. 4.5, corresponding, as we shall see shortly, to the first-order Born approximation for impurity scattering (1BA) $(v(-\mathbf{p}) = v^*(\mathbf{p}))$:

$$\Sigma^{1BA}(\varepsilon_n \mathbf{p}) = \rho \sum_{\mathbf{q}} |v(\mathbf{q})|^2 \frac{1}{i\varepsilon_n - \xi(\mathbf{p} - \mathbf{q})} = \rho \sum_{\mathbf{p}'} |v(\mathbf{p} - \mathbf{p}')|^2 \frac{1}{i\varepsilon_n - \xi(\mathbf{p}')}$$

$$(4.23)$$

where, as usual, $\xi(\mathbf{p}) = \varepsilon_p - \mu \approx v_F(|\mathbf{p}| - p_F)$.

---

[4]The averaged Green's function $< G(\mathbf{r}\mathbf{r}'\varepsilon_n) > = G(\mathbf{r} - \mathbf{r}'\varepsilon_n)$ depends only on $|\mathbf{r} - \mathbf{r}'|$ — the averaging "restores" translational invariance.

In metals we have typically $E_F \sim 7eV \sim 80000K$, and in most cases we are interested in studying electrons close enough to the Fermi level $E_F \approx \mu$. For example, at temperatures $T < 800K$ we have $\frac{T}{E_F} < 10^{-2}$. Thus, we only need to know $\Sigma^{1BA}(\varepsilon_n \mathbf{p})$ for $|\mathbf{p}| \sim p_F$ and[5] $|i\varepsilon_n \to \varepsilon + isign(\varepsilon_n)\delta| \ll E_F$. From the previous analysis of screening, it is clear that the impurity potential is also screened, so that in fact $v(\mathbf{p} - \mathbf{p}')$ is rather smooth function on the interval of $0 < |\mathbf{p} - \mathbf{p}'| < 2p_F$. These facts will help us in calculations to follow.

We have:

$$\Sigma^{1BA}(\mathbf{p}, \varepsilon + isign(\varepsilon_n)\delta) = \rho \sum_{\mathbf{p}'} |v(\mathbf{p} - \mathbf{p}')|^2 \frac{1}{\varepsilon - \xi(\mathbf{p}') + isign(\varepsilon_n)\delta}$$

$$= \rho \sum_{\mathbf{p}'} |v(\mathbf{p} - \mathbf{p}')|^2 \left\{ \frac{\varepsilon - \xi(\mathbf{p}')}{(\varepsilon - \xi(\mathbf{p}'))^2 + \delta^2} - isign(\varepsilon_n)\pi\delta(\varepsilon - \xi(\mathbf{p}')) \right\}$$

$$(4.24)$$

As $|v(\mathbf{p} - \mathbf{p}')|^2$ changes rather slowly and we are interested in $|\varepsilon - \xi(p')| \ll E_F \approx \mu$, we have qualitative picture shown in Fig. 4.6. Due to the fact that $\frac{\varepsilon - \xi(p')}{(\varepsilon - \xi(p'))^2 + \delta^2}$ is an odd function of $\varepsilon - \xi(p')$, we have $Re\Sigma^{1BA}(\mathbf{p}\varepsilon_n) \approx 0$.[6] The (4.24) is reduced to purely imaginary contribution:

$$\Sigma^{1BA}(\mathbf{p}\varepsilon) = -i\pi sign(\varepsilon_n) \sum_{\mathbf{p}'} \rho |v(\mathbf{p} - \mathbf{p}')|^2 \delta(\varepsilon - \xi(\mathbf{p}')) \approx -i\frac{\varepsilon_n}{|\varepsilon_n|} \frac{1}{2\tau_p}$$

$$\equiv -i\frac{\varepsilon_n}{|\varepsilon_n|}\gamma_p \qquad\qquad (4.25)$$

where, putting $\varepsilon \approx \xi(p)$ (close to the pole!), we have introduced

$$\frac{1}{\tau_p} = 2\gamma_p = 2\pi \sum_{\mathbf{p}'} \rho |v(\mathbf{p} - \mathbf{p}')|^2 \delta(\xi(\mathbf{p}) - \xi(\mathbf{p}')) \qquad (4.26)$$

— the scattering rate of electrons due to impurities, calculated in Born approximation (Fermi "golden rule").

---

[5]Note that finally we have to perform analytic continuation to the real axis from the upper half-plane, where $\varepsilon_n > 0$, i.e. $i\varepsilon_n \to \varepsilon + i\delta$, or from the lower half-plane, where $\varepsilon_n < 0$, i.e. $i\varepsilon_n \to \varepsilon - i\delta$.

[6]Strictly speaking, the integral over p' in the first term of (4.24) can be split in two: one over p', which are from p_F, and the other, over p' close to p_F. The limits in the second integral can be taken symmetric in $p' - p_F$, leading to the integral being zero (if we neglect the deviations from $v(\mathbf{p} - \mathbf{p}')$ — behavior of the spectrum close to p_F). The first integral gives just a real constant, which again can be included into renormalized chemical potential.

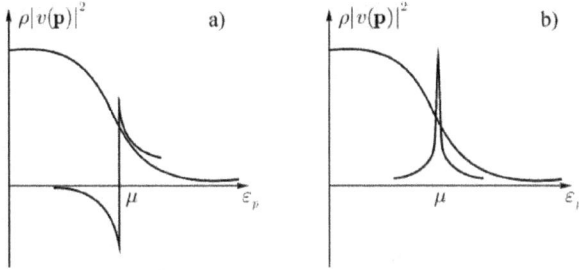

Fig. 4.6 (a) Comparison of $\rho|v(p)|^2$ and $\frac{\varepsilon-\varepsilon_p+\mu}{(\varepsilon-\varepsilon_p+\mu)^2+\delta^2}$, entering $Re\Sigma^{1BA}(p\varepsilon)$.
(b) Comparison of $\rho|v(p)|^2$ and $\frac{\delta}{(\varepsilon-\varepsilon_p+\mu)^2+\delta^2}$, entering $Im\Sigma^{1BA}(p\varepsilon)$.

If, for simplicity, from the very beginning we introduce point-like impurity potential $v(\mathbf{p}) = v$, and linearized spectrum for electrons close to the Fermi surface, all calculations become much simpler and we immediately obtain:

$$\Sigma^{1BA}(\mathbf{p}\varepsilon_n) = \rho v^2 \sum_{\mathbf{p}'} \frac{1}{i\varepsilon_n - \xi(\mathbf{p}')} \approx -\rho v^2 \nu_F \int_{-\infty}^{\infty} d\xi \frac{i\varepsilon_n + \xi}{\varepsilon_n^2 + \xi^2}$$

$$= -i\rho v^2 \nu_F arctg \frac{\xi}{\varepsilon_n}\Big|_{-\infty}^{\infty} \frac{\varepsilon_n}{|\varepsilon_n|} = -i\frac{\varepsilon_n}{|\varepsilon_n|}\pi\rho v^2 \nu_F \quad (4.27)$$

which, in fact, coincides with (4.25), and

$$\gamma_p = \pi\rho v^2 \nu_F \quad (4.28)$$

is a constant, determined by the impurity potential and electron density of states at the Fermi level.

Finally, in this approximation, the averaged single-electron Green's function can be written as:

$$G^{1BA}(\mathbf{p}\varepsilon_n) = \frac{1}{i\varepsilon_n - \xi(p) + i\gamma_p sign\varepsilon_n} \quad (4.29)$$

which, after the continuation $i\varepsilon_n \to z$, gives:

$$G^{1BA}(\mathbf{p}z) = \begin{cases} \frac{1}{z-\xi(p)+i\gamma_p} & Imz > 0 \\ \frac{1}{z-\xi(p)-i\gamma_p} & Imz < 0 \end{cases} \quad (4.30)$$

which for $z \to \varepsilon \pm i\delta$ (where $\delta \to +0$) defines $G^{R(A)}(\mathbf{p}\varepsilon)$. According to general analyticity properties [Abrikosov A.A., Gorkov L.P., Dzyaloshinskii I.E. (1963)], $G^{1BA}(\mathbf{p}z)$ possess a cut along the real axis of z.

After an elementary Fourier transformation, we obtain:

$$G^R(\mathbf{p}t) = \int \frac{d\varepsilon}{2\pi} \frac{e^{-i(\varepsilon+i\delta)t}}{\varepsilon - \xi(p) + i\gamma_p} = -i\theta(t)e^{-i\xi(p)t}e^{-\gamma_p t} \qquad (4.31)$$

and similarly:

$$G^R(\mathbf{r}\varepsilon) = \int \frac{d^3p}{(2\pi)^3} \frac{e^{i\mathbf{p}\mathbf{r}}}{\varepsilon - \xi(p) + i\gamma_p} = -\frac{\pi\nu_F}{p_F r} e^{ip_F r} e^{-r/2l_p} \qquad (4.32)$$

where $l_p = v_F \tau_p$.[7] Thus, $\gamma_p = \frac{1}{2\tau_p}$ determines "damping" of the averaged Green's function both in time and space (on the length l_p, similar to the mean free path).

Spectral density, corresponding to (4.30), has the form of a simple Lorentzian with the width γ_p:

$$A(\mathbf{p}\varepsilon) = -\frac{1}{\pi} ImG^R(\mathbf{p}\varepsilon) = \frac{1}{\pi} \frac{\gamma_p}{(\varepsilon - \xi(p))^2 + \gamma_p^2} \qquad (4.34)$$

which naturally transforms to $A(\mathbf{p}\varepsilon) = \delta(\varepsilon - \xi(p))$ for the gas of free electrons for $\gamma_p \to 0$.

Let us analyze now the role of neglected diagrams and possible generalizations.

We can introduce the "full" Born approximation, which is exact in the lowest order in impurity concentration ρ, and taking into account the multiple-scattering of an electron by a single impurity. Appropriate diagrams for the self-energy are shown in Fig. 4.7(a). Analytically:

$$\Sigma^{FBA}(\mathbf{p}\varepsilon_n) = t_{\mathbf{pp}}(\varepsilon_n) \qquad (4.35)$$

---

[7]We can write (4.32) as (use $p - p_F = \xi/v_F$):

$$G^R(\mathbf{r}\varepsilon) = \frac{\nu_F}{p_F r} \int_{-\infty}^{\infty} d\xi \frac{\sin pr}{\varepsilon - \xi + \frac{i}{2\tau_p}}$$

$$= \frac{\nu_F}{2ip_F r} \int_{-\infty}^{\infty} d\xi \frac{\exp(ip_F r + i\frac{\xi}{v_F}r) - \exp(-ip_F r - i\frac{\xi}{v_F}r)}{\varepsilon - \xi + \frac{i}{2\tau_p}} = -\frac{\pi\nu_F}{p_F r} e^{ip_F r} e^{-\frac{r}{2v_F \tau_p}}$$

$$(4.33)$$

which reduces to the diagonal element of the scattering matrix $t_{\mathbf{pp'}}$, which is determined by the equation, shown diagrammatically in Fig. 4.7(b), or analytically:

$$t_{\mathbf{pp'}}(\varepsilon_n) = \rho v(0)\delta_{\mathbf{pp'}} + \sum_{\mathbf{p''}} v(\mathbf{p} - \mathbf{p''})G_0(\mathbf{p''})t_{\mathbf{p''p}}(\varepsilon_n) \qquad (4.36)$$

Here again we have great simplifications for electrons close to the Fermi surface. The real part of the diagonal element of t-matrix $t_{\mathbf{pp}}(i\varepsilon_n)$ is practically constant for $|\mathbf{p}| \sim p_F$ and can be included in μ. Then we only have to analyze $Imt_{\mathbf{pp}}(i\varepsilon_n)$. Using the optical theorem of quantum theory of scattering[8]

$$Imt_{\mathbf{pp}} = Im\sum_{\mathbf{p'}} t_{\mathbf{pp'}}^+ G_0(\mathbf{p'})t_{\mathbf{p'p}} \qquad (4.38)$$

we have:

$$Im\Sigma^{FBA}(\mathbf{p}\varepsilon_n) = Imt_{\mathbf{pp}}(\varepsilon_n) = Im\sum_{\mathbf{p'}} \frac{|t_{\mathbf{pp'}}|^2}{i\varepsilon_n - \xi(\mathbf{p'})}$$

$$= -sign\varepsilon_n\pi\sum_{\mathbf{p'}} |t_{\mathbf{pp'}}|^2\delta(\varepsilon - \xi(\mathbf{p'})) \qquad (4.39)$$

where in the last equality $i\varepsilon_n \to \varepsilon + i\delta sign\varepsilon_n$. Eq. (4.39) coincides with (4.25), where we substitute $\rho|v(\mathbf{p} - \mathbf{p'})|^2$ by $|t_{\mathbf{pp'}}|^2$:

$$\Sigma^{FBA}(\mathbf{p}\varepsilon_n) = -isign\varepsilon_n\frac{1}{2\tau_p} = -isign\varepsilon_n\gamma_p \qquad (4.40)$$

where

$$\frac{1}{\tau_p} = 2\gamma_p = 2\pi\sum_{\mathbf{p'}} |t_{\mathbf{pp'}}|^2\delta(\xi(\mathbf{p}) - \xi(\mathbf{p'})) \qquad (4.41)$$

If in Eq. (4.26) we replace $\rho|v(\mathbf{p} - \mathbf{p'})|^2$ by $|t_{\mathbf{pp'}}|^2$, we get precisely this result. In this sense, we may limit ourselves by the second diagram of Fig. 4.7(a) only, as was done above, but assume that $v(\mathbf{p} - \mathbf{p'})$ is just the matrix element of the single-impurity scattering matrix.

Now let us consider the self-consistent Born approximation, which is achieved by "dressing" internal electronic lines in self-energy diagrams, as shown in Fig. 4.7(c). Analytically:

$$\Sigma^{SCBA}(\mathbf{p}\varepsilon_n) = \rho v(0)\delta_{\mathbf{pp}} + \sum_{\mathbf{p'}} v(\mathbf{p} - \mathbf{p'})G(\mathbf{p'}\varepsilon_n)t_{\mathbf{p'p}} \qquad (4.42)$$

---

[8]From (4.36) we have: $t = v + vG_0t$, $v^+ = v$, $v = -t^+G_0^+v + t^+$ $t = v + (t^+G_0t - t^+G_0^+vG_0t)$, so that due to Hermiticity of v and $t^+G_0^+vG_0t$, we get:

$$Imt_{\mathbf{pp}} = Im < \mathbf{p}|t^+G_0t|\mathbf{p} >= Im\sum_{\mathbf{p'}} t_{\mathbf{pp'}}^+ G_0t_{\mathbf{p'p}} \qquad (4.37)$$

which reduces to (4.38).

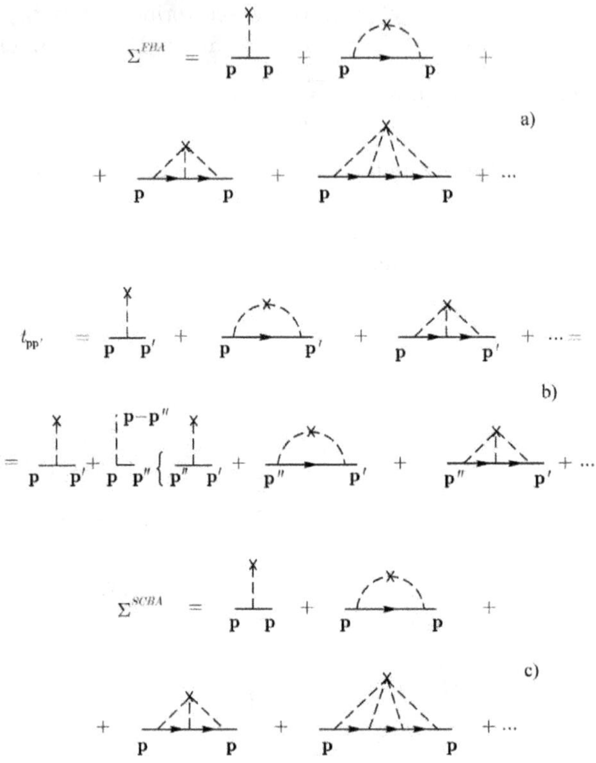

$$\Sigma^{FBA} = \quad \underset{p \quad p}{} + \underset{p \qquad p}{} +$$

a)

$$+ \underset{p \qquad p}{} + \underset{p \qquad p}{} + \ldots$$

$$t_{pp'} = \underset{p \quad p'}{} + \underset{p \qquad p'}{} + \underset{p \qquad p'}{} + \ldots =$$

b)

$$= \underset{p \quad p'}{} + \underset{p \quad p''}{} \left\{ \underset{p'' \quad p'}{} + \underset{p'' \qquad p'}{} + \underset{p'' \qquad p'}{} + \ldots \right.$$

$$\Sigma^{SCBA} = \quad \underset{p \quad p}{} + \underset{p \qquad p}{} +$$

c)

$$+ \underset{p \qquad p}{} + \underset{p \qquad p}{} + \ldots$$

Fig. 4.7 (a) Diagrams for the self-energy, accounting for the multiple scattering by a single impurity. (b) Equation for t-matrix. (c) Diagrams for self-energy in self-consistent approximation, accounting for multiple scattering.

where the difference with (4.36) is in replacement of Green's function G_0 by:

$$G(\mathbf{p}\varepsilon_n) = \frac{1}{i\varepsilon_n - \xi(p) - \Sigma(\mathbf{p}\varepsilon_n)} \tag{4.43}$$

so that, in fact, we obtain self-consistency procedure, determining self-energy part (Green's function).

After that, we can repeat our arguments. Using the weak energy dependence of $t_{\mathbf{pp}}$ for $|\mathbf{p}| \approx p_F$ and $\varepsilon \ll E_F$, and assuming weak enough scattering, so that $|\Sigma^{SCBA}| \ll E_F$, we again obtain the result of the type of (4.39). Again only $Im\Sigma$

is relevant, as $Re\Sigma$ can be "hidden" in the chemical potential μ. Finally we get:

$$Im\Sigma^{SCBA}(\mathbf{p}\varepsilon_n) = Imt_{\mathbf{pp}} = Im \sum_{\mathbf{p'}} \frac{|t_{\mathbf{pp'}}|^2}{i\varepsilon_n - \xi(\mathbf{p'}) - iIm\Sigma^{SCBA}(\mathbf{p'}\varepsilon_n)}$$

$$\approx -sign(\varepsilon_n - Im\Sigma^{SCBA})\pi \sum_{\mathbf{p'}} |t_{\mathbf{pp'}}|^2 \delta(\varepsilon - \xi(\mathbf{p'})) \qquad (4.44)$$

where the approximate equality is valid for small $Im\Sigma^{SCBA}$. For self-consistency it is sufficient to take $Im\Sigma^{SCBA}(i\varepsilon_n) \sim -sign(\varepsilon_n)$, which is checked by direct substitution. The difference between the Born approximation discussed above and its self-consistent variant appears only in the case of strong enough scattering, when δ-function in (4.44) is replaced by Lorentzian of the finite width, which in the model with point-like scattering (cf. (4.27)) changes nothing at all.

Finally we again get the well known result:

$$\Sigma^{SCBA}(\mathbf{p}\varepsilon_n) = -isign(\varepsilon_n)\frac{1}{2\tau_p} = -isign(\varepsilon_n)\gamma_p \qquad (4.45)$$

where τ_p and γ_p are defined as in (4.41), or (4.28) for the case of point-like impurities.[9]

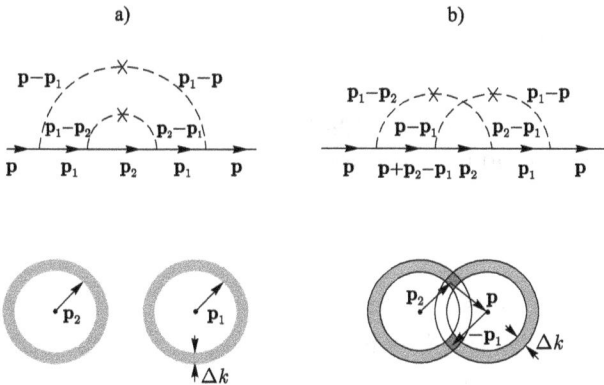

Fig. 4.8 Diagram without intersections of interaction lines (a) and "crossing" diagram (b). Shown also are corresponding regions of integration in momentum space.

In diagram expansion for the Green's function our approximation corresponds to the account of only "non-crossing" diagrams, shown in Fig. 4.4, *without intersections* of interaction lines. Why and under what conditions we can neglect "crossing" diagrams? Let us compare two diagrams, shown in Fig. 4.8. Calculating the contribution of the diagram

---

[9]If we consider the band of the finite width, ν_F denotes the density of states at the Fermi level, with the account of scattering effects.

of Fig. 4.8(a) we note, that integration momenta $\mathbf{p}_1$ and $\mathbf{p}_2$ can take any values in the spherical layer of the width $\Delta k \sim 1/l$, so that this contribution is proportional to the appropriate phase space volume of the order of $\Omega_a \approx (4\pi p_F^2 \hbar \Delta k)^2$. In the case of diagram shown in Fig. 4.8(b) the same limitations apply to $\mathbf{p}_1$ and $\mathbf{p}_2$, but in addition we have to satisfy $|\mathbf{p} + \mathbf{p}_2 - \mathbf{p}_1| \approx p_F$. For fixed $\mathbf{p}_2$, the change of $\mathbf{p}_1$ is limited to the region of intersection of its layer and the appropriate layer for $\mathbf{p} + \mathbf{p}_2 - \mathbf{p}_1$, as shown by doubly dashed region in Fig. 4.8(b). The phase space of the "ring", formed by intersection of two spherical layers, is of the order of $\Omega_b \approx (4\pi p_F^2 \hbar \Delta k)(2\pi p_F \hbar^2 \Delta k^2)$. Then we understand that the ratio of the contributions of "crossing" and "non-crossing" diagrams is of the order of $\frac{\Omega_b}{\Omega_a} \sim \frac{\hbar \Delta k}{p_F} = \frac{\hbar}{p_F l} \ll 1$ ("weak" disorder corresponds to large enough mean free path, i.e. $p_F l / \hbar \gg 1$). Thus, the dimensionless "small parameter" of our perturbation theory is given by $\frac{\hbar}{p_F l} \ll 1$, which is equivalent to $\frac{\hbar}{E_F \tau_p} \ll 1$. Taking into account $p_F \sim \hbar/a$ (where a is interatomic spacing), we see that the smallness of this parameter corresponds to the condition $l \gg a$, so that the mean free path must be significantly larger than interatomic spacing (lattice constant). In fact, this is the usual condition of applicability of kinetic (Boltzmann-like) equation of the standard transport theory.

4.3   Keldysh model

Condition of $p_F l / \hbar \gg 1 (l \gg a)$ allows us to limit ourselves to the subseries of "dominating" diagrams (diagrams without intersections of interaction lines) of Feynman perturbation series. In majority of problems solved by diagram technique we act precisely in this way, i.e. we are looking (using some physical criteria) for some infinite subseries of diagrams, which we are *able* to sum. From "mathematical" point of view it is not very well defined procedure. Neglected diagrams (though small over some physical parameter) also constitute an infinite subseries of the full perturbation expansion and their contribution, strictly speaking, remains unknown. In some (very rare!) cases we can actually sum the *whole* Feynman series and obtain an exact solution of the problem. Unfortunately, it is usually possible only for some oversimplified models. However, conclusions, obtained via the analysis of such model may be very instructive.

As an example of such a model, we shall consider an electron, moving in Gaussian random field (when perturbation series is given by diagrams, shown in Fig. 4.4) with special form of pair correlator, defined in momentum representation as $W^2(\mathbf{q}) = (2\pi)^3 W^2 \delta(\mathbf{q})$. In this case, the momentum transferred by each interaction line is equal to zero. In coordinate space this gives the pair correlator for the random field of the form:

$$< V(\mathbf{r})V(\mathbf{r}') > = \int \frac{d^3 q}{(2\pi)^3} e^{i\mathbf{q}(\mathbf{r}-\mathbf{r}')}(2\pi)^3 W^2 \delta(\mathbf{q}) = W^2 \qquad (4.46)$$

corresponding to the case of "infinite range" correlations of the random field $V(\mathbf{r})$. It is the case just opposite to the "white noise" correlator (4.20).[10]

In this model we can easily sum the whole Feynman series (L.V. Keldysh, 1965).[11] Let us return to the series for non averaged Green's function, shown in Fig. 4.1. After we perform averaging over the Gaussian random field, interaction lines, corresponding to external field, are joined "pairwise" in all possible combinations, and we obtain diagrammatic expansion shown in Fig. 4.4. Accordingly, in the n-th order over the correlator of Gaussian field, in each term of the expansion for the averaged Green's function we have $2n$ vertices, joined pairwise by interaction lines (all diagrams in this order are obtained if we perform all possible pairwise connections of $2n$ vertices by interaction lines in $2n$-th order term of the expansion shown in Fig. 4.1). If each interaction line transfers zero momentum to the electron, it is easily seen that all diagrams in the given order of perturbation theory (including those with "crossing" interaction lines!) give just *equal* contributions. Then, the complete perturbation series is written as:

$$G(\varepsilon\mathbf{p}) = G_0(\varepsilon\mathbf{p}) \left\{ 1 + \sum_{n=1}^{\infty} G_n W^{2n} G_0^{2n}(\varepsilon\mathbf{p}) \right\} \qquad (4.47)$$

where A_n is the total number of diagrams in the n-th order of this series (in $2n$-th order in interaction amplitude W). The contribution of the arbitrary diagram is equal to:

$$W^{2n} G_0^{2n+1}(\varepsilon\mathbf{p}) \qquad (4.48)$$

---

[10] In general case, correlator $< V(\mathbf{r})V(\mathbf{r}') >$ can be characterized by some *correlation length* ξ of fluctuations of the random field. "White noise" corresponds to the limit of $\xi \to 0$, in the model under consideration we have $\xi \to \infty$.

[11] Doctoral Thesis, P.N. Lebedev Physical Institute, Moscow, 1965.

which corresponds to $2n$ vertices (factor of W^{2n}), connected pairwise by dashed (interaction) lines, and also the product of $2n + 1$ free electron Green's functions. The factor of A_n is directly determined by combinatorics — this is just the number of possible ways, in which we can join pairwise $2n$ vertices by dashed (interaction) lines. It is easily seen that:

$$G_n = (2n - 1)!! \tag{4.49}$$

There are $2n$ vertices and $2n + 1$ electronic lines in each diagram. Take an arbitrary vertex. It can be joined in $2n - 1$ ways with each of remaining $2n - 1$ vertices. After that we have $2n - 2$ "unjoined" vertices at our disposal. Again, take one. It can be joined with the others in $2n - 3$ ways. Then, there remain $2n - 4$ "unjoined". Any of these can be joined with the remaining in $2n - 5$ ways, etc. The total number of ways we can join $2n$ vertices in the given order is equal to $(2n-1)(2n-3)(2n-5)... = (2n-1)!!$, which gives us (4.49).

Now we may use rather well known integral representation:[12]

$$(2n - 1)!! = \frac{1}{\sqrt{2\pi}} \int_{-\infty}^{\infty} dt\, t^{2n} e^{-\frac{t^2}{2}} \tag{4.54}$$

---

[12] By definition: $(2n)!! = 2.4.6...(2n) = 2^n n!$. Similarly:

$$(2n - 1)!! = 1.3.5...(2n - 1) = 2^n \frac{1}{\sqrt{\pi}} \Gamma\left(n + \frac{1}{2}\right) \tag{4.50}$$

Using the integral representation of Γ-function:

$$\Gamma(z) = \int_0^{\infty} dx\, x^{z-1} e^{-x} \tag{4.51}$$

it is easy to obtain quite useful relation:

$$n! = \Gamma(n + 1) = \int_0^{\infty} dx\, x^n e^{-x} \tag{4.52}$$

and also

$$\Gamma\left(n + \frac{1}{2}\right) = \int_0^{\infty} dx\, x^{n-1/2} e^{-x} = \int_0^{\infty} dx\, x^{\frac{2n-1}{2}} e^{-x} \tag{4.53}$$

which, after the substitution $x \to t^2/2$ and with the account of (4.50), gives (4.54).

Then (4.47) reduces to:

$$
\begin{aligned}
G(\varepsilon\mathbf{p}) &= G_0(\varepsilon\mathbf{p})\left\{1 + \sum_{n=1}^{\infty} \frac{1}{\sqrt{2\pi}} \int_{-\infty}^{\infty} dt\, t^{2n} e^{-\frac{t^2}{2}} W^{2n} G_0(\varepsilon\mathbf{p})^{2n}\right\} \\
&= \frac{1}{\sqrt{2\pi}} \int_{-\infty}^{\infty} dt\, e^{-\frac{t^2}{2}} G_0(\varepsilon\mathbf{p})\left\{1 + \sum_{n=1}^{\infty} t^{2n} W^{2n} G_0^{2n}(\varepsilon\mathbf{p})\right\} \\
&= \frac{1}{\sqrt{2\pi}} \int_{-\infty}^{\infty} dt\, e^{-\frac{t^2}{2}} G_0(\varepsilon\mathbf{p})\left\{1 + \frac{t^2 G_0^2(\varepsilon\mathbf{p}) W^2}{1 - t^2 G_0^2(\varepsilon\mathbf{p}) W^2}\right\} \quad (4.55)
\end{aligned}
$$

where during the calculations we have changed the order of summation and integration, and summed the simple progression.[13] After the elementary transformations we have:

$$
\begin{aligned}
G(\varepsilon\mathbf{p}) &= \frac{1}{\sqrt{2\pi}} \int_{-\infty}^{\infty} dt\, e^{-\frac{t^2}{2}} G_0(\varepsilon\mathbf{p}) \frac{1}{1 - t^2 W^2 G_0^2(\varepsilon\mathbf{p})} \\
&= \frac{1}{\sqrt{2\pi}} \int_{-\infty}^{\infty} dt\, e^{-\frac{t^2}{2}} G_0(\varepsilon\mathbf{p}) \frac{1}{2}\left\{\frac{1}{1 - tW G_0(\varepsilon\mathbf{p})} + \frac{1}{1 + tW G_0(\varepsilon\mathbf{p})}\right\}
\end{aligned}
$$
$$(4.56)$$

so that finally we get:

$$
\begin{aligned}
G^R(\varepsilon\mathbf{p}) &= \frac{1}{\sqrt{2\pi}} \int_{-\infty}^{\infty} dt\, e^{-\frac{t^2}{2}} \frac{G_0(\varepsilon\mathbf{p})}{1 - tW G_0(\varepsilon\mathbf{p})} \\
&= \frac{1}{\sqrt{2\pi}} \int_{-\infty}^{\infty} dt\, e^{-\frac{t^2}{2}} \frac{1}{\varepsilon - \varepsilon_p - tW + i\delta} \quad (4.57)
\end{aligned}
$$

where we have used the explicit form of $G_0^R(\varepsilon\mathbf{p}) = \frac{1}{\varepsilon - \varepsilon_p + i\delta}$.[14] In the following we take $\varepsilon_p = \frac{p^2}{2m}$. Introducing $tW = V$ we rewrite (4.57) in a more descriptive form:

$$
G^R(\varepsilon\mathbf{p}) = \frac{1}{\sqrt{2\pi}W} \int_{-\infty}^{\infty} dV\, e^{-\frac{V^2}{2W^2}} \frac{1}{\varepsilon - \frac{p^2}{2m} - V + i\delta} \quad (4.60)
$$

---

[13]This approach in mathematics is called *Borel summation*.

[14]Define the function of complex variable z:

$$
\Psi(z) = \frac{1}{\sqrt{2\pi}} \int_{-\infty}^{\infty} dt\, e^{-\frac{t^2}{2}} \frac{1}{z - t} \quad (4.58)
$$

Then (4.57) can be written as:

$$
G(\varepsilon\mathbf{p}) = \frac{1}{W} \Psi\left(\frac{1}{W G_0(\varepsilon\mathbf{p})}\right) \quad (4.59)
$$

The physical meaning of this result is obvious — we have an electron, moving in spatially homogeneous random field V, with Gaussian distribution of the width W. The averaged Green's function describes an ensemble of "samples", with field V being constant along each "sample", but having random values in different "samples" (elements of am ensemble).

Let us give another derivation of this elegant result (L.V. Keldysh, 1965, A.L. Efros, 1970). Consider Dyson equation:

$$G^{-1}(\varepsilon\mathbf{p}) = \varepsilon - \frac{p^2}{2m} - \Sigma(\varepsilon\mathbf{p}) \qquad (4.61)$$

where the self-energy part can be represented by the diagram shown in Fig. 4.9(a). In analytic form we have:

$$\Sigma(\varepsilon\mathbf{p}) = \int \frac{d^3q}{(2\pi)^3}\Gamma(\mathbf{p}, \mathbf{p}-\mathbf{q}, \mathbf{q})G(\varepsilon\mathbf{p}-\mathbf{q})W^2(\mathbf{q}) \qquad (4.62)$$

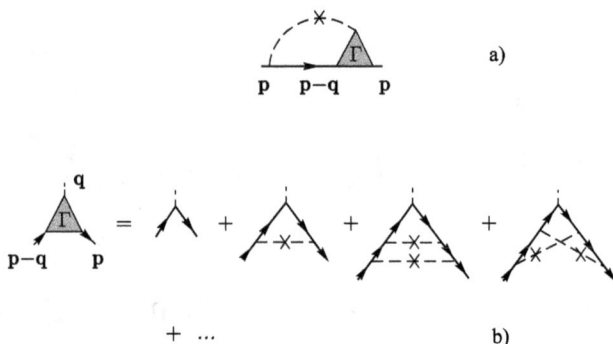

Fig. 4.9   Diagrams for an exact self-energy part of an electron in Gaussian random field (a) and for the vertex-part, determining this self-energy (b).

Here $\Gamma(\mathbf{p}, \mathbf{p}-\mathbf{q}, \mathbf{q})$ is an exact vertex-part, defined by diagrams shown in Fig. 4.9(b).[15] Taking into account $W^2(\mathbf{q}) = (2\pi)^3 W^2\delta(\mathbf{q})$, equations (4.61) and (4.62) are reduced to:

$$\varepsilon - \frac{p^2}{2m} - G^{-1}(\varepsilon\mathbf{p}) = W^2 G(\varepsilon\mathbf{p})\Gamma(\mathbf{p}, \mathbf{p}, 0) \qquad (4.63)$$

[15]Note that both electronic "legs" of this vertex correspond to retarded or advanced Green's function $G^{R(A)}$, depending on which of them is obtained from Eq. (4.61). It is precisely because of this fact we can use Ward identity (4.64) to find $\Gamma(\mathbf{p}, \mathbf{p}, 0)$.

The vertex-part $\Gamma(\mathbf{p}, \mathbf{p}, 0)$ satisfies the following Ward identity:

$$\Gamma(\mathbf{p}, \mathbf{p}, 0) = \frac{dG^{-1}(\varepsilon\mathbf{p})}{d\varepsilon} \tag{4.64}$$

which is easily derived by direct differentiation of diagrammatic series for self-energy $\Sigma(\varepsilon\mathbf{p})$. Then, from (4.63) we obtain the differential equation, determining the Green's function:

$$W^2\frac{dG}{dx} + Gx - 1 = 0 \tag{4.65}$$

where we introduced $x = \varepsilon - \frac{p^2}{2m} + i\delta$. Solving Eq. (4.65) with the boundary condition $G(x) = \frac{1}{x}$ for $x \to \infty$ immediately leads to (4.60).

The main result given by Eq. (4.60) and obtained by an *exact* summation of the whole Feynman series (4.47) is rather instructive. For example, the spectral density, corresponding to (4.60), has the form:

$$A(\varepsilon\mathbf{p}) = \frac{1}{\pi}\frac{1}{\sqrt{2\pi}W}\int_{-\infty}^{\infty} dV e^{-\frac{V^2}{2W^2}}\delta\left(\varepsilon - \frac{p^2}{2m} - V\right)$$

$$= \frac{1}{\pi\sqrt{2\pi}W}\exp\left(-\frac{(\varepsilon - \frac{p^2}{2m})^2}{2W^2}\right) \tag{4.66}$$

i.e. is given by wide Gaussian peak. Note also, that Eq. (4.60) does not reduce to something similar to Eq. (4.30), i.e. to the Green's function with smeared "quasiparticle" pole, it does not possess poles at all.

From Eqs. (4.60) and (4.66) we can easily calculate the density of (electronic) states:

$$N(\varepsilon) = -\frac{2}{\pi}\int\frac{d^3p}{(2\pi)^3}ImG^R(\varepsilon\mathbf{p}) = 2\int\frac{d^3p}{(2\pi)^3}A(\varepsilon\mathbf{p})$$

$$= \frac{2}{\sqrt{2\pi}W}\int_{-\infty}^{\infty} dV e^{-\frac{V^2}{2W^2}}\int\frac{d^3p}{(2\pi)^3}\delta\left(\varepsilon - \frac{p^2}{2m} - V\right) \tag{4.67}$$

where we have added 2 to account for two spin projections. Accordingly:

$$N(\varepsilon) = \frac{2^{3/4}m^{3/2}W^{1/2}}{\pi^2\hbar^3}G_0\left(\frac{\varepsilon}{\sqrt{2}W}\right) \tag{4.68}$$

where the dimensionless function $G_0(x)$ is defined as:

$$G_0(x) = \frac{1}{\sqrt{\pi}}\int_{-\infty}^{x} dy e^{-y^2}(x - y)^{1/2} \tag{4.69}$$

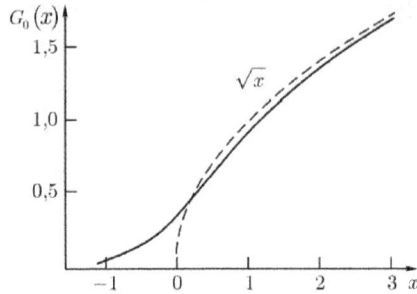

Fig. 4.10   Dimensionless function $G(x)$, determining the density of states in Keldysh model.

and shown in Fig. 4.10. For $\varepsilon > 0$ and $\varepsilon \gg W$ we have:

$$N(\varepsilon) = N_0(\varepsilon) - \frac{(2m)^{3/2}W^2}{16\pi^2\hbar^3\varepsilon^{3/2}} \qquad (4.70)$$

where the second term represents a small correction to the density of states of free electrons (dashed line in Fig. 4.10):

$$N_0(\varepsilon) = \frac{(2m)^{3/2}}{2\pi^2\hbar^3}\sqrt{\varepsilon} \qquad (4.71)$$

Most important result following from an exact solution is the appearance of the "tail" of the density of states in the region of $\varepsilon < 0$. For $\varepsilon < 0$ and $|\varepsilon| \gg W$ we obtain Gaussian asymptotic behavior for the "tail":

$$N(\varepsilon) = \frac{2^{1/4}m^{3/2}\sqrt{W}}{4\pi^2\hbar^3}\left(\frac{\sqrt{2W}}{\varepsilon}\right)^{3/2}\exp\left(-\frac{\varepsilon^2}{2W^2}\right) \qquad (4.72)$$

Formation of the "tail" in the density of states within the band gap is the general result of electronic theory of disordered systems [Lifshits I.M., Gredeskul S.A., Pastur L.A. (1988)]. Of course, the specific energy dependence of the "tail" is not universal and depends on the model of the random field (disorder).

In particular, the "tail" of the density of states appears also in the "white noise" model, discussed above. It can be shown that in this model (for three-dimensional case, $d = 3$) at $\varepsilon < 0$ [Lifshits I.M., Gredeskul S.A., Pastur L.A. (1988)]:

$$N(\varepsilon) \sim \exp\left\{-const\frac{|\varepsilon|^{1/2}\hbar^3}{m^{3/2}\rho v^2}\right\} \qquad (4.73)$$

for $|\varepsilon| \gg E_{sc} \sim m^3 (\rho v^2)/\hbar^6$, which corresponds to $|\varepsilon| \gg \gamma(\varepsilon)$, where $\gamma(\varepsilon) = \pi \rho v^2 N_0(\varepsilon) \sim \rho v^2 \frac{m^{3/2}}{\hbar^3} \sqrt{\varepsilon}$.

For the model of Gaussian random field with correlator, characterized by some finite correlation length ξ, there appear characteristic energy regions shown in Fig. 4.11. For the spatial dimensions $d < 4$ the energy E_{sc}, determining the size of the "strong coupling" region (following from $\gamma(\varepsilon) \sim \varepsilon$) is defined as (M.V. Sadovskii, 1977):

$$E_{sc} = \frac{m^{\frac{d}{4-d}}}{\hbar^{2d}} (\rho v^2)^{\frac{2}{4-d}} \tag{4.74}$$

Besides that, one more characteristic energy scale appears in this problem:

$$E_0 \sim \frac{\hbar^2}{m \xi^2} \tag{4.75}$$

The energy dependence of the "tail" in the density of states in the region of $E_{sc} \ll |\varepsilon| \ll E_0$ is determined by the following expression, directly generalizing (4.73) (M.V. Sadovskii, 1979):

$$N(\varepsilon) \sim \exp \left\{ -A_d \frac{|\varepsilon|^{2-\frac{d}{2}} \hbar^d}{\rho v^2} \right\} = \exp \left\{ -A_d \left(\frac{|\varepsilon|}{E_{sc}} \right)^{2-\frac{d}{2}} \right\} \tag{4.76}$$

where $A_d = const$, depending only on d. In the region of $|\varepsilon| \gg E_0$ the "tail" asymptotics becomes Gaussian:

$$N(\varepsilon) \sim \exp \left\{ -\frac{\xi^d}{\rho V^2} E^2 \right\} \tag{4.77}$$

which is the same as (4.72).

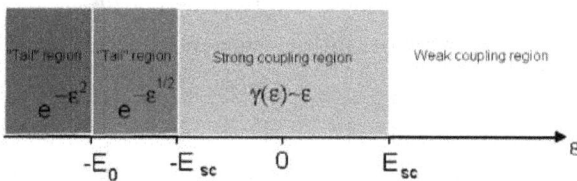

Fig. 4.11 Characteristic energy regions in the problem of an electron in the random field.

4.4 Conductivity and two-particle Green's function

Let us return to the problem of an electron in the field of random impurities (Gaussian random field with "white noise" correlation). The major task is to formulate the general method to calculate conductivity of such a system. We have already seen above that to calculate conductivity we need to know the density–density response function $\chi(\mathbf{q}\omega)$. Then we can use (2.111) and (2.116). We also convinced ourselves, that (up to a sign) this response function may be obtained via analytic continuation ($i\omega_m \to \omega + i\delta$) of Matsubara polarization operator $\Pi(i\omega_m\mathbf{q})$ of an electron gas:

$$\sigma(\omega) = -\lim_{q\to 0}\frac{ie^2}{q^2}\omega\chi(\mathbf{q}\omega) = \lim_{q\to 0}\frac{ie^2}{q^2}\omega\Pi(\mathbf{q}i\omega_m \to \omega + i\delta) \qquad (4.78)$$

We consider free (noninteracting with each other) electrons in the field of random impurities.[16] From the general quantum mechanical point of view we are dealing with single-particle problem. For such an electron there always exist some (in general unknown!) *exact* eigenfunctions and eigenenergies in the potential field, determined by the given (fixed) configuration of impurities:[17]

$$H\varphi_n(\mathbf{r}) = \varepsilon_n\varphi_n(\mathbf{r}) \qquad (4.79)$$

where

$$H = \int d\mathbf{r}\psi^+(\mathbf{r})\left\{-\frac{\nabla^2}{2m} + \sum_i v(\mathbf{r} - \mathbf{R}_i)\right\}\psi(\mathbf{r}) \qquad (4.80)$$

Accordingly, we can introduce (non averaged!) retarded and advanced Green's functions of the Schroedinger equation (4.79) as:

$$G^{R,A}(\mathbf{r}\mathbf{r}',\varepsilon) = \sum_n \frac{\varphi_n(\mathbf{r})\varphi_n^*(\mathbf{r}')}{\varepsilon - \varepsilon_n \pm i\delta} \qquad (4.81)$$

or, after the Fourier transformation over each of the coordinates:

$$G^{R,A}(\mathbf{p}\mathbf{p}',\varepsilon) = \sum_n \frac{\varphi_n(\mathbf{p})\varphi_n^*(\mathbf{p}')}{\varepsilon - \varepsilon_n \pm i\delta} \qquad (4.82)$$

---

[16]General discussion of the approach proposed below can also be found in [Sadovskii M.V. (2000); Altshuler B.L., Aronov A.G. (1985)].

[17]The averaging procedure discussed above leads to formally "multi-particle" structure of perturbation theory, but the averaged Green's function is no longer Green's function (propagators) of any *quantum mechanical problem*. They give only the effective picture averaged over the statistical ensemble of "samples" with all possible configurations of scatterers (impurities).

Expanding electron operators over exact eigenfunctions:

$$\psi(\mathbf{r}) = \sum_n a_n \varphi_n(\mathbf{r}) \qquad \psi^+(\mathbf{r}) = \sum_n a_n^+ \varphi_n^*(\mathbf{r}) \qquad (4.83)$$

we define the density operator as:

$$\rho(\mathbf{r}) = \psi^+(\mathbf{r})\psi(\mathbf{r}) = \sum_{mn} \varphi_n^*(\mathbf{r})\varphi_m(\mathbf{r}) a_n^+ a_m \qquad (4.84)$$

Then the density–density response function can be calculated using the general scheme, described above in (2.82)–(2.88), so that for $T = 0$ we get:

$$\chi(\mathbf{rr}', \omega) = i \int_0^\infty dt e^{i(\omega + i\delta)t} < 0|[\rho(\mathbf{r}, t), \rho(\mathbf{r}', 0)]|0 >$$

$$= 2 \sum_{mn} \varphi_n^*(\mathbf{r})\varphi_m(\mathbf{r})\varphi_m^*(\mathbf{r}')\varphi_n(\mathbf{r}') \frac{n(\varepsilon_m) - n(\varepsilon_n)}{\omega + \varepsilon_n - \varepsilon_m + i\delta} \qquad (4.85)$$

where the averaging is performed over the ground state, and the factor of 2 accounts for both spin projections. Eq. (4.85) gives an explicit expression for (2.82). In fact, we are interested in *averaged* (over impurity configurations) response function $\chi(\mathbf{r} - \mathbf{r}', \omega) = < \chi(\mathbf{rr}', \omega) >$, or its spatial Fourier-transform:

$$\chi(\mathbf{q}\omega) = 2 \sum_{\mathbf{pp}'} \sum_{nm} \left\langle \varphi_m(\mathbf{p}_+)\varphi_m^*(\mathbf{p}'_+)\varphi_n(\mathbf{p}'_-)\varphi_n^*(\mathbf{p}_-) \frac{n(\varepsilon_m) - n(\varepsilon_n)}{\omega + \varepsilon_n - \varepsilon_m + i\delta} \right\rangle$$

$$= 2 \sum_{\mathbf{pp}'} \int_{-\infty}^\infty \frac{d\varepsilon}{2\pi i} \{ [n(\varepsilon + \omega) - n(\varepsilon)] < G^R(\mathbf{p}_+ \mathbf{p}'_+, \varepsilon + \omega) G^A(\mathbf{p}'_- \mathbf{p}_-, \varepsilon) >$$

$$+ n(\varepsilon) < G^R(\mathbf{p}_+ \mathbf{p}'_+, \varepsilon + \omega) G^R(\mathbf{p}'_- \mathbf{p}_+, \varepsilon) >$$

$$- n(\varepsilon + \omega) < G^A(\mathbf{p}_+ \mathbf{p}'_+, \varepsilon + \omega) G^A(\mathbf{p}'_- \mathbf{p}_-, \varepsilon) > \} \qquad (4.86)$$

where the second expression is directly checked using (4.82), and we introduced $\mathbf{p}_\pm = \mathbf{p} \pm \frac{1}{2}\mathbf{q}$. Eq. (4.86) can be rewritten as:

$$\chi(\mathbf{q}\omega) = - \int_{-\infty}^\infty d\varepsilon \{ [n(\varepsilon + \omega) - n(\varepsilon)] \Phi^{RA}(\varepsilon\omega\mathbf{q}) + n(\varepsilon) \Phi^{RR}(\varepsilon\omega\mathbf{q})$$

$$- n(\varepsilon + \omega) \Phi^{AA}(\varepsilon\omega\mathbf{q}) \} \qquad (4.87)$$

where we have introduced:

$$\Phi^{RA}(\varepsilon\omega\mathbf{q}) = -\frac{1}{2\pi i} 2 \sum_{\mathbf{pp}'} < G^R(\mathbf{p}_+ \mathbf{p}'_+, \varepsilon + \omega) G^A(\mathbf{p}'_- \mathbf{p}_-, \varepsilon) > \qquad (4.88)$$

and similar expressions for Φ^{RR} and Φ^{AA}. Note that we included here a factor of 2 due to spin.

Previous discussion was concerned with the case of $T = 0$. For $T \neq 0$ similar expressions are obtained after analytic continuation of appropriate Matsubara functions.

Using the rules of "impurity" diagram technique, we can obtain Matsubara polarization operator as shown in Fig. 4.12(a). Analytically:

$$\Pi(\mathbf{q}\omega_m) = 2T \sum_n \sum_{\mathbf{p}} G(\mathbf{p}\varepsilon_n)G(\mathbf{p}+\mathbf{q}\varepsilon_n+\omega_m)\mathcal{T}(\mathbf{p},\mathbf{p}+\mathbf{q},\varepsilon_n,\varepsilon_n+\omega_m)$$

$$(4.89)$$

where equations for vertices Γ and $\mathcal{T}$ are shown graphically in Fig. 4.12(b),(c). the sum over frequencies in (4.89) can be calculated using the general approach, described above in connection with (3.40)–(3.44). The presence in (4.89) of the pair of Green's functions (with frequencies, differing by $i\omega_m$) leads to the appearance of two cuts in

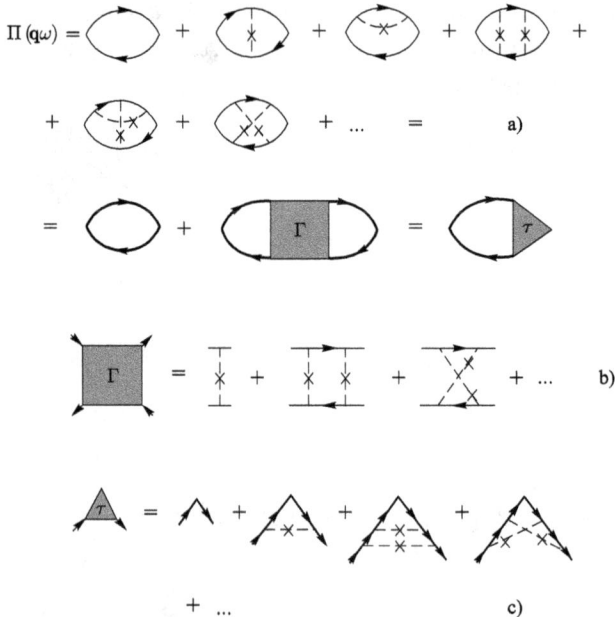

Fig. 4.12 Diagrammatic expansion for polarization operator in impure system (a) and appropriate vertices Γ (b) and $\mathcal{T}$ (c).

their product[18] in the complex plane of frequency — one along $z = \varepsilon$, and the other along $z = \varepsilon - i\omega_m$. Then, using Eq. (3.38) we may write the sum in (4.89) as:

$$S(i\omega_m) = T\sum_n f(i\varepsilon_n, i\varepsilon_n + i\omega_m) = -\int_C \frac{dz}{2\pi i} n(z) f(z, z + i\omega_m) \quad (4.90)$$

where $f(z, z + i\omega_m) = G(z)G(z + i\omega_m)T(z, z + i\omega_m)$ (we drop momenta arguments for shortness!), and $n(z)$ is the Fermi distribution. The contour of integration C is shown in Fig. 4.13. Rewriting (4.90) via integrals over four horizontal lines of C (integrals over infinitely far away arcs vanish!), we obtain:

$$S(i\omega_m) = -\int_{-\infty}^{\infty} \frac{d\varepsilon}{2\pi i} n(\varepsilon)[f(\varepsilon + i\delta, \varepsilon + i\omega_m) - f(\varepsilon - i\delta, \varepsilon + i\omega_m)]$$

$$-\int_{-\infty}^{\infty} \frac{d\varepsilon}{2\pi i} n(\varepsilon - i\omega_m)[f(\varepsilon - i\omega_m, \varepsilon + i\delta) - f(\varepsilon - i\omega_m, \varepsilon - i\delta)]$$

$$(4.91)$$

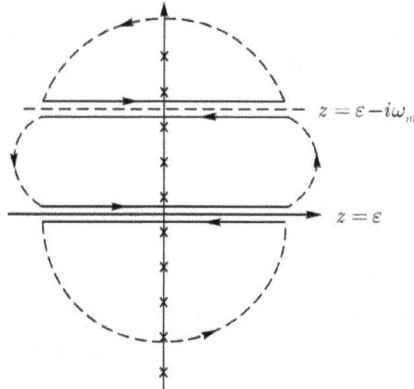

Fig. 4.13   Integration contour used in calculations of the sum over Matsubara frequencies.

---

[18]Remember, that Green's function $G(\mathbf{p}z)$ possess a cut in the complex plane of frequency z along the line defined by $Im z = 0$.

After the substitution $\omega_m \to \omega + i\delta$, we get:

$$S(\omega) = - \int_{-\infty}^{\infty} \frac{d\varepsilon}{2\pi i} n(\varepsilon)[f^{RR}(\varepsilon, \varepsilon + \omega) - f^{RA}(\varepsilon, \varepsilon + \omega)$$
$$+ f^{RA}(\varepsilon - \omega, \varepsilon) - f^{AA}(\varepsilon - \omega, \varepsilon)] \qquad (4.92)$$

where notations of the type $f^{RA}(\varepsilon, \varepsilon')$ correspond to $f(\varepsilon - i\delta, \varepsilon + i\delta)$, etc. The shift of the summation variable $\varepsilon \to \varepsilon + \omega$ in the last two terms gives:

$$S(\omega) = \int_{-\infty}^{\infty} \frac{d\varepsilon}{2\pi i} [n(\varepsilon) - n(\varepsilon + \omega)] f^{RA}(\varepsilon, \varepsilon + \omega)$$
$$- \int_{-\infty}^{\infty} \frac{d\varepsilon}{2\pi i} [n(\varepsilon) f^{RR}(\varepsilon, \varepsilon + \omega) - n(\varepsilon + \omega) f^{AA}(\varepsilon, \varepsilon + \omega)]$$

$$(4.93)$$

This, in fact, coincides with (4.87), if we introduce:

$$\Phi^{RA}(\varepsilon \omega \mathbf{q}) = \frac{1}{2\pi i} f^{RA}(\varepsilon, \varepsilon + \omega) =$$
$$- \frac{1}{2\pi i} 2 \sum_{\mathbf{p}} G^R(\varepsilon + \omega \mathbf{p} + \mathbf{q}) G^A(\varepsilon \mathbf{p}) \mathcal{T}(\mathbf{p}, \mathbf{p} + \mathbf{q}, \varepsilon, \varepsilon + \omega) \quad (4.94)$$

which is just another form of (4.88), where everything is expressed via averaged Green's functions $G^R(\varepsilon + \omega \mathbf{p} + \mathbf{q})$ and $G^A(\varepsilon \mathbf{p})$, and the averaged vertex $\mathcal{T}(\mathbf{p}, \mathbf{p} + \mathbf{q}, \varepsilon, \varepsilon + \omega)$. As all calculations were made in Matsubara technique, Fermi functions in (4.93) are taken at $T > 0$. Thus, we have:

$$\chi(\mathbf{q}\omega) = -\Pi(\mathbf{q}i\omega_m \to \omega + i\delta) = -\int_{-\infty}^{\infty} d\varepsilon \left\{ [n(\varepsilon + \omega) - n(\varepsilon)] \Phi^{RA}(\varepsilon \omega \mathbf{q}) \right.$$
$$\left. + n(\varepsilon) \Phi^{RR}(\varepsilon \omega \mathbf{q}) - n(\varepsilon + \omega) \Phi^{AA}(\varepsilon \omega \mathbf{q}) \right\} \qquad (4.95)$$

which equivalent to (4.87), as expected!

Let us return to Eq. (4.88). Using $G^A(\mathbf{p}\mathbf{p}', \varepsilon) = [G^R(\mathbf{p}'\mathbf{p}, \varepsilon)]^*$, we obtain $\Phi^{RR}(\varepsilon 00) = -[\Phi^{AA}(\varepsilon 00)]^*$. Then, we easily see that:

$$\Phi^{RR}(\varepsilon 00) = \frac{1}{2\pi i} 2 \sum_{\mathbf{p}\mathbf{p}'} \frac{\partial}{\partial \varepsilon} < G^R(\mathbf{p}\mathbf{p}', \varepsilon) > \qquad (4.96)$$

where $< G^{R,A}(\mathbf{p}\mathbf{p}', \varepsilon) > \equiv G^{R,A}(\mathbf{p}\varepsilon)\delta(\mathbf{p} - \mathbf{p}')$ and

$$G^{R,A}(\mathbf{p}\varepsilon) = \frac{1}{\varepsilon - \frac{p^2}{2m} + \mu - \Sigma^{R,A}(\mathbf{p}\varepsilon)} \qquad (4.97)$$

is the retarded (advanced) Green's function, determined by "impurity" diagram technique.[19] It is not difficult to check the following identity:

$$\int_{-\infty}^{\infty} d\varepsilon n(\varepsilon) Im\left\{-2\pi i \Phi^{RR}(\varepsilon 00)\right\} = \pi N(E_F) \qquad (4.98)$$

which is obtained by direct substitution of (4.96) into (4.98) and partial integration, using the definition of the density of states:

$$N(\varepsilon) = -\frac{2}{\pi} \sum_{\mathbf{p}} Im G^R(\mathbf{p}\varepsilon) \qquad (4.99)$$

For $T \ll E_F$ we have:[20]

$$\left(-\frac{\partial n(\varepsilon)}{\partial \varepsilon}\right) \approx \delta(\varepsilon) \qquad (4.100)$$

and accordingly:

$$n(\varepsilon + \omega) - n(\varepsilon) = -\omega\delta(\varepsilon) \qquad (4.101)$$

Then, from (4.87) or (4.95), using (4.98), and for small $q \ll p_F$ and $\omega \ll E_F$, we have:

$$\chi(\mathbf{q}\omega) = \omega\Phi^{RA}(\mathbf{q}\omega) + N(E_F) \qquad (4.102)$$

where we have introduced the notation:

$$\Phi^{RA}(\mathbf{q}\omega) = \Phi^{RA}(\mathbf{q}\omega\varepsilon = 0) \qquad (4.103)$$

For $q = 0$ the density–density response function must be zero for arbitrary values of ω. This is a general property [Nozieres P., Pines D. (1966)], which is clear e.g. from comparison of (2.114) and (2.115). Then, from (4.102) we obtain:

$$\Phi^{RA}(0\omega) = -\frac{N(E_F)}{\omega} \qquad (4.104)$$

which, in fact, is directly related to Ward identity, connected with charge conservation. Then we can rewrite (4.102) as:

$$\chi(\mathbf{q}\omega) = \omega\left\{\Phi^{RA}(\mathbf{q}\omega) - \Phi^{RA}(0\omega)\right\} \qquad (4.105)$$

---

[19]Using Matsubara technique we obtain R or A functions via analytic continuation $i\varepsilon_n \to \varepsilon \pm i\delta$.

[20]Remember, that for us E_F is an origin of energy scale, so that in all expressions here formally we have $E_F = 0$.

and general relation for conductivity (2.111) reduces to:

$$\sigma(\omega) = -\lim_{q \to 0} ie^2 \frac{\omega^2}{q^2} \left\{ \Phi^{RA}(\mathbf{q}\omega) - \Phi^{RA}(0\omega) \right\} \qquad (4.106)$$

This expression (D. Vollhardt, P. Wölfle, 1980) is quite convenient for direct calculations, as we can effectively calculate two-particle Green's function (loop) $\Phi^{RA}(\mathbf{q}\omega)$ using "impurity" diagram technique.

Often to study conductivity diagrammatically another approach is used, based on the calculation of the response to an external vector-potential. The general expression for conductivity tensor is then given by:[21]

$$\sigma_{\mu\nu}(\mathbf{q}\omega) = \frac{1}{i\omega} \left\{ \Phi_{\mu\nu}(\mathbf{q}\omega + i\delta) - \Phi_{\mu\nu}(\mathbf{q}0 + i\delta) \right\} \qquad (4.107)$$

where

$$\Phi_{\mu\nu}(\mathbf{q}i\omega_m) = T \int_0^\beta d\tau \int_0^\beta d\tau' e^{i\omega_m(\tau - \tau')} < T_\tau J_\mu(\mathbf{q}\tau) J_\nu(-\mathbf{q}, \tau') > \qquad (4.108)$$

and $\mathbf{J}(\mathbf{p})$ is the Fourier transform of the (so-called "paramagnetic") current operator:

$$\mathbf{J}(\mathbf{q}) = -\frac{ie}{2m} \int d\mathbf{r} e^{-i\mathbf{q}\mathbf{r}} \left[\psi^+(\mathbf{r})\nabla\psi(\mathbf{r}) - \nabla\psi^+(\mathbf{r})\psi(\mathbf{r}) \right] \qquad (4.109)$$

Schematic derivation of these expressions is as follows. Consider the response of a system to an external vector-potential, leading to the following perturbation term in the Hamiltonian:

$$\mathcal{H}_{ext} = -\mathbf{J}(\mathbf{r})\mathbf{A}(\mathbf{r}t) \qquad (4.110)$$

and take $\mathbf{A}(\mathbf{r}t)$ in the form of a plane-wave:

$$\mathbf{A}(\mathbf{r}t) = \mathbf{A}(\mathbf{q}\omega)e^{i\mathbf{q}\mathbf{r} - i\omega t} \qquad (4.111)$$

so that the appropriate electric field is given by:

$$\mathbf{E}(\mathbf{r}t) = -\frac{\partial \mathbf{A}(\mathbf{r}t)}{\partial t} = i\omega \mathbf{A}(\mathbf{q}\omega)e^{i\mathbf{q}\mathbf{r} - i\omega t} \qquad (4.112)$$

Then, according to the general Kubo formalism [Sadovskii M.V. (2019a)], we obtain ("paramagnetic") response as:

$$J_\mu(\mathbf{q}\omega) = \chi_{\mu\nu}^p(\mathbf{q}\omega)A^\nu(\mathbf{q}\omega) = \frac{\chi_{\mu\nu}^p(\mathbf{q}\omega)}{i\omega}i\omega A^\nu(\mathbf{q}\omega) = \frac{\chi_{\mu\nu}^p(\mathbf{q}\omega)}{i\omega}E^\nu(\mathbf{q}\omega) \qquad (4.113)$$

---

[21]Obviously, in isotropic system and in the absence of an external magnetic field we have $\sigma_{\mu\nu} = \sigma\delta_{\mu\nu}$.

where

$$\chi^p_{\mu\nu}(\mathbf{q}\omega) = i \int_0^\infty dt e^{i\omega t} < [J_\mu(\mathbf{q}t), J_\nu(-\mathbf{q}0)] > \qquad (4.114)$$

From the general discussion of the connection between linear response and Matsubara formalism given above, it is clear that $\chi^p_{\mu\nu}(\mathbf{q}\omega) = \Phi_{\mu\nu}(\mathbf{q}\omega+i\delta)$, i.e. it can be obtained via analytic continuation of (4.108).

The full electric current in the presence of an external vector-potential is given by:

$$\mathbf{J}^{tot} = \mathbf{J} - \frac{ne^2}{m}\mathbf{A} \qquad (4.115)$$

where the second term represents "diamagnetic" current, appearing because of electron velocity having now the form: $\mathbf{v} = \frac{1}{m}(\mathbf{p} - e\mathbf{A})$. On the other hand, for $\omega = 0$ (static vector-potential (magnetic field)) an electric current in the system (normal metal) is just absent, so that "diamagnetic" part of (4.115) practically cancels "paramagnetic" one.[22] Then we have:

$$\chi_{\mu\nu}(\mathbf{q}0) = \Phi_{\mu\nu}(\mathbf{q}0) = \frac{ne^2}{m}\delta_{\mu\nu} \qquad (4.116)$$

and we immediately obtain the general expression (4.107) for conductivity. In fact, Eq. (4.116), similarly to (4.104), is also some version of the Wars identity.

Now it is clear that the diagonal element of conductivity tensor at $\mathbf{q} = \mathbf{0}$ can be written as:

$$\sigma_{xx}(\omega) = \frac{1}{i\omega}\{\Phi_{xx}(\omega + i\delta) - \Phi_{xx}(0 + i\delta)\} \qquad (4.117)$$

where, similarly to (4.89), we can write:

$$\Phi_{xx}(i\omega_m) = -2eT\sum_n\sum_{\mathbf{p}}\frac{p_x}{m}J_x(\mathbf{p}, \mathbf{p}, \varepsilon_n, \varepsilon_n + \omega_m)G(\varepsilon_n\mathbf{p})G(\varepsilon_n + \omega_m\mathbf{p})$$

$$(4.118)$$

Here we have introduced the "current" vertex:

$$J_\mu(\mathbf{p}, \mathbf{p}, \varepsilon_n, \varepsilon_n + \omega_m) \equiv \frac{e}{m}p_\mu\Xi(\mathbf{p}, \varepsilon_n, \varepsilon_n + \omega_m) \qquad (4.119)$$

which can be defined diagrammatically as shown in Fig. 4.12(c), where the "bare" vertex is given by $\frac{e}{m}p_\mu$.

---

[22]Up to a small contribution due to Landau diamagnetism! More detailed discussion of this situation will be given below in the Chapter on superconductivity.

Accordingly, rewriting (4.118) as:

$$\Phi_{xx}(i\omega_m) = -2e^2 T \sum_n \sum_{\mathbf{p}} \frac{p_x^2}{m^2} \Xi(\mathbf{p}, \varepsilon_n, \varepsilon_n + \omega_m) G(\varepsilon_n \mathbf{p}) G(\varepsilon_n + \omega_m \mathbf{p})$$

(4.120)

and performing summation over n and analytic continuation $i\omega_m \to \omega + i\delta$ as it was done above, we obtain the following expression for static conductivity ($\omega \to 0$):

$$\sigma_{xx} = \frac{e^2}{2\pi} \sum_{\mathbf{p}} \frac{p_x^2}{m^2} \Xi(\mathbf{p}) G^R(\mathbf{p}, 0) G^A(\mathbf{p}, 0)$$

(4.121)

where we have introduced the static limit $\Xi(\mathbf{p}) = \Xi(\mathbf{p}, 0 - i\delta, 0 + i\delta)$.

Let us give also, just for reference and without derivation, the general expression for non-diagonal (Hall) conductivity in the presence of the weak external magnetic field H (H. Fukuyama, H. Ebisawa, Y. Wada, 1969):

$$\sigma_{xy} = \frac{eH}{m} \frac{e^2}{4\pi i} \sum_{\mathbf{p}} \frac{p_x}{m} \Xi^2(\mathbf{p}) \left\{ G^R(\mathbf{p}, 0) \frac{\partial}{\partial p_x} G^A(\mathbf{p}0) - \frac{\partial}{\partial p_x} G^R(\mathbf{p}, 0) G^A(\mathbf{p}, 0) \right\}$$

(4.122)

which gives us diagrammatic method to calculate the Hall effect.

4.5 Bethe–Salpeter equation, "diffuson" and "Cooperon"

Thus, according to (4.105), to calculate the density–density response function $\chi(\mathbf{q}\omega)$ and conductivity of a system (4.106), we have to find the way to calculate:[23]

$$\Phi^{RA}(\omega\mathbf{q}) = -\frac{1}{2\pi i} 2 \sum_{\mathbf{p}\mathbf{p}'} < G^R(\mathbf{p}_+\mathbf{p}'_+, E + \omega) G^A(\mathbf{p}'_-\mathbf{p}_-, E) >$$

(4.123)

which, in turn, is defined via two-particle Green's function:

$$\Phi^{RA}_{\mathbf{p}\mathbf{p}'}(E\omega\mathbf{q}) = -\frac{1}{2\pi i} < G^R(\mathbf{p}_+\mathbf{p}'_+, E + \omega) G^A(\mathbf{p}'_-\mathbf{p}_-, E) > \quad (4.124)$$

which is determined by diagrams, shown in Fig. 4.14(a). It is convenient to introduce the vertex function $\Gamma_{\mathbf{p}\mathbf{p}'}(\mathbf{q}\omega)$, defined by:[24]

$$\Phi^{RA}_{\mathbf{p}\mathbf{p}'}(E\omega\mathbf{q}) = -\frac{1}{2\pi i} G^R(\mathbf{p}_+ E + \omega) G^A(\mathbf{p}_- E) \{\delta_{\mathbf{p}\mathbf{p}'}$$
$$+ \Gamma_{\mathbf{p}\mathbf{p}'}(\mathbf{q}\omega) G^R(\mathbf{p}'_+ E + \omega) G^A(\mathbf{p}'_- E)\} \quad (4.125)$$

and diagrams of Fig. 4.14(b).

---

[23] In the following, we always assume $\mu = E_F \equiv E$, $T = 0$.

[24] Here and below we mainly follow [Sadovskii M.V. (2000)] and [Altshuler B.L., Aronov A.G. (1985)].

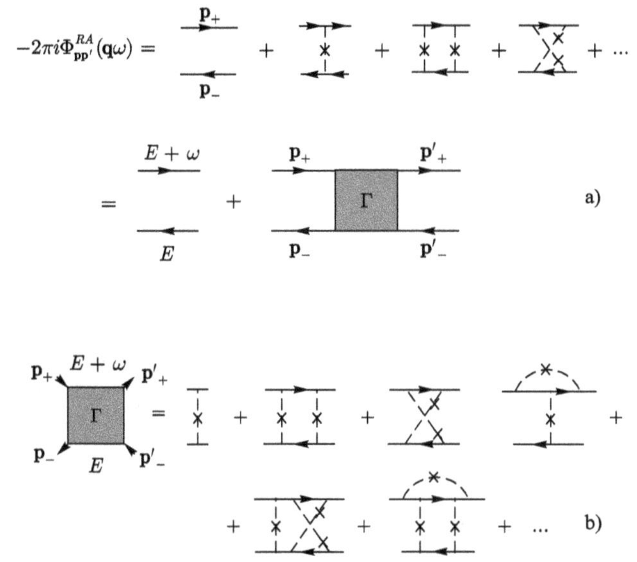

Fig. 4.14 Diagrams for two-particle Green's function (a) and vertex part $\Gamma_{\mathbf{pp'}}(\mathbf{q}\omega)$ (b). Upper electron line corresponds to retarded Green's function $G^R(\mathbf{p}_+ E+\omega)$ ("particle"), while lower to advanced $G^A(\mathbf{p}_- E)$ ("hole").

Now it is convenient to classify diagrams for the vertex as reducible (i.e. those which can be "cut" over two R and A lines) and irreducible in "$R-A$-channel" (or "particle–hole" channel). For example, in Fig. 4.14, the second and the fifth diagrams are reducible, while the rest are irreducible. Then it is clear that the full vertex $\Gamma_{\mathbf{pp'}}(\mathbf{q}\omega)$ is described by Bethe–Salpeter integral equation, shown diagrammatically in Fig. 4.15, or analytically:

$$\Gamma_{\mathbf{pp'}}(\mathbf{q}\omega)=U_{\mathbf{pp'}}(\mathbf{q}\omega)+\sum_{\mathbf{p''}}U_{\mathbf{pp''}}(\mathbf{q}\omega)G^R(E+\omega\mathbf{p''}_+)G^A(E\mathbf{p''}_-)\Gamma_{\mathbf{p''p'}}(\mathbf{q}\omega)$$

(4.126)

where $U_{\mathbf{pp'}}(\mathbf{q}\omega)$ denotes the sum of all diagrams, irreducible in $R-A$-channel, of the type shown in Fig. 4.15(b).

In the simplest approximation, we can take for $U_{\mathbf{pp'}}(\mathbf{q}\omega)$ only the first diagram from the r.h.s. of Fig. 4.15(b), i.e. just put:

$$U_0(\mathbf{p}-\mathbf{p'})=\rho|v(\mathbf{p}-\mathbf{p'})|^2 \qquad U_0=\rho v^2 \qquad (4.127)$$

where the second equality is valid for point-like impurities. Then Eq. (4.126) takes the form, shown diagrammatically in Fig. 4.16(a).

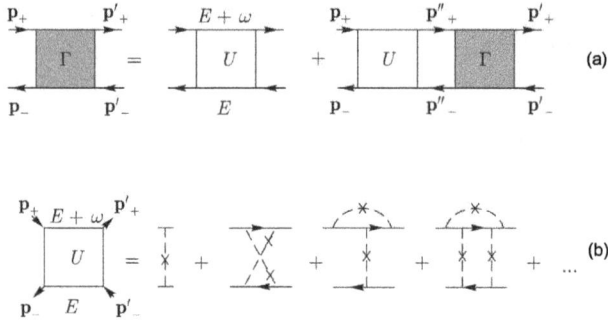

Fig. 4.15 Bethe–Salpeter equation for the vertex part $\Gamma_{\mathbf{pp}'}(\mathbf{q}\omega)$ (a) and diagrams of the lowest orders for irreducible (in $R - A$-channel) vertex $U_{\mathbf{pp}'}(\mathbf{q}\omega)$ (b).

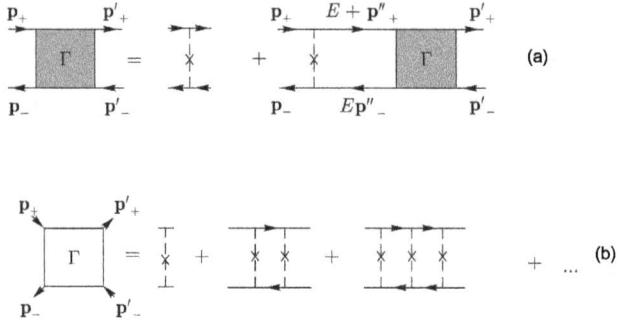

Fig. 4.16 Bethe–Salpeter equation for the vertex part $\Gamma_{\mathbf{pp}'}(\mathbf{q}\omega)$ in "ladder" approximation (a) and appropriate diagrams of lowest orders (b).

Its solution can be written as:

$$\Gamma^0_{\mathbf{pp}'}(\mathbf{q}\omega) = \frac{U_0}{1 - U_0 \sum_{\mathbf{p}} G^R(E + \omega \mathbf{p}_+) G^A(E \mathbf{p}_-)} \qquad (4.128)$$

which defines $\Gamma_{\mathbf{pp}'}(\mathbf{q}\omega)$ via the sum of "ladder" diagrams, shown in Fig. 4.16(b).

Now let us perform explicit calculations for the case of arbitrary spatial dimensionality d. Basic element, determining (4.128), can be written as:

$$
\begin{aligned}
I_d(\mathbf{q}\omega) &= \sum_{\mathbf{p}} G^R(E + \omega \mathbf{p}_+) G^A(E\mathbf{p}_-) \\
&= \int \frac{d^d p}{(2\pi)^d} G^R\left(E + \omega \mathbf{p} + \frac{\mathbf{q}}{2}\right) G^A\left(E\mathbf{p} - \frac{\mathbf{q}}{2}\right) \\
&= \int \frac{d^d p}{(2\pi)^d} G^R(E + \omega \mathbf{p} - \mathbf{q}) G^A(E\mathbf{p}) \\
&= \int \frac{d^d p}{(2\pi)^d} \frac{1}{E + \omega - \xi(p) + v_F q \cos\theta + i\gamma} \frac{1}{E - \xi(p) - i\gamma}
\end{aligned}
$$
$$(4.129)$$

where we have taken the averaged Green's functions in simplest approximation (4.29), (4.30). Here $v_F = \sqrt{\frac{2E}{m}}$ is electron velocity at the Fermi level, while damping (for point-like impurities) $\gamma = \pi \rho v^2 \nu(E) = \pi U_0 \nu(E)$, where $\nu(E)$ is density of states at the Fermi level E in d-dimensional space and for a single spin direction.

Integrals over the polar angle θ in d dimensions, are calculated using the following rules:

$$
\begin{aligned}
\int \frac{d^d p}{(2\pi)^d} f(p,\theta) &= \frac{1}{(2\pi)^d} \int_0^\infty dp\, p^{d-1} \Omega_{d-1} \int_0^\pi d\theta \sin^{d-2}\theta f(p,\theta) \\
&= \frac{\Omega_d}{(2\pi)^d} \int_0^\infty dp\, p^{d-1} \frac{\Omega_{d-1}}{\Omega_d} \int_0^\pi d\theta \sin^{d-2}\theta f(p,\theta) \\
&\approx \nu(E) \int_{-\infty}^\infty d\xi \frac{\Omega_{d-1}}{\Omega_d} \int_0^\pi d\theta \sin^{d-2}\theta f(\xi,\theta)
\end{aligned}
$$
$$(4.130)$$

where $\Omega_d = \frac{2\pi^{d/2}}{\Gamma(d/2)}$ is the surface of the sphere with radius unity in d-dimensional space.

For small ω and q we can write:[25]

$$
\begin{aligned}
&\frac{1}{E + \omega - \xi(p) \pm v_F q \cos\theta + i\gamma} \\
&\approx \frac{1}{E - \xi(p) + i\gamma} - \frac{\omega \pm v_F q \cos\theta}{(E - \xi(p) + i\gamma)^2} + \frac{(\omega \pm v_F q \cos\theta)^2}{(E - \xi(p) + i\gamma)^3} + \cdots
\end{aligned}
$$
$$(4.131)$$

---

[25] In the following it will become clear, that we are speaking about $\omega \ll \gamma$ and $v_F q \ll \gamma$.

so that:

$$I_d(q\omega) \approx \nu(E) \int_{-\infty}^{\infty} d\xi \frac{\Omega_{d-1}}{\Omega_d} \int_0^\pi d\theta \sin^{d-2}\theta \frac{1}{E - \xi - i\gamma}$$

$$\times \left\{ \frac{1}{E - \xi + i\gamma} - \frac{\omega + v_F q \cos\theta}{(E - \xi + i\gamma)^2} + \frac{(\omega + v_F q \cos\theta)^2}{(E - \xi + i\gamma)^3} + \cdots \right\}$$

$$(4.132)$$

Integrals over ξ are elementary (calculate residues!), so that:

$$I_d(q\omega) \approx \nu(E) \frac{\Omega_{d-1}}{\Omega_d} 2\pi i \int_0^\pi d\theta \sin^{d-2}\theta$$

$$\times \left\{ \frac{1}{2i\gamma} - \frac{\omega + v_F q \cos\theta}{(2i\gamma)^2} + \frac{(\omega + v_F q \cos\theta)^2}{(2i\gamma)^3} + \cdots \right\}$$

$$= \frac{\pi}{\gamma} \nu(E) \frac{\Omega_{d-1}}{\Omega_d} \int_0^\pi d\theta \sin^{d-2}\theta$$

$$\times \left\{ \left[1 + \frac{i\omega}{2\gamma} - \frac{\omega^2}{4\gamma^2} \right] + \left[\frac{iv_F q}{2\gamma} - \frac{v_F q \omega}{2\gamma^2} \right] \cos\theta - \frac{v_F^2 q^2}{4\gamma^2} \cos^2\theta \right\}$$

$$(4.133)$$

Now θ integrations give:

$$\frac{\Omega_{d-1}}{\Omega_d} \int_0^\pi d\theta \sin^{d-2}\theta = 1 \qquad (4.134)$$

$$\frac{\Omega_{d-1}}{\Omega_d} \int_0^\pi d\theta \sin^{d-2}\theta \cos\theta = 0 \qquad (4.135)$$

$$\frac{\Omega_{d-1}}{\Omega_d} \int_0^\pi d\theta \sin^{d-2}\theta \cos^2\theta = \frac{1}{d} \qquad (4.136)$$

Then, taking into account $\frac{\pi}{\gamma}\nu(E) = U_0$ and $\gamma = \frac{1}{2\tau}$, we finally obtain the denominator of (4.128) as:

$$1 - U_0 I_d(q\omega) = -\frac{i\omega}{2\gamma} + \frac{\omega^2}{4\gamma^2} + D_0 \frac{q^2}{2\gamma} \approx -i\omega\tau + D_0\tau q^2 \quad (4.137)$$

$$1 - U_0 I_d^*(q\omega) \approx i\omega\tau + D_0\tau q^2 \quad (4.138)$$

where

$$D_0 = \frac{1}{d} v_F^2 \tau = \frac{1}{d} \frac{2E}{m} \tau = \frac{1}{d} \frac{E}{m\gamma} \qquad (4.139)$$

is the usual Drude diffusion coefficient in d dimensions. Accordingly, from (4.128) we obtain the following typical (and very important!) expression for the vertex with *diffusion pole*:

$$\Gamma^0_{\mathbf{pp'}}(\mathbf{q}\omega) = \frac{U_0\tau^{-1}}{-i\omega + D_0 q^2} = \frac{2U_0\gamma}{-i\omega + D_0 q^2} \qquad (4.140)$$

or the so-called *diffuson*.[26]

It is very important to note, that in the case of time-reversal symmetry, the full vertex $\Gamma_{\mathbf{p,p'}}(\mathbf{q},\omega)$ possesses the following general property:

$$\Gamma_{\mathbf{p,p'}}(\mathbf{q},\omega) = \Gamma_{\frac{1}{2}(\mathbf{p}-\mathbf{p'}+\mathbf{q}),\frac{1}{2}(\mathbf{p'}-\mathbf{p}+\mathbf{q})}(\mathbf{p}+\mathbf{p'},\omega) \qquad (4.141)$$

To prove this, consider the general vertex part shown in Fig. 4.17(a). Here we performed the "ordering" of momenta: on the lower electronic line the numbers attributed to momenta are even, while on the upper — odd. The smaller number corresponds to incoming line, while the larger — to outgoing. We have the general conservation law:

$$\mathbf{p}_1 + \mathbf{p}_2 = \mathbf{p}_3 + \mathbf{p}_4 = \mathbf{p} + \mathbf{p'} \qquad (4.142)$$

so that

$$\mathbf{q} = \mathbf{p}_1 - \mathbf{p}_4 = \mathbf{p}_3 - \mathbf{p}_2 \qquad \mathbf{p} = \frac{1}{2}(\mathbf{p}_1 + \mathbf{p}_4) \qquad \mathbf{p'} = \frac{1}{2}(\mathbf{p}_2 + \mathbf{p}_3)$$
$$(4.143)$$

In the case of time-reversal symmetry the single-particle eigenstates $\mathbf{p}$ and $-\mathbf{p}$ are just equivalent! Then we can reverse the direction of e.g. lower ("hole"-like) line of the diagram, changing the sign of its momentum. Then the diagram of Fig. 4.17(a) is transformed into the second diagram of Fig. 4.17(b), and Eqs. (4.142) and (4.143) are transformed to:

$$\mathbf{p}_1 + \mathbf{p}_2 = \mathbf{p}_3 + \mathbf{p}_4 = \mathbf{q} \qquad (4.144)$$

and

$$\mathbf{p} + \mathbf{p'} = \mathbf{p}_1 - \mathbf{p}_4 = \mathbf{p}_3 - \mathbf{p}_2,$$
$$\frac{1}{2}(\mathbf{p}_1 + \mathbf{p}_4) = \frac{1}{2}(\mathbf{p} - \mathbf{p'} + \mathbf{q}), \quad \frac{1}{2}(\mathbf{p}_2 + \mathbf{p}_3) = \frac{1}{2}(\mathbf{p'} - \mathbf{p} + \mathbf{q}) \quad (4.145)$$

---

[26]Note that the cancellation of contributions independent of ω and q in the denominator of (4.128), leading to the appearance of diffusion pole in (4.140), follows, in general, from particle conservation (Ward identity)(S.V. Maleev, B.P. Toperverg, 1975).

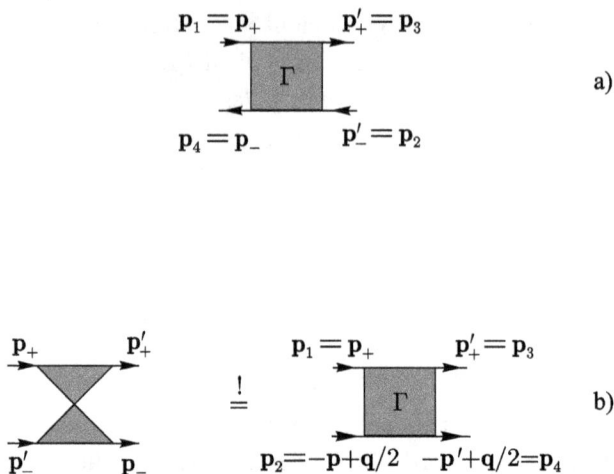

Fig. 4.17 The general vertex part (a) and the vertex obtained from it by reversal of one of electronic lines (b). First diagram (b) is obtained by a simple "unwrapping" of diagram (a), and obviously it is the same as (a). The second diagram (b) is obtained from the first one by the reversal of the direction of the lower electronic line, which is possible in the case of time-reversal symmetry.

Thus, the reversal of the direction of lower electronic line of the diagram of Fig. 4.17(a), acceptable in the case of $t \to -t$ symmetry, is equivalent to the following change of variables of the vertex:

$$\mathbf{q} \to \mathbf{p} + \mathbf{p}' \qquad \mathbf{p} \to \frac{1}{2}(\mathbf{p} - \mathbf{p}' + \mathbf{q}) \qquad \mathbf{p}' \to \frac{1}{2}(\mathbf{p}' - \mathbf{p} + \mathbf{q}) \quad (4.146)$$

which reduces to (4.141).

Let us apply this procedure to diffuson (4.140) and diagrams of Fig. 4.16(b). We can see that the reversal of the lower electronic line on these diagrams leads to the "ladder" in "particle–particle" channel, or, equivalently, to "maximally crossed" diagrams in "particle–hole" channel (first introduced by Langer and Neal (J.S. Langer, T. Neal, 1966)), as shown in Fig. 4.18. Analytic expression for the sum of these diagrams is easily obtained by the simple change of variables (4.146) in the expression for diffuson (4.140):[27]

$$U_{\mathbf{pp}'}^{C}(\mathbf{q}\omega) = \frac{2\gamma U_0}{-i\omega + D_0(\mathbf{p} + \mathbf{p}')^2}; \qquad \mathbf{p} \approx -\mathbf{p}' \qquad (4.147)$$

---

[27]Of course, the same result can be obtained by direct diagram summation!

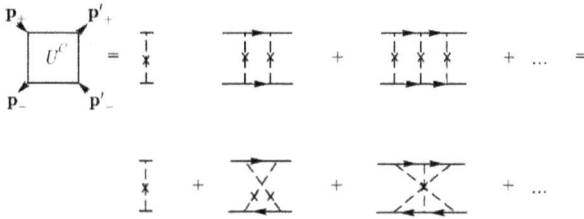

Fig. 4.18 "Cooperon" — reversal of electron line in diffuson gives "ladder" in "particle–particle" channel, or to "maximally-crossed" diagrams in "particle–hole" channel.

Due to the obvious analogy with diagrams, appearing during the analysis of Cooper instability in the theory of superconductivity (cf. Ch. 4), this sum and the result of summation (4.147) is usually called *"Cooperon"* (L.P. Gorkov, A.I. Larkin, D.E. Khmelnitskii, 1979). The necessity of $\mathbf{p} \approx -\mathbf{p}'$ directly follows from the criteria of applicability of diffusion approximation (smallness of q and ω in (4.140)) and corresponds, in the case of Cooperon, to the scattering of particles with almost opposite momenta (nearly "backward" scattering).

Now note, that diagrams of Fig. 4.18 are in fact *irreducible* in RA-channel ("particle–hole"). Thus, we can use Eq. (4.147) as an irreducible vertex $U_{\mathbf{pp}'}(\mathbf{q}\omega)$ in Bethe–Salpeter equation (4.126), shown diagrammatically in Fig. 4.15. In this way we, of course, obtain much more complicated approximation, than those given by (4.127), which takes into account, as will be shown below, quantum (localization) corrections to electron propagation in the field of random impurities (and also to conductivity).

Returning to our general analysis, let us use (4.126) in (4.125) and write down Bethe–Salpeter equation for the two-particle Green's function in the following form:

$$\Phi_{\mathbf{pp}'}^{RA}(\mathbf{q}\omega) = G^R(E + \omega \mathbf{p}_+) G^A(E \mathbf{p}_-) \left\{ -\frac{1}{2\pi i} \delta(\mathbf{p} - \mathbf{p}') \right.$$

$$\left. + \sum_{\mathbf{p}'} U_{\mathbf{pp}''}(\mathbf{q}\omega) \Phi_{\mathbf{p}''\mathbf{p}'}^{RA}(\mathbf{q}\omega) \right\} \qquad (4.148)$$

which is shown diagrammatically in Fig. 4.19. The product of two

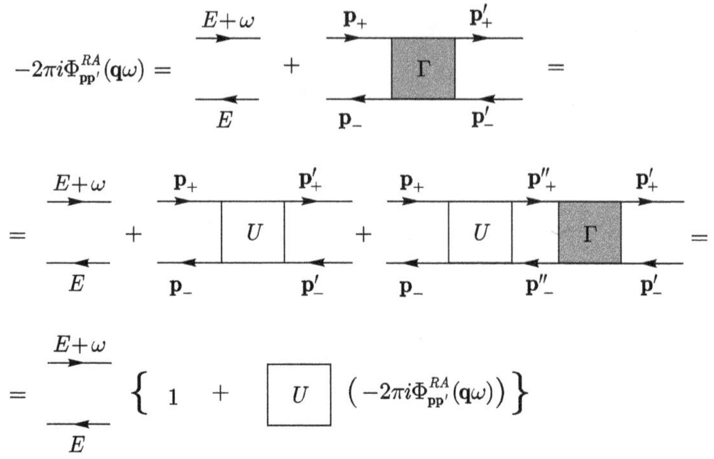

Fig. 4.19 Bethe–Salpeter equation for the two-particle Green's function.

Green's functions entering this equation can be rewritten as:

$$G^R\left(E+\omega\mathbf{p}+\frac{\mathbf{q}}{2}\right)G^A\left(E\mathbf{p}-\frac{\mathbf{q}}{2}\right)$$

$$=\frac{1}{(E+\omega-\xi(\mathbf{p}_+)-\Sigma^R(E+\omega,\mathbf{p}_+))(E+\xi(\mathbf{p}_-)-\Sigma^A(E,\mathbf{p}_-))}$$

$$=\left\{\frac{1}{E-\xi(\mathbf{p}_-)-\Sigma^A(E,\mathbf{p}_-)}-\frac{1}{E+\omega-\xi(\mathbf{p}_+)-\Sigma^R(E+\omega,\mathbf{p}_+)}\right\}$$

$$\times\frac{1}{\omega+(\xi(\mathbf{p}_-)-\xi(\mathbf{p}_+))-\Sigma^R+\Sigma^A}$$

$$=-\frac{\Delta G_\mathbf{p}}{\omega+(\xi(\mathbf{p}_-)-\xi(\mathbf{p}_+))-\Sigma^R+\Sigma^A}\equiv-\frac{G^R-G^A}{(G^R)^{-1}-(G^A)^{-1}}\quad(4.149)$$

where $\Delta G_\mathbf{p}=G^R(E\mathbf{p}_-)-G^A(E+\omega\mathbf{p}_+)$. Then, taking into account
(4.149) and $\xi(\mathbf{p}_-)-\xi(\mathbf{p}_+)\approx-\frac{1}{m}\mathbf{p}\mathbf{q}$, we can rewrite (4.148) as:

$$\left\{\omega-\frac{1}{m}\mathbf{p}\mathbf{q}-\Sigma^R(E+\omega\mathbf{p}_+)+\Sigma^A(E\mathbf{p}_-)\right\}\Phi^{RA}_{\mathbf{pp}'}(\mathbf{q}\omega)$$

$$=\Delta G_\mathbf{p}\left\{\frac{1}{2\pi i}\delta(\mathbf{p}-\mathbf{p}')-\sum_{\mathbf{p}''}U_{\mathbf{pp}''}\Phi^{RA}_{\mathbf{p}''\mathbf{p}'}(\mathbf{q}\omega)\right\}\quad(4.150)$$

Sometimes Eq. (4.150) is called the generalized kinetic (transport) equation.

Let us now introduce (without a complete proof) an important Ward identity, which gives an *exact* relation between the self-energy and irreducible vertex part for our impurity scattering problem (D. Vollhardt, P. Wölfle, 1980):

$$\Sigma^R(E + \omega \mathbf{p}_+) - \Sigma^A(E\mathbf{p}_-) = \sum_{\mathbf{p}'} U_{\mathbf{p}\mathbf{p}'}(\mathbf{q}\omega)\Delta G_{\mathbf{p}'} \qquad (4.151)$$

This identity can be used as an important check of self-consistency of different diagrammatic approximations. It will also be used during the derivation of general equations of self-consistent theory of localization.

We shall give here only an idea of the proof of the Ward identity (4.151). In fact, this identity follows from the simple fact, that all diagrams for the irreducible vertex $U_{\mathbf{p}\mathbf{p}'}(\mathbf{q}\omega)$ can be obtained by "cutting" (internal) electronic line in all diagrams for the self-energy in all possible ways. Consider as an example the typical diagram of the second order for the self-energy $\Sigma_{\mathbf{p}}^{(2)}$, shown in Fig. 4.20(a). Direct calculations give:

$$\Delta\Sigma_{\mathbf{p}}^{(2)} = \Sigma_{\mathbf{p}_+}^{R(2)} - \Sigma_{\mathbf{p}_-}^{A(2)}$$

$$= \sum_{\mathbf{p}_1\mathbf{p}_2} U(\mathbf{p}_1)U(\mathbf{p}_2)\left\{G_{\mathbf{p}_+-\mathbf{p}_1}^R G_{\mathbf{p}_+-\mathbf{p}_1-\mathbf{p}_2}^R G_{\mathbf{p}_+-\mathbf{p}_2}^R - G_{\mathbf{p}_--\mathbf{p}_1}^A G_{\mathbf{p}_--\mathbf{p}_1-\mathbf{p}_2}^A G_{\mathbf{p}_--\mathbf{p}_2}^A\right\}$$

$$= \sum_{\mathbf{p}_1\mathbf{p}_2} U(\mathbf{p}_1)U(\mathbf{p}_2)\left\{G_{\mathbf{p}_+-\mathbf{p}_1}^R G_{\mathbf{p}_+-\mathbf{p}_1-\mathbf{p}_2}^R \Delta G_{\mathbf{p}-\mathbf{p}_2} + G_{\mathbf{p}_+-\mathbf{p}_1}^R \Delta G_{\mathbf{p}-\mathbf{p}_1-\mathbf{p}_2}^R G_{\mathbf{p}_--\mathbf{p}_2}^A\right.$$

$$\left. + \Delta G_{\mathbf{p}-\mathbf{p}_1} G_{\mathbf{p}_--\mathbf{p}_1-\mathbf{p}_2}^A G_{\mathbf{p}_--\mathbf{p}_2}^A\right\}$$

$$= \sum_{\mathbf{p}_1\mathbf{p}_2}\left\{U(\mathbf{p}_1)U(\mathbf{p}-\mathbf{p}')G_{\mathbf{p}_+-\mathbf{p}_1}^R G_{\mathbf{p}'_+-\mathbf{p}_1}^R - U(\mathbf{p}_1)U(\mathbf{p}-\mathbf{p}'-\mathbf{p}_1)G_{\mathbf{p}_+-\mathbf{p}_1}^R G_{\mathbf{p}'_-+\mathbf{p}_1}^A\right.$$

$$\left. + U(\mathbf{p}-\mathbf{p}')U(\mathbf{p}_2)G_{\mathbf{p}'_--\mathbf{p}_2}^A G_{\mathbf{p}_--\mathbf{p}_2}^A\right\}\Delta G_{\mathbf{p}'} = \sum_{\mathbf{p}'} U_{\mathbf{p}\mathbf{p}'}^{(2)}(\mathbf{q}\omega)\Delta G_{\mathbf{p}'} \qquad (4.152)$$

Now we can convince ourselves that an expression for $U_{\mathbf{p}\mathbf{p}'}^{(2)}(\mathbf{q}\omega)$ (defined by the figure bracket in an expression before the last formula) is given by diagrams, shown in Fig. 4.20(b), which are obtained by all possible "cuts" of internal electronic lines in $\Sigma^{(2)}(\mathbf{p})$ (and by the reversal of one of those remaining uncut). The generalization of this analysis to the case of an arbitrary diagram of higher order gives us the complete proof of the Ward identity given in (4.151).

4.6 Combinatorics of diagrams

In any problem, dealing with summation of Feynman diagrams, any kind of information on combinatorics of graphs, i.e. on the number of diagrams of different

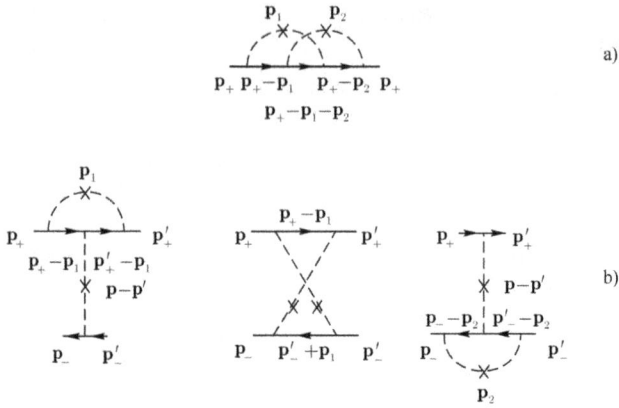

Fig. 4.20 Typical diagram of the second order for electron self-energy (a) and corresponding diagrams for the irreducible vertex $U_{\mathbf{pp}'}(\mathbf{q}\omega)$ in the Ward identity (b).

types for the given order of perturbation theory, is quite useful. In the problem of an electron in Gaussian random field the complete analysis of this problem can be performed using Keldysh model (E.Z. Kuchinskii, M.V. Sadovskii, 1998). We have seen above, that the total number of diagrams in n-th order of perturbation theory for the Green's function is determined by (4.49), which corresponds to the number of ways to connect $2n$ vertices by n impurity lines. Naturally, this result is not specific to Keldysh model, but is always valid within impurity technique for a general case of the Gaussian random field. The similar problem for the number of diagrams in the expansion of self-energy Σ_n is much more difficult. However it can also be solved exactly. Let us return to an exact solution for the Green's function in Keldysh model, written as (4.59):

$$G(E,p) = \frac{1}{W}\Psi\left(\frac{1}{WG_0(E,p)}\right) \qquad (4.153)$$

where the function (4.58):

$$\Psi(z) = -\frac{1}{\sqrt{2\pi}}\int\limits_{-\infty}^{\infty} dt e^{-\frac{t^2}{2}}\frac{1}{t-z} \qquad (4.154)$$

Consider the self-energy part corresponding to the Green's function (4.153), dealing with its expansion over "skeleton" graphs, containing only Green's functions "dressed" by impurity scattering. Addition of a new impurity line to any diagram in this problem leads just to the appearance of an additional factor W^2G^2, so that the self-energy part can be written as:

$$\Sigma = Q(W^2G^2)W^2G \qquad (4.155)$$

where $Q(x)$ is some function. We shall see, that this function is the generating function for the number of "skeleton" graphs for the self-energy part, i.e. the coefficients of its Taylor expansion determine the required numbers Σ_n.

The Dyson's equation for the problem under consideration is written as:

$$G = G_0 + G_0\Sigma G = G_0\left(1 + Q(W^2G^2)W^2G^2\right) \tag{4.156}$$

Introducing $z = (WG_0)^{-1}$ and $y = W^2G^2$, from Eqs. (4.153) and (4.156) we obtain the following parametric representation for $Q(y)$:

$$1 + yQ(y) = z\Psi(z) = z\sqrt{y}$$
$$y = \Psi^2(z) \tag{4.157}$$

This representation of Q is rather inconvenient. Below we show that this function obeys certain differential equation. It is easily seen that previously introduced function $\Psi(z)$, satisfies the usual dispersion relation:[28]

$$Re\Psi(z) = \frac{1}{\pi}\int\limits_{-\infty}^{\infty}dt\frac{Im\Psi(t)}{t-z}; \qquad \frac{1}{\pi}Im\Psi(t) = \mp\frac{1}{\sqrt{2\pi}}e^{-\frac{t^2}{2}} \tag{4.158}$$

from which it follows immediately that $\Psi(z)$ satisfies the differential equation:

$$\frac{d\Psi}{dz} = 1 - z\Psi \tag{4.159}$$

with the initial condition:

$$\Psi(z = \pm i0) = \mp i\sqrt{\frac{\pi}{2}} \tag{4.160}$$

Differentiating the first equation in (4.157) over y, we obtain:

$$\frac{dz}{dy} = \frac{1}{2}y^{-\frac{3}{2}}\left\{2y^2\frac{dQ(y)}{dy} + yQ(y) - 1\right\} \tag{4.161}$$

Differentiating the second-equation in (4.157) over z and using (4.159) we obtain:

$$\frac{dy}{dz} = 2\Psi(z)\frac{d\Psi(z)}{dz} = 2\Psi(z)(1 - z\Psi(z)) = -2y^{\frac{3}{2}}Q(y) \tag{4.162}$$

Comparing (4.161) and (4.162) we obtain the following non-linear differential equation for $Q(y)$:

$$\frac{dQ(y)}{dy} = \frac{1}{2y^2}\left\{1 - Q^{-1}(y) + yQ(y)\right\} \tag{4.163}$$

Using (4.157) and (4.160) we get $y = \Psi^2(z)\big|_{z=\pm i0} = -\frac{\pi}{2}$, so that

$$Q(-\frac{\pi}{2}) = \frac{z\Psi(z) - 1}{y}\bigg|_{z=\pm i0} = \frac{2}{\pi} \tag{4.164}$$

---

[28]The sign of imaginary part is defined as we consider either retarded or advanced Green's function.

which gives the initial condition for Eq. (4.163). Note that the condition $Q(0) = 1$, which obviously follows from diagram expansion for Σ, is a special one for Eq. (4.163) and can not be used as an initial condition. Eq. (4.163) can be expressed in more convenient form:

$$Q(y) = 1 + y\frac{d}{dy}yQ^2(y) \tag{4.165}$$

We are interested in Taylor expansion for $Q(y)$:

$$Q(y) = \sum_{n=0}^{\infty} a_n y^n \tag{4.166}$$

As the number of "skeleton" diagrams of the n-th order for the self-energy part is just the coefficient of W^{2n} in the expansion of Σ over the powers of W^2, it is easily seen that Eq. (4.155) gives the required value of Σ_n as:

$$\Sigma_n = a_{n-1} \tag{4.167}$$

This means that function $Q(y)$ is the generating function for combinatorial factors Σ_n.

Substitution of Eq. (4.166) into Eq. (4.165) leads to the following recurrence relation for the coefficients a_n:

$$a_n = n\sum_{m=0}^{n-1} a_m a_{n-1-m} \tag{4.168}$$

where $a_0 = 1$. From $a_0 = 1$ it follows that $Q(0) = 1$. It is due to this fact this point is special — equation $Q(0) = 1$ is satisfied for any initial conditions, for which Eq. (4.165) has a solution. From Eq. (4.168) it is easy to find the values of a_n for small values of n, the appropriate results are presented in Table below.

The knowledge of combinatorics for the self-energy part allows to find also the combinatorics for the two-particle Green's function — both for the full vertex-part Γ and for the irreducible vertex U. The appropriate diagram representations of these vertices is shown in Figs. 4.14, 4.15. The self-energy part Σ is connected with the vertex-part Γ by the equation shown graphically in Fig. 4.9(a). For the problem with zero transferred momentum this equation has the following form:

$$\Sigma = W^2 G(1 + G^2\Gamma) \tag{4.169}$$

Thus, for the number of diagrams in the n-th order of the full vertex Γ_n, we obtain immediately:

$$\Gamma_N = \Sigma_{n+1} = a_n \tag{4.170}$$

In this sense $Q(y)$ is also the generating function for the number of diagrams for the full vertex-part.

The number of diagrams of the n-th order for the irreducible vertex-part U_n can be easily obtained if we note, that the cut of any of $2n - 1$ internal Green's function lines in the diagram for the self-energy part of N-th order produces the appropriate diagram for the n-th order contribution to U. Thus:

$$U_n = (2n - 1)\Sigma_n = (2n - 1)a_{n-1} \tag{4.171}$$

In the limit of large orders of perturbation theory $n \gg 1$ the use of the recurrence relation (4.168) becomes inconvenient due to the factorial growth of the number of diagrams. At the same time the fact of this factorial growth itself can be used for significant simplification of the problem. Let us rewrite (4.168) as:

$$a_n = 2na_0a_{n-1} + 2na_1a_{n-2} + 2na_2a_{n-3} + \cdots \tag{4.172}$$

where $a_0 = 1$, $a_1 = 1$, $a_2 = 4$. It is natural to assume that in the limit of large n we have $a_n \approx (2n + \beta)a_{n-1}$, then $a_{n-2} \approx \frac{a_{n-1}}{2n-2+\beta}$ etc. Substituting these expressions into (4.172) immediately leads to $\beta = 1$ and

$$a_n = (2n + 1 + O(\frac{1}{n}))a_{n-1} \tag{4.173}$$

This means that in the limit of large n we have $a_n \sim (2n + 1)!!$. Let us define b_n as:

$$b_n = \frac{a_n}{(2n + 1)!!} \tag{4.174}$$

Substituting (4.174) into (4.168), we obtain the recurrence relation for b_n:

$$b_n = n \sum_{m=0}^{n-1} \frac{(2m + 1)!!(2n - 2m - 1)!!}{(2n + 1)!!} b_m b_{n-1-m} \tag{4.175}$$

with $b_0 = 1$. In the limit of large n, taking into account $b_1 = \frac{1}{3}$, $b_2 = \frac{4}{15}$, and limiting ourselves to the accuracy of the order of b/n^2 (where $b \sim b_n \sim b_{n-1} \sim b_{n-2} \sim b_{n-3}$), we get:

$$\Delta b_n = b_n - b_{n-1} = \frac{5}{4}\frac{b_{n-1}}{n^2} + O(\frac{b}{n^3}) \tag{4.176}$$

Thus, in the limit of large n we can write down the following differential equation for b_n:

$$\frac{db_n}{dn} = \frac{5}{4}\frac{b_n}{n^2} + O(\frac{b}{n^3}) \tag{4.177}$$

from which it follows immediately:

$$b_n = b \cdot exp\left(-\frac{5}{4}\frac{1}{n} + O(\frac{1}{n^2})\right) = b\left\{1 - \frac{5}{4}\frac{1}{n} + O(\frac{1}{n^2})\right\} \tag{4.178}$$

It is natural, that this analysis can not provide us with the value of the constant $b = \lim_{n \to \infty} b_n$. Numerical study of b_n using the recurrence relation (4.175) completely supports the dependence defined by Eq. (4.178) and leads to the value of $b = \frac{1}{e} = 0.36787944\cdots$. This funny result can also be derived analytically (I.M. Suslov, 2007), but we shall not go into further details here.

Finally, the asymptotic expressions for the number of diagrams of different types for large n have the following form:

$$\Sigma_n = a_{n-1} = b_{n-1}(2n-1)!! = \frac{1}{e}\left\{1 - \frac{5}{4}\frac{1}{n} + O(\frac{1}{n^2})\right\}(2n-1)!!$$

$$= \frac{1}{\sqrt{\pi e}}\left\{1 - \frac{5}{4}\frac{1}{n} + O(\frac{1}{n^2})\right\}2^n\Gamma(n+\frac{1}{2}) \qquad (4.179)$$

$$\Gamma_n = a_n = \frac{1}{e}\left\{1 - \frac{5}{4}\frac{1}{n} + O(\frac{1}{n^2})\right\}(2n+1)!!$$

$$= \frac{1}{\sqrt{\pi e}}\left\{1 - \frac{5}{4}\frac{1}{n} + O(\frac{1}{n^2})\right\}2^{n+1}\Gamma(n+\frac{3}{2}) \qquad (4.180)$$

$$U_n = (2n-1)a_{n-1} = \frac{1}{e}\left\{1 - \frac{5}{4}\frac{1}{n} + O(\frac{1}{n^2})\right\}(2n-1)(2n-1)!!$$

$$= \frac{1}{e}\left\{1 - \frac{9}{4}\frac{1}{n} + O(\frac{1}{n^2})\right\}(2n+1)!!$$

$$= \frac{1}{\sqrt{\pi e}}\left\{1 - \frac{9}{4}\frac{1}{n} + O(\frac{1}{n^2})\right\}2^{n+1}\Gamma(n+\frac{3}{2}) \qquad (4.181)$$

It is interesting to note that:

$$\frac{\Sigma_n}{G_n} = b_{n-1} = \frac{1}{e}\left\{1 - \frac{5}{4}\frac{1}{n} + O(\frac{1}{n^2})\right\} \to \frac{1}{e} \qquad (4.182)$$

$$\frac{U_n}{\Gamma_n} = 1 - \frac{1}{n} + O(\frac{1}{n^2}) \to 1 \qquad (4.183)$$

In Table below we present the summary of the main results for the number of diagrams of different types.

n	$\Gamma_n = a_n$	$b_n = a_n/(2n+1)!!$	$\Sigma_n = a_{n-1}$	$U_n = (2n-1)a_{n-1}$
1	1	0.3333	1	1
2	4	0.2667	1	3
3	27	0.2571	4	20
4	248	0.2624	27	189
5	2830	0.2722	248	2232
6	38232	0.2829	2830	3130
7	593859	0.2930	38232	497016
8	10401712	0.3019	593859	8907885
9	202601898	0.3158	10401712	176829104
10	4342263000	0.3211	202601898	3849436062
$n \gg 1$	$\frac{1}{e}[1-\frac{5}{4n}](2n+1)!!$	$\frac{1}{e}[1-\frac{5}{4n}]$	$\frac{1}{e}[1-\frac{5}{4n}](2n-1)!!$	$\frac{1}{e}[1-\frac{9}{4n}](2n+1)!!$

4.7 Quantum corrections, self-consistent theory of localization and Anderson transition

4.7.1 *Quantum corrections to conductivity*

Information contained in two-particle Green's function (4.124) is, in some sense, excessive. To calculate conductivity we need only to know (4.123), i.e. two-particle Green's function summed over the momenta $\mathbf{p}$ and $\mathbf{p}'$. From (4.150) we can directly obtain an approximate system of equations determining this function. Let us sum over $\mathbf{p}$ and $\mathbf{p}'$ both sides of Eq. (4.150) using also the Ward identity (4.151). Then we immediately get:

$$\omega \Phi^{RA}(\omega \mathbf{q}) - v_F q \Phi_1^{RA}(\omega \mathbf{q}) = -N(E) \qquad (4.184)$$

where we have introduced

$$\Phi_1^{RA}(\omega \mathbf{q}) = 2 \sum_{\mathbf{pp}'} (\hat{\mathbf{p}}\hat{\mathbf{q}}) \Phi_{\mathbf{pp}'}^{RA}(\omega \mathbf{q}) \qquad (4.185)$$

Here $\hat{\mathbf{p}}$ and $\hat{\mathbf{q}}$ are the unit vectors along the directions of $\mathbf{p}$ and $\mathbf{q}$, $|\mathbf{p}| \approx |\mathbf{p}'| \approx p_F$, $v_F = \frac{p_F}{m} = \sqrt{\frac{2E}{m}}$, while $N(E)$ is the density of states at the Fermi level for both spin projections:

$$N(E) = 2\nu(E) = -\frac{2}{\pi} \sum_{\mathbf{p}} Im G^R(E\mathbf{p}) = - \lim_{\omega \to 0, q \to 0} \frac{2}{2\pi i} \sum_{\mathbf{p}} \Delta G_{\mathbf{p}} \qquad (4.186)$$

The appearance of the spin factor of 2 here is due to its presence in our definitions (4.88), (4.185). In the r.h.s. of (4.184) we can limit ourselves to the limit used in (4.186) because $N(E)$ is practically constant on the energy interval of $\omega \ll E$, $v_F q \ll E$.

But now we have a new function $\Phi_1^{RA}(\omega \mathbf{q})$ defined by (4.185). For this function we also derive an equation using (4.150). Let us multiply (4.150) by $(\hat{\mathbf{p}}\hat{\mathbf{q}})$, sum both sides of the equation over $\mathbf{p}$ and $\mathbf{p}'$, use the Ward identity (4.151) and also the following approximate representation:[29]

$$\sum_{\mathbf{p'}} \Phi^{RA}_{\mathbf{pp'}}(\omega\mathbf{q}) \approx -\frac{\Delta G_{\mathbf{p}}}{2\pi i \nu(E)} \sum_{\mathbf{p'p''}} \{1 + d(\hat{\mathbf{p}}\hat{\mathbf{q}})(\hat{\mathbf{p}}''\hat{\mathbf{q}})\} \Phi^{RA}_{\mathbf{p''p'}}(\omega\mathbf{q})$$

(4.192)

we obtain:

$$[\omega + M(\mathbf{q}\omega)] \Phi^{RA}_1(\omega\mathbf{q}) - \frac{1}{d}v_F q \Phi^{RA}(\omega\mathbf{q}) = 0 \qquad (4.193)$$

where we have introduced the so-called *relaxation kernel*:

$$M(\mathbf{q}\omega) = 2i\gamma + \frac{id}{2\pi\nu(E)} \sum_{\mathbf{pp'}} (\hat{\mathbf{p}}\hat{\mathbf{q}})\Delta G_{\mathbf{p}} U_{\mathbf{pp'}}(\mathbf{q}\omega)\Delta G_{\mathbf{p'}}(\hat{\mathbf{p}}'\hat{\mathbf{q}}) \quad (4.194)$$

with $\gamma = \pi\rho v^2 \nu(E)$ the usual Born frequency of impurity scattering. Now Eqs. (4.184) and (4.193) form the closed system, allowing us to express two-particle function $\Phi^{RA}(\mathbf{q}\omega)$ via $M(\mathbf{q}\omega)$, which, in turn, is expressed via the irreducible vertex part $U_{\mathbf{pp'}}(\mathbf{q}\omega)$ with the help of Eq. (4.194).

4.7.1.1 *Technical details*

Let us present the detailed derivation of Eqs. (4.193) and (4.194). Multiplying both sides of (4.150) by $\frac{\mathbf{p}\hat{\mathbf{q}}}{m}$ and performing the summation over $\mathbf{p}$ and $\mathbf{p'}$ we

---

[29]Note that (4.192) reduces to the first two terms of the following expansion over Legendre polynomials:

$$\sum_{\mathbf{p'}} \Phi^{RA}_{\mathbf{pp'}} = \sum_{l=0}^{\infty} P_l(\cos\theta_{\mathbf{pq}})\Phi^l_p \approx \Phi^0_{\mathbf{p}} + \cos\theta_{\mathbf{pq}}\Phi^1_{\mathbf{p}} + \cdot \qquad (4.187)$$

where $\theta_{\mathbf{pq}}$ is an angle between vectors $\mathbf{p}$ and $\mathbf{q}$. Assuming $\Phi^0_{\mathbf{p}} \sim \Phi^1_{\mathbf{p}} \sim \Delta G_{\mathbf{p}}$, we can write:

$$\Phi^0_{\mathbf{p}} = -[2\pi i\nu(E)]^{-1}\Delta G_{\mathbf{p}} \sum_{\mathbf{p'p''}} \Phi^{RA}_{\mathbf{p''p'}}(\omega\mathbf{q}) \qquad (4.188)$$

$$\Phi^1_{\mathbf{p}} = -[2\pi i\nu(E)]^{-1}\Delta G_{\mathbf{p}} \sum_{\mathbf{p'p''}} \frac{(\mathbf{p}''\hat{\mathbf{q}})}{p_F}\Phi_{\mathbf{p''p'}}(\omega\mathbf{q}) \qquad (4.189)$$

which was, in fact, done in (4.192). For $\omega \to 0$, $q \to 0$, with the account of (4.186) and $|\mathbf{p}| \approx |\mathbf{p'}| \approx p_F$, we have:

$$\sum_{\mathbf{pp'}} \Phi^{RA}_{\mathbf{pp'}} = -\frac{1}{2\pi i\nu(E)}\sum_{\mathbf{p}} \Delta G_{\mathbf{p}} \sum_{\mathbf{p'p''}} \Phi^{RA}_{\mathbf{p''p'}} = \sum_{\mathbf{p'p''}} \Phi^{RA}_{\mathbf{p''p'}} \qquad (4.190)$$

$$\sum_{\mathbf{pp'}} (\mathbf{p}\hat{\mathbf{q}})\Phi^{RA}_{\mathbf{pp'}} = -\frac{d}{2\pi i\nu(E)}\sum_{\mathbf{p}} \Delta G_{\mathbf{p}} \frac{(\mathbf{p}\hat{\mathbf{q}})^2}{p_F^2} \sum_{\mathbf{p'p''}} (\mathbf{p}''\hat{\mathbf{q}})\Phi^{RA}_{\mathbf{p''p'}} = \sum_{\mathbf{p'p''}} (\mathbf{p}''\hat{\mathbf{q}})\Phi^{RA}_{\mathbf{p''p'}}$$

(4.191)

where d is the number of space dimensions.

obtain:

$$\sum_{\mathbf{pp'}} \frac{\mathbf{p\hat{q}}}{m} \left[\omega - \frac{\mathbf{pq}}{m} - \underbrace{\Sigma^R(E+\omega, \mathbf{p_+}) + \Sigma^A(E\mathbf{p_-})}_{-\sum_{\mathbf{p''}} U_{\mathbf{pp''}}(\mathbf{q}\omega)\Delta G_{\mathbf{p''}}} \right] \Phi_{\mathbf{pp'}}^{RA}(\mathbf{q}\omega)$$

$$= \sum_{\mathbf{pp'}} \frac{\mathbf{p\hat{q}}}{m} \Delta G_{\mathbf{p}} \frac{1}{2\pi i} \delta(\mathbf{p} - \mathbf{p'}) - \sum_{\mathbf{pp'}} \frac{\mathbf{p\hat{q}}}{m} \Delta G_{\mathbf{p}} \sum_{\mathbf{p''}} U_{\mathbf{pp''}} \Phi_{\mathbf{p''p'}}^{RA}(\mathbf{q}\omega) \quad (4.195)$$

where the difference of self-energies in the l.h.s. is rewritten with the use of the Ward identity (4.151). Then we get:

$$\omega \sum_{\mathbf{pp'}} \frac{\mathbf{p\hat{q}}}{m} \Phi_{\mathbf{pp'}}^{RA}(\mathbf{q}\omega) - \sum_{\mathbf{pp'}} \underbrace{\frac{(\mathbf{p\hat{q}})^2}{m^2}}_{\frac{1}{d}\frac{p^2}{m^2} = \frac{2E}{dm}} \Phi_{\mathbf{pp'}}^{RA}(\mathbf{q}\omega) - \sum_{\mathbf{pp'}} \frac{(\mathbf{p\hat{q}})}{m} \sum_{\mathbf{p''}} U_{\mathbf{pp''}} \Delta G_{\mathbf{p''}} \Phi_{\mathbf{pp'}}^{RA}(\mathbf{q}\omega)$$

$$= \underbrace{\sum_{\mathbf{p}} \frac{\mathbf{p\hat{q}}}{m} \frac{\Delta G_{\mathbf{p}}}{2\pi i}}_{\text{Zero after the angular integration!}} - \sum_{\mathbf{pp'}} \frac{\mathbf{p\hat{q}}}{m} \Delta G_{\mathbf{p}} \sum_{\mathbf{p''}} U_{\mathbf{pp''}} \Phi_{\mathbf{p''p'}}^{RA}(\mathbf{q}\omega)$$

$$(4.196)$$

which gives:

$$\omega \sum_{\mathbf{pp'}} \frac{\mathbf{p\hat{q}}}{m} \Phi_{\mathbf{pp'}}^{RA}(\mathbf{q}\omega) + \underbrace{\sum_{\mathbf{pp'}} \frac{\mathbf{p\hat{q}}}{m} \Delta G_{\mathbf{p}} \sum_{\mathbf{p''}} U_{\mathbf{pp''}} \Phi_{\mathbf{p''p'}}^{RA}(\mathbf{q}\omega)}_{(I)}$$

$$= \underbrace{\sum_{\mathbf{pp'}} \frac{\mathbf{p\hat{q}}}{m} \sum_{\mathbf{p''}} U_{\mathbf{pp''}} \Delta G_{\mathbf{p''}} \Phi_{\mathbf{pp'}}^{RA}(\mathbf{q}\omega)}_{(II)} + q \frac{2E}{dm} \sum_{\mathbf{pp'}} \Phi_{\mathbf{pp'}}^{RA}(\mathbf{q}\omega) \quad (4.197)$$

Let us consider the contributions (I) and (II) separately. Using (4.192) we have:

$$(I) = \sum_{\mathbf{pp''}} \frac{\mathbf{p\hat{q}}}{m} \Delta G_{\mathbf{p}} U_{\mathbf{pp''}} \sum_{\mathbf{p'}} \Phi_{\mathbf{pp'}}^{RA}$$

$$= -\sum_{\mathbf{pp''}} \frac{\mathbf{p\hat{q}}}{m} \Delta G_{\mathbf{p}} U_{\mathbf{pp''}} \frac{\Delta G_{\mathbf{p''}}}{2\pi i \nu(E)} \sum_{\mathbf{p'p'''}} \left[1 + d(\mathbf{p''\hat{q}})(\mathbf{p'''\hat{q}})/p_F^2 \right] \Phi_{\mathbf{p'''p'}}^{RA}$$

$$= \underbrace{\frac{i}{2\pi\nu(E)} \sum_{\mathbf{pp''}} \frac{\mathbf{p\hat{q}}}{m} \Delta G_{\mathbf{p}} U_{\mathbf{pp''}} \Delta G_{\mathbf{p''}} \sum_{\mathbf{p'p'''}} \Phi_{\mathbf{p'''p}}^{RA}}_{(III)}$$

$$+ \frac{id}{2\pi\nu(E)p_F^2} \sum_{\mathbf{pp''}} (\mathbf{p\hat{q}}) \Delta G_{\mathbf{p}} U_{\mathbf{pp''}} \Delta G_{\mathbf{p''}} (\mathbf{p''\hat{q}}) \sum_{\mathbf{p'p'''}} \frac{\mathbf{p'''\hat{q}}}{m} \Phi_{\mathbf{p'''p'}}^{RA} \quad (4.198)$$

Similarly:

$$(II) = \sum_{\mathbf{p}} \frac{\mathbf{p}\hat{\mathbf{q}}}{m} \sum_{\mathbf{p}''} U_{\mathbf{p}\mathbf{p}''} \Delta G_{\mathbf{p}''} \sum_{\mathbf{p}'} \Phi_{\mathbf{p}\mathbf{p}'}^{RA}$$

$$= -\sum_{\mathbf{p}} \frac{\mathbf{p}\hat{\mathbf{q}}}{m} \sum_{\mathbf{p}''} U_{\mathbf{p}\mathbf{p}''} \Delta G_{\mathbf{p}''} \frac{\Delta G_{\mathbf{p}}}{2\pi i \nu(E)} \sum_{\mathbf{p}'\mathbf{p}'''} \{1 + d(\mathbf{p}\hat{\mathbf{q}})(\mathbf{p}'''\hat{\mathbf{q}})/p_F^2\} \, \Phi_{\mathbf{p}'''\mathbf{p}'}^{RA}$$

$$= (III) + \frac{id}{2\pi\nu(E)} \sum_{\mathbf{p}} \underbrace{\frac{(\mathbf{p}\hat{\mathbf{q}})^2}{p_F^2}}_{1/d} \Delta G_{\mathbf{p}} \sum_{\mathbf{p}''} U_{\mathbf{p}\mathbf{p}''} \Delta G_{\mathbf{p}''} \sum_{\mathbf{p}'\mathbf{p}'''} \frac{\mathbf{p}'''\hat{\mathbf{q}}}{m} \Phi_{\mathbf{p}'''\mathbf{p}'}^{RA}$$

$$= (III) + \sum_{\mathbf{p}''} U_{\mathbf{p}\mathbf{p}''} \Delta G_{\mathbf{p}''} \sum_{\mathbf{p}'\mathbf{p}'''} \frac{\mathbf{p}'''\hat{\mathbf{q}}}{m} \Phi_{\mathbf{p}'''\mathbf{p}'}^{RA} = (III) - 2i\gamma \sum_{\mathbf{p}\mathbf{p}'} \frac{\mathbf{p}\mathbf{q}}{m} \Phi_{\mathbf{p}\mathbf{p}'}^{RA}$$

$$(4.199)$$

where to obtain the last equality we again used the Ward identity (4.151) and rewritten the difference of self-energies using the simplest approximation (4.25) as $\Sigma^A - \Sigma^R = 2i Im \Sigma^A = 2i\gamma$. Then from (4.198) and (4.199) we obtain:

$$(I) - (II) = 2i\gamma \sum_{\mathbf{p}\mathbf{p}'} \frac{\mathbf{p}\hat{\mathbf{q}}}{m} \Phi_{\mathbf{p}\mathbf{p}'}^{RA}$$

$$+ \frac{id}{2\pi\nu(E)p_F^2} \sum_{\mathbf{p}\mathbf{p}''} (\mathbf{p}\hat{\mathbf{q}}) \Delta G_{\mathbf{p}} U_{\mathbf{p}\mathbf{p}''} \Delta G_{\mathbf{p}''} (\mathbf{p}''\hat{\mathbf{q}}) \sum_{\mathbf{p}\mathbf{p}'} \frac{\mathbf{p}\hat{\mathbf{q}}}{m} \Phi_{\mathbf{p}\mathbf{p}'}^{RA}$$

$$= M(\mathbf{q}\omega) \sum_{\mathbf{p}\mathbf{p}'} \frac{\mathbf{p}\hat{\mathbf{q}}}{m} \Phi_{\mathbf{p}\mathbf{p}'}^{RA} \qquad (4.200)$$

where $M(\mathbf{q}\omega)$ is defined in (4.194). As a result Eq. (4.197) reduces to (4.193), completing our derivation.

In principle, all these manipulations are "almost exact". Most serious limitation of our analysis is the use of the simplest approximation for the self-energy (4.25) in (4.199). If we do not use this simplification, we obtain more general (compared to (4.194)) expression for the relaxation kernel:

$$M(\mathbf{q}\omega) = \frac{1}{2\pi i \nu(E)} \sum_{\mathbf{p}} \Delta G_{\mathbf{p}} \left[\Sigma^R(E + \omega \mathbf{p}_+) - \Sigma^A(E\mathbf{p}_-) \right]$$

$$+ \frac{id}{2\pi\nu(E)p_F^2} \sum_{\mathbf{p}\mathbf{p}'} (\mathbf{p}\hat{\mathbf{q}}) \Delta G_{\mathbf{p}} U_{\mathbf{p}\mathbf{p}'} \Delta G_{\mathbf{p}'} (\mathbf{p}'\hat{\mathbf{q}}) \qquad (4.201)$$

Next everything depends on the approximation we use for the irreducible vertex $U_{\mathbf{p}\mathbf{p}'}$, but explicit results, up to now, are only obtained using the approximate expressions given above and valid, strictly speaking, in the limit of weak enough disorder: $\gamma \ll E$. However, as we shall see below, some kind of self-consistency procedure allows to overcome these limitations.

Solving now the system of equations (4.184), (4.193) we get:

$$\Phi^{RA}(\mathbf{q}\omega) = -N(E)\frac{\omega + M(\mathbf{q}\omega)}{\omega^2 + \omega M(\mathbf{q}\omega) - \frac{1}{d}v_F^2 q^2} \tag{4.202}$$

Using (4.102) we immediately obtain the density–density response function in the following form:

$$\chi(\mathbf{q}\omega) = \omega \Phi^{RA}(\mathbf{q}\omega) + N(E_F) = v_F q \Phi_1^{RA}(\mathbf{q}\omega)$$

$$= -N(E)\frac{\frac{1}{d}v_F^2 q^2}{\omega^2 + \omega M(\mathbf{q}\omega) - \frac{1}{d}v_F^2 q^2} \tag{4.203}$$

For small ω, neglecting ω^2 in the denominator of (4.202) or (4.203), we can write:

$$\Phi^{RA}(\mathbf{q}\omega) = -N(E)\frac{1}{\omega + iD_E(\mathbf{q}\omega)q^2} \tag{4.204}$$

$$\chi(\mathbf{q}\omega) = N(E)\frac{iD_E(\mathbf{q}\omega)q^2}{\omega + iD_E(\mathbf{q}\omega)q^2} \tag{4.205}$$

where we have introduced, by definition, the *generalized diffusion coefficient*:

$$D_E(\mathbf{q}\omega) = i\frac{2E}{dm}\frac{1}{M(\mathbf{q}\omega)} = \frac{v_F^2}{d}\frac{i}{M(\mathbf{q}\omega)} \tag{4.206}$$

which is directly expressed via the relaxation kernel.

Using the general expression (4.106) for conductivity, we obtain:

$$\sigma(\omega) = \frac{ne^2}{m}\frac{i}{\omega + M(0\omega)} \to e^2 D_E(00)N(E) \qquad \omega \to 0 \quad (4.207)$$

where we have used $\frac{n}{N(E)} = \frac{2E}{d}$ (n is electron density).

Thus, all important characteristics of the system are, in fact, expressed via the relaxation kernel (4.194). The question is in what approximation we can calculate it! If we take $U_{\mathbf{p}\mathbf{p}'}(\mathbf{q}\omega)$ in (4.194) in the simplest possible approximation (4.127), Eq. (4.194) in the limit of small disorder (precisely when this approximation is valid!) reduces to:[30]

$$M(00) = \frac{i}{\tau_{tr}} = 2\pi i\rho \sum_{\mathbf{p}'}\delta\left(E - \frac{\mathbf{p}'^2}{2m}\right)|v(\mathbf{p} - \mathbf{p}')|^2(1 - \hat{\mathbf{p}}\hat{\mathbf{p}}') \quad (4.208)$$

---

[30]In ΔG_p we just take here the limit of $\gamma \to 0$.

i.e. the usual expression for the *transport* relaxation time τ_{tr} due to impurity scattering in metals, determining the value of the residual resistivity (remember that here $E = E_F$ and $\hat{\mathbf{p}}\hat{\mathbf{p}}' = \cos\theta_{\mathbf{p}\mathbf{p}'}$)) and (4.207) gives the standard Drude expression:

$$\sigma = \frac{ne^2}{m}\tau_{tr} \qquad (4.209)$$

for $\omega = 0$, or

$$\sigma(\omega) = \frac{ne^2}{m}\frac{\tau_{tr}}{1 - i\omega\tau_{tr}} \qquad (4.210)$$

for the finite frequencies of external field.

Let us now take $U_{\mathbf{p}\mathbf{p}'}(\mathbf{q}\omega)$ as given by Cooperon (4.147). Then it is clear that the presence of diffusion pole can, in general case, lead to *divergence* of relaxation kernel $M(0\omega)$ for $\omega \to 0$, as in this limit:

$$\int dk \frac{k^{d-1}}{\omega + iD_0 k^2} \sim \begin{cases} \frac{1}{\sqrt{\omega}} & d = 1 \\ \ln\omega & d = 2 \end{cases} \qquad (4.211)$$

This ("infrared") divergence appears for $d \leq 2$.[31] Thus, besides the usual Drude contribution (finite for $\omega \to 0$), we can get the singular (in the "infrared" limit of $\omega \to 0$) contribution to $M(0\omega)$, which leads to important physical effects (such as Anderson localization). Our next task is to perform an accurate separation and analysis of such contributions.

Substituting (4.147) into (4.194) we have ($q \to 0,\ \omega \to 0$):

$$M(0\omega) = 2i\gamma - \frac{2id}{\pi\nu(E)p_F^2}\sum_{\mathbf{p}\mathbf{p}'}(\mathbf{p}\hat{\mathbf{q}})ImG^R(E\mathbf{p})U_{\mathbf{p}\mathbf{p}'}(0\omega)ImG^R(E\mathbf{p}')(\mathbf{p}'\hat{\mathbf{q}})$$

$$= 2i\gamma + \frac{4dU_0\gamma}{\pi\nu(E)p_F^2}\sum_{\mathbf{p}\mathbf{p}'}(\mathbf{p}\hat{\mathbf{q}})ImG^R(E\mathbf{p})\frac{1}{\omega + iD_0(\mathbf{p}+\mathbf{p}')^2}ImG^R(E\mathbf{p}')(\mathbf{p}'\hat{\mathbf{q}})$$

$$(4.212)$$

---

[31]For $d \geq 2$ in (4.211) we have to introduce the cut-off at the upper limit for $k \sim l^{-1} \sim v_F^{-1}\gamma$ (which is connected with the applicability limit of diffusion approximation (cf. footnote after (4.131)).

After the variable change $\mathbf{k} = \mathbf{p} + \mathbf{p}'$; $\mathbf{p}' = \mathbf{k} - \mathbf{p}$ we rewrite (4.212) as:

$$M(0\omega) = 2i\gamma + \frac{4dU_0\gamma}{\pi\nu(E)p_F^2}\sum_{\mathbf{pk}}(\mathbf{p}\hat{\mathbf{q}})ImG^R(Ep)\frac{1}{\omega + iD_0\mathbf{k}^2}ImG^R(E\mathbf{p} - \mathbf{k})(\mathbf{k}\hat{\mathbf{q}})$$

$$- \frac{4dU_0\gamma}{\pi\nu(E)p_F^2}\sum_{\substack{\mathbf{pk}\\ \frac{1}{d}p_F^2}}(\mathbf{p}\hat{\mathbf{q}})^2 ImG^R(Ep)\frac{1}{\omega + iD_0\mathbf{k}^2}ImG^R(E\mathbf{p} - \mathbf{k}) \quad (4.213)$$

Note that the second term here is finite for $\omega \to 0$ (we can say that it just "renormalize" $2i\gamma$), and in the third the singular at $\omega \to 0$ contribution appears from the zeroth-order term in the expansion of the integrand over $k \to 0$ (then integral of the type of (4.211) appears). So finally we write:

$$M(0\omega) \approx 2i\gamma - 4U_0^2\sum_{\mathbf{p}}(ImG^R(E\mathbf{p}))^2\sum_{\mathbf{k}}\frac{1}{\omega + iD_0\mathbf{k}^2} \quad (4.214)$$

Now take into account that in our approximation:

$$\sum_{\mathbf{p}}(ImG^R(E\mathbf{p}))^2 \approx \nu(E)\int_{-\infty}^{\infty}d\xi\frac{\gamma^2}{[(E - \xi)^2 + \gamma^2]^2} = \frac{\pi}{2\gamma}\nu(E)$$

$$(4.215)$$

so that:

$$2U_0\sum_{\mathbf{p}}(ImG^R(E\mathbf{p}))^2 = \pi U_0\nu(E)\gamma^{-1} = 1 \quad (4.216)$$

Accordingly, (4.214) reduces to:

$$M(0\omega) = 2i\gamma - 2U_0\sum_{\mathbf{k}}\frac{1}{\omega + iD_0\mathbf{k}^2} \quad (4.217)$$

Consider now in details the case of two-dimensional system $(d = 2)$. We have to calculate the integral:

$$I = \sum_{\mathbf{k}}\frac{1}{\omega + iD_0\mathbf{k}^2} = \frac{\Omega_2}{(2\pi)^2}\int_0^{\Lambda}dkk\frac{1}{\omega + iD_0k^2} = \frac{1}{2\pi}\int_0^{\Lambda}dkk\frac{1}{\omega + iD_0k^2}$$

$$= \frac{1}{4\pi}\int_0^{\Lambda^2}dx\frac{1}{\omega + iD_0x} = \frac{1}{4\pi iD_0}\int_0^{\Lambda^2}\frac{1}{x + \frac{\omega}{iD_0}} = \frac{1}{4\pi iD_0}\int_0^{\Lambda^2}dx\frac{x + i\frac{\omega}{D_0}}{x^2 + \frac{\omega^2}{D_0^2}}$$

$$(4.218)$$

Now we see that:

$$ReI = \frac{1}{4\pi D_0}\frac{\omega}{D_0}\int_0^{\Lambda^2}dx\frac{1}{x^2 + \frac{\omega^2}{D_0^2}} \to 0 \qquad \omega \to 0 \quad (4.219)$$

due to convergence of the integral. On the other hand:

$$ImI = -\frac{1}{4\pi D_0}\int_0^{\Lambda^2} dx\,\frac{x}{x^2 + \frac{\omega^2}{D_0^2}} \tag{4.220}$$

giving the logarithmic divergence for $\omega \to 0$:

$$ImI = -\frac{1}{8\pi D_0}\int_0^{\Lambda^4} dz\,\frac{1}{z + \frac{\omega^2}{D_0^2}} = -\frac{1}{8\pi D_0}\ln\left(1 + \frac{D_0^2\Lambda^4}{\omega^2}\right) \tag{4.221}$$

Accordingly, for $\omega \to 0$ we have:

$$ImI \approx -\frac{1}{8\pi D_0}\ln\frac{D_0^2\Lambda^4}{\omega^2} = -\frac{1}{4\pi D_0}\ln\frac{D_0\Lambda^2}{\omega}$$

$$= -\frac{1}{4\pi D_0}\ln\frac{1}{\omega\tau} = -\frac{m}{2}\frac{1}{2\pi E\tau}\ln\frac{1}{\omega\tau} \tag{4.222}$$

where we have chosen the cut-off Λ such that:[32]

$$D_0\Lambda^2 \sim \frac{1}{\tau} = 2\gamma \tag{4.223}$$

which gives:

$$\Lambda \sim \sqrt{\frac{2m}{E}\frac{1}{\tau}} = 2\sqrt{\frac{2m}{E}}\gamma \sim l^{-1} \tag{4.224}$$

where $l = v_F\tau$ is the mean-free path, or

$$v_F\Lambda \sim \frac{1}{\tau} = 2\gamma \tag{4.225}$$

with the account of $v_F = \sqrt{\frac{2E}{m}}$ and $D_0 = \frac{E\tau}{m}$ for $d = 2$. Then we obtain:

$$- 2U_0I \approx imU_0\frac{1}{2\pi E\tau}\ln\frac{1}{\omega\tau} \tag{4.226}$$

For $d = 2$ $\nu(E) = \frac{m}{2\pi}$, so that $mU_0 = \frac{m}{2\pi\nu(E)\tau} = \frac{1}{\tau}$.
Finally (4.217) reduces to:

$$M(0\omega) = \frac{i}{\tau} + \frac{i}{\tau}\frac{1}{2\pi E\tau}\ln\frac{1}{\omega\tau} \tag{4.227}$$

Using this expression in (4.206), we find the generalized diffusion coefficient of two-dimensional system as:

$$D(\omega) = D_0\frac{i}{\tau}\frac{1}{M(\omega)} = \frac{D_0}{1 + \frac{1}{2\pi E\tau}\ln\frac{1}{\omega\tau}}$$

$$\approx D_0\left\{1 - \frac{1}{2\pi E\tau}\ln\frac{1}{\omega\tau}\right\} \tag{4.228}$$

---

[32]Later we shall discuss this choice in detail!

where in the last equality we have taken into account the fact that all our procedures are valid only in the limit of weak disorder, when $2\pi E\tau \gg 1$. The second term in (4.228) describes quantum corrections to diffusion (conductivity) in two-dimensional system of electrons and impurities (L.P. Gorkov, A.I. Larkin, D.E. Khmelnitskii, 1979). As the sign of this correction (in this simple case of potential scattering) is negative (diminishing diffusion coefficient compared to its classical Drude value D_0), this phenomenon is often called "weak localization".[33]

All the previous analysis was performed for the case of $T = 0$. For finite temperatures, inelastic scattering of electrons becomes important, leading to "phase decoherence" of wave functions, with characteristic time $\tau_\phi = AT^{-p}$ (the power p depends on the type of inelastic scattering). Thus, the expression (4.228) is changed to:

$$D(\omega) = D_0 \left\{ 1 - \frac{1}{2\pi E\tau} \ln \frac{1}{Max \left[\omega, \frac{1}{\tau_\phi}\right]\tau} \right\} \qquad (4.229)$$

In particular, for $\omega = 0$ we obtain the static conductivity in the following form:

$$\sigma = \sigma_0 \left\{ 1 - \frac{1}{2\pi E\tau} \ln \frac{\tau_\phi}{\tau} \right\} = \sigma_0 \left\{ 1 - \frac{p}{2\pi E\tau} \ln \frac{T_0}{T} \right\} \qquad (4.230)$$

where T_0 is some temperature (energy) scale. Using $\frac{n}{N(E)} = 2\pi \frac{n}{m} = E$ (which is valid for $d = 2$), we obtain $\sigma_0 = \frac{ne^2}{m}\tau = \frac{e^2}{2\pi}E\tau$, so that (4.230) can be rewritten as:

$$\sigma = \frac{e^2}{2\pi}E\tau \left\{ 1 - \frac{1}{2\pi E\tau} \ln \frac{\tau_\phi}{\tau} \right\} \qquad (4.231)$$

Note that $\frac{e^2}{2\pi} = \frac{e^2}{2\pi\hbar} = \frac{e^2}{h}$ defines the quantum scale of conductivity: $\frac{e^2}{\hbar} = 2.5 \cdot 10^{-4} \mathrm{Ohm}^{-1}$.

Logarithmic temperature dependence of the type of (4.230), (4.231) is experimentally observed for low enough temperatures practically in all two-dimensional metallic systems (such as thin films, two-dimensional electron gas in MOSFETs, etc.) [Alshuler B.L., Aronov A.G., Khmelnitskii D.E., Larkin A.I. (1982); Lee P.A., Ramakrishnan T.V. (1985)].

---

[33] If we formally consider in (4.228) the limit of $\omega \to 0$, the logarithmic divergence leads to quantum correction becoming of the order and greater than the classical contribution, signifying the possibility of static conductivity of the system at $T = 0$ becoming zero (metal–insulator or Anderson transition). Of course, in this case we are already outside the limits of applicability of our expressions and special analysis is needed, which will be given later.

4.7.1.2 *"Poor man" interpretation of quantum corrections*

Let us consider now the physical meaning of quantum corrections to conductivity (A.I. Larkin, D.E. Khmelnitskii, 1980). Consider (for the moment!) the case of weak disorder with mean-free path $l \gg \frac{\hbar}{p_F}$ or ($\hbar = 1$) $p_F l \gg 1$, $E\tau \gg 1$. Starting with the usual "metallic" regime of conductivity, we are looking for small corrections due to "weak localization":

$$\sigma = \sigma_0 + \delta\sigma; \qquad |\delta\sigma| \ll \sigma_0 \qquad (4.232)$$

The usual Drude conductivity (the result of the standard transport theory), as we have just seen, is given by $\sigma_0 \sim \frac{e^2}{\hbar^2} E\tau$. In this (classic) theory the different acts of scattering by impurities are considered as independent (uncorrelated). Electrons are moving by (classical) diffusion, so that for the particle which at the moment $t = 0$ is at the point $\mathbf{r}_0$, the probability to arrive at some moment $t > 0$ at the point $\mathbf{r}$ is given by the solution of diffusion equation, which for space of d dimensions is given by:

$$P(\mathbf{r}t) = \frac{e^{-\frac{|\mathbf{r}-\mathbf{r}_0|^2}{4D_0 t}}}{(4\pi D_0 t)^{d/2}} \qquad (4.233)$$

where the classical diffusion coefficient $D_0 = v_F^2 \tau/d$. This probability is essentially nonzero within the volume V_{diff}, which is defined by $|\mathbf{r} - \mathbf{r}_0|^2 \ll 4D_0 t$, when:

$$P(\mathbf{r}t) \sim \frac{1}{V_{diff}} = \frac{1}{(D_0 t)^{d/2}} \qquad (4.234)$$

This is all just a classical picture. Consider now *quantum* propagation of an electron described by Feynman trajectories [Sadovskii M.V. (2019b)] going from some point A to point B, as shown in Fig. 4.21(a). Due to Heisenberg indeterminacy principle each trajectory can be, in fact, represented by a "tube" with finite width of the order of:

$$\lambda_F = \frac{\hbar}{mv_F} \qquad (4.235)$$

Then, the effective crossection of this "tube" is of the order of λ_F^{d-1}. In the classical limit we have $\hbar \to 0$ and $\lambda_F = 0$. In case of weak disorder $\lambda_F/l = \frac{\hbar}{p_F l} \ll 1$ and our "tubes" are thin enough. Let the temperature be low, so that acts of inelastic scattering, characterized by τ_ϕ are rare enough and $\tau_\phi \gg \tau$.

Probability of $A \to B$ transition according to Feynman is given by:

$$W = \left| \sum_i A_i \right|^2 = \sum_i |A_i|^2 + \sum_{i \neq j} A_i A_j^* \qquad (4.236)$$

where A_i is the transition amplitude of $A \to B$ transition along the i-th trajectory. The usual Boltzmann transport theory naturally neglects quantum interference contribution in (4.236). In most cases this is well justified — the

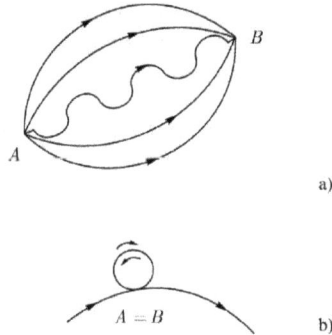

Fig. 4.21 Feynman trajectories of an electron propagating from the point A to point B (a) and an example of self-intersecting trajectory (b).

trajectories (paths) have different lengths and amplitudes A_i have different (essentially random!) phases. However, there is one special case, that is when points A and B just coincide, i.e. the case of self-intersecting paths of the type shown in Fig. 4.21(b).[34] Such a closed path may be traversed in both (opposite) directions 1 and 2. Now, as paths 1 and 2 coincide, the phases of amplitudes A_1 and A_2 are coherent and the second term in (4.236) gives a finite contribution! Appropriate classical transition probability in the case of $A_1 = A_2 = A$ is $W_{cl} = 2|A|^2$ (the first term in (4.236)), but in quantum case we have an additional contribution due to interference (second term in (4.236)), so that:

$$W_{qm} = 2|A|^2 + A_1^* A_2 + A_1 A_2^* = 4|A|^2 \qquad (4.237)$$

and

$$W_{qm} = 2W_{cl} \qquad (4.238)$$

Thus, the probability of *return* for the quantum particle is twice that of the classical particle — quantum diffusion is "slower" than classical.[35] This leads to the *suppression* of conductivity σ and tendency to localization!

Let us estimate now the value of $\delta\sigma/\sigma_0$. From the previous discussion it is clear that the sign of $\delta\sigma/\sigma_0$ is negative and the change of conductivity is proportional to probability of the self-intersecting trajectory appearance (in the process of diffusion). Consider the path ("tube" with the cross-section λ_F^{d-1}) in d-dimensional space. During the time interval dt an electron passes a distance of the order of $dl = v_F dt$ and the corresponding volume of the "tube" is $dV = v_F dt \lambda_F^{d-1}$.

---

[34]The case of backward scattering, leading to return to the initial point.

[35]This is due to the *wave-like* nature of particles in quantum mechanics. All these conclusions are also valid and well known for the classical *waves*. In particular, this leads to the enhancement of the usual echo in the forest!

On the other hand, the maximal available volume for diffusing particle is V_{diff} (4.234). Then, the probability for an electron to find itself inside the closed tube can be roughly estimated to be determined by the ratio of these volumes:

$$W = \int_\tau^{\tau_{in}} \frac{dV}{V_{diff}} = v_F \lambda_F^{d-1} \int_\tau^{\tau_\phi} \frac{dt}{(D_0 t)^{d/2}} \qquad (4.239)$$

The lower limit of integration here is determined by the applicability of diffusion approximation, while the upper — by the limit of applicability of the picture of coherent quantum propagation. Remembering that $\sigma_0 \sim \frac{e^2}{\hbar^2} E\tau$, we obtain the simple estimate:

$$\frac{\delta\sigma}{\sigma_0} \sim \frac{\hbar}{E\tau} \begin{cases} \left(\frac{\tau_\phi}{\tau}\right)^{1/2} & d = 1 \\ \ln\left(\frac{\tau_\phi}{\tau}\right) & d = 2 \\ \left(\frac{\tau_\phi}{\tau}\right)^{-1/2} & d = 3 \end{cases} \qquad (4.240)$$

If $\frac{1}{\tau_\phi} \sim T^p$, we have from (4.240):

$$\frac{\delta\sigma}{\sigma_0} \sim \frac{\hbar}{E\tau} \begin{cases} T^{-p/2} & d = 1 \\ p \ln \frac{T_0}{T} & d = 2 \\ T^{p/2} & d = 3 \end{cases} \qquad (4.241)$$

Above we have used the equality $A_1 = A_2$ assuming the equivalence of electronic states with momenta $\mathbf{p}$ and $-\mathbf{p}$, i.e. time reversal invariance ($t \to -t$). This is valid in the absence of an external magnetic field and magnetic impurities.[36] In the presence of external magnetic field $\mathbf{H}$ these transition amplitudes acquire phases of opposite sign:

$$A_1 \to A_1 e^{i\varphi} \qquad A_2 \to A_2 e^{-i\varphi} \qquad (4.242)$$

with

$$\varphi = \frac{e}{\hbar c} \oint d\mathbf{l} \mathbf{A} = \frac{2\pi\phi}{hc/e} \qquad (4.243)$$

where $\phi = \mathbf{H}\mathbf{S}$ is magnetic flux through the cross-section $\mathbf{S}$ of the closed path (trajectory). As our electron moves by diffusion, we have $S \sim D_0 t$, so that $\phi \sim H D_0 t$. As a result, instead of (4.237), we obtain the probability of return as:

$$W_H = 2|A|^2 \left\{ 1 + \cos\left(2\pi \frac{\phi}{\phi_0}\right) \right\} \qquad (4.244)$$

where $\phi_0 = \frac{ch}{2e}$ is magnetic flux quantum appearing in the theory of superconductivity [Sadovskii M.V. (2019a)] and corresponding to electric charge 2e! Charge doubling appears here due to the interference of contributions of a *pair* of electrons with $\mathbf{p}$ and $-\mathbf{p}$ (similarly, to the case of superconductivity, where the Cooper pair is formed by electrons with opposite momenta). For $H = 0$ we obviously obtain

---

[36] Very clear discussion of time reversal invariance and its absence for the systems in an external magnetic field or containing magnetic impurities can be found in Ch. VIII of [De Gennes P.G. (1966)].

$W_{H=0} = 4|A|^2$. Now, we can conclude that external magnetic field leads to the effect of *negative* magnetoresistance given by:

$$\Delta\sigma(H) = \delta\sigma(H) - \delta\sigma(0) \sim W_H - W_{H=0}$$

$$\sim v_F \lambda_F^{d-1} \int_\tau^{\tau_\phi} \frac{dt}{(D_0 t)^{d/2}} \left\{ 1 - \cos\left(2\pi\frac{\phi}{\phi_0}\right) \right\} > 0 \quad (4.245)$$

and connected with suppression of quantum (localization) corrections by an external magnetic field. It is clear that $\Delta\sigma(H)$ is a function of $\frac{H D_0 \tau_\phi}{\phi_0}$. "Critical" magnetic field, leading to almost complete suppression of localization corrections, is determined from $\frac{H_c D_0 \tau_\phi}{\phi_0} \sim 1$, which gives (for typical values of D_0 and τ_ϕ) $H_c \sim 100 \div 500 \text{Gs}$. This effect of negative magnetoresistance in weak magnetic fields is widely observed in disordered systems and gives a practical method for investigation of quantum corrections and characteristic time τ_ϕ [Alshuler B.L., Aronov A.G., Khmelnitskii D.E., Larkin A.I. (1982); Lee P.A., Ramakrishnan T.V. (1985)].[37]

4.7.2 *Self-consistent theory of localization*

Let us return to general discussion. Below we shall present some self-consistent approach, which apparently allows to analyze the case of strong enough disorder. Consider again Eq. (4.217). The basic idea of *self-consistent theory of localization* (W. Götze, 1979, D. Vollhardt, P. Wölfle, 1980) is to substitute the classical diffusion coefficient D_0 in the *denominator* of Eq. (4.217) by the generalized diffusion coefficient (4.206), which is expressed via the relaxation kernel in its turn determined by Eq. (4.217). As a result we obtain the following *self-consistency equation* determining the relaxation kernel $M(0\omega)$:[38]

$$M(\omega) = 2i\gamma \left\{ 1 + \frac{1}{\pi\nu(E)} \sum_{|\mathbf{k}| < k_0} \frac{i}{\omega - \frac{2E}{dm}\frac{k^2}{M(\omega)}} \right\} \quad (4.246)$$

or equivalent (due to (4.206)) equation for the generalized diffusion coefficient:

$$\frac{D_0}{D_E(\omega)} = 1 + \frac{1}{\pi\nu(E)} \sum_{|\mathbf{k}| < k_0} \frac{1}{-i\omega + D_E(\omega)k^2} \quad (4.247)$$

---

[37] Note that this effect is completely different from the classical magnetoresistance where $\frac{\Delta\sigma(H)}{\sigma_0} \sim -(\omega_H \tau)^2$, with $\omega_H = \frac{eH}{mc}$. The value of classical magnetoresistance for magnetic fields studied here is in orders of magnitude smaller than (4.245) and have the opposite sign!

[38] Note that here we neglect the possible spatial dispersion: $M(\mathbf{q}\omega) \to M(0\omega)$, $D_E(\mathbf{q}\omega) \to D_E(0\omega)$.

The choice of cut-off momentum k_0 will be discussed later.[39]

Before we go to the detailed analysis of solutions of Eqs. (4.246) and (4.247), let us see what can be said from "general considerations". Consider again expression (4.204) for $\Phi^{RA}(\mathbf{q}\omega)$. We are expecting that $M(\omega)$ may be singular for $\omega \to 0$. Assume that the existence of the following limit:

$$R_{loc}^2(E) = -\frac{2E}{dm} \lim_{\omega \to 0} \frac{1}{\omega M(\omega)} \tag{4.248}$$

defining characteristic length R_{loc}. Then the following singular contribution appears in (4.204)[40] (for simplicity we also assume that $q \to 0$):

$$\Phi^{RA}(\mathbf{q}\omega) = -\frac{N(E)}{\omega} \frac{1}{1 + R_{loc}^2 q^2} \tag{4.249}$$

In this case it is convenient to introduce also the characteristic frequency $\omega_0(E)$:

$$\omega_0^2(E) = -\lim_{\omega \to 0} \omega M(\omega) = \frac{2E}{dm} \frac{1}{R_{loc}^2(E)} > 0 \tag{4.250}$$

so that

$$R_{loc}(E) = \sqrt{\frac{2E}{dm} \frac{1}{\omega_0(E)}} \tag{4.251}$$

The length R_{loc} has the meaning of *localization radius (length)* of electronic states in the field of random potential of impurities [Sadovskii M.V. (2000)]. If these limits exist electronic states at the Fermi level E are localized and our system becomes the *Anderson insulator*. Localization of electrons in disordered systems or Anderson transition is one of the basic concepts of the modern theory of disordered systems [Mott N.F. (1974)]. Below we shall see that the self-consistent theory localization gives rather satisfactory description of this transition.

---

[39]Surely, the proposed self-consistency scheme introduces into our theory some "uncontrollable" elements, which is typical for many other "self-consistent" approximations. In fact, there exists more rigorous derivation of Eqs. (4.246), (4.247) (D. Vollhardt, P. Wölfle, 1982), based upon the general diagrammatic analysis and the drop of "less singular" (for $\omega \to 0$) contributions, but we shall not discuss it here. As a matter of fact, the fruitfulness of self-consistent approach is justified mainly by physical results obtained below.

[40]The presence of such contribution in $\Phi^{RA}(\mathbf{q}\omega)$ corresponds to the general criterion of electron localization in the random field of impurities, i.e. to (Anderson) transition to insulating state [Sadovskii M.V. (2000)].

Appearance of localized states (Anderson insulator) is related here to the existence of the limit in (4.248), i.e. with the existence of finite frequency $\omega_0(E)$, defined in (4.250). In another words, it corresponds to the appearance of divergent contribution to relaxation kernel: $ReM(0\omega) = -\frac{\omega_0^2(E)}{\omega}$ for $\omega \to 0$. Then, the relaxation kernel (at $q \to 0$ and $\omega \to 0$) can be written as:

$$M(0\omega) = \begin{cases} \frac{i}{\tau_E} & \text{(metal)} \\ \frac{i}{\tau_E} - \frac{\omega_0^2(E)}{\omega} & \text{(Anderson insulator)} \end{cases} \quad (4.252)$$

The problem is whether such solutions of Eq. (4.246) really exist?

Returning to discussion of the cut-off momentum in Eqs. (4.246) and (4.247) we note that from Eq. (4.194) and the simple estimate (for the case of weak disorder!) $\Delta G_{\mathbf{p}} \sim ImG^R(E\mathbf{p}) \sim \delta\left(E - \frac{\mathbf{p}^2}{2m}\right)$, it is clear that the modulus of the sum of momenta $\mathbf{k} = \mathbf{p} + \mathbf{p}'$ in $U_{\mathbf{pp}'}(\mathbf{q}\omega)$ can change from 0 to $2p_F$. At the same time, our expression for $U_{\mathbf{pp}'}(\mathbf{q}\omega)$ given by (4.147) ("Cooperon") is valid only for $|\mathbf{p} + \mathbf{p}'| \leq l^{-1}$ (criterion of validity of diffusion approximation). Then it is clear that the cut-off momentum in (4.246) and (4.247) can be estimated as:

$$k_0 \sim Min\{p_F, \; l^{-1}\} \quad (4.253)$$

In fact, the Anderson transition takes place for the mean-free path given by the simple estimate $p_F l \sim 1$ (so-called Ioffe–Regel criterion) [Mott N.F. (1974)], so that we may write:

$$k_0 = x_0 p_F = x_0\sqrt{2mE} \quad (4.254)$$

where $x_0 = const \sim 1 \div 2$.

Let us now turn to the solution of self-consistency equation (4.246) (A.V. Myasnikov, M.V. Sadovskii, 1982). Introduce the dimensionless integration variable $y = \frac{k}{x_0\sqrt{2mE}}$ and rewrite Eq. (4.246) as:

$$M(\omega) = 2i\gamma + d\lambda x_0^{d-2} M(\omega) \int_0^1 dy y^{d-1} \frac{1}{y^2 - \frac{d\omega}{4x_0^2 E^2}M(\omega)} \quad (4.255)$$

where λ is (dimensionless) perturbation theory parameter:

$$\lambda = \frac{1}{2\pi E\tau} = \frac{\gamma}{\pi E} = \left(\frac{m}{2\pi}\right)^{d/2} \frac{E^{\frac{d}{2}-2}}{\Gamma\left(\frac{d}{2}\right)}\rho v^2 \quad (4.256)$$

and we have taken into account $\gamma = \pi\rho v^2 \nu(E)$ and the form of the density of states of free electrons in d-dimensional space:

$$\nu(E) = \left(\frac{m}{2\pi}\right)^{d/2} \frac{E^{\frac{d}{2}-1}}{\Gamma\left(\frac{d}{2}\right)} \quad (4.257)$$

4.7.2.1   *Metallic phase*

Putting in (4.255) $\omega = 0$ and considering the metallic regime of (4.252), when $ReM(\omega = 0) = 0$ and $ImM(\omega = 0) = 1/\tau_E$, we immediately obtain from (4.255):

$$\tau_E = \frac{1}{2\gamma}\left\{1 - \frac{d}{d-2}\lambda x_0^{d-2}\right\} \tag{4.258}$$

Then, from (4.207) we get the metallic conductivity in the following form:

$$\sigma = \frac{ne^2}{m}\frac{1}{2\gamma}\left\{1 - \left(\frac{E_c}{E}\right)^{\frac{4-d}{2}}\right\}; \qquad 2 < d < 4 \tag{4.259}$$

where

$$E_c = \left\{\frac{d}{d-2}\frac{x_0^{d-2}}{\Gamma\left(\frac{d}{2}\right)}(2\pi)^{-\frac{d}{2}}\right\}^{\frac{2}{4-d}}E_{sc} \tag{4.260}$$

and we introduced characteristic energy:

$$E_{sc} = m^{\frac{d}{4-d}}(\rho v^2)^{\frac{2}{4-d}} \tag{4.261}$$

which already appeared in (4.74) and defines the width of "strong coupling" (strong scattering) region on the energy axis for an electron in the random field of impurities. In fact it follows from the simplest estimate of $\gamma \sim \rho v^2 \nu(E) \sim E$ (with the account of (4.257)), which corresponds to Ioffe–Regel criterion (and limit of validity of our perturbation theory!).[41] It is easily seen that energy $E_c \sim E_{sc}$ (4.260) plays the role of *mobility edge* [Mott N.F. (1974)]. In fact, for $E > E_c$ we obtain from (4.259):

$$\sigma \approx \frac{ne^2}{m}\frac{1}{2\gamma(E_c)}\left(\frac{4-d}{2}\right)\left(\frac{E - E_c}{E_c}\right) \sim \frac{E - E_c}{E_c} \tag{4.262}$$

so that static conductivity of the system at $T = 0$ tends to zero as Fermi energy $E \to E_c$. This corresponds to Anderson metal–insulator transition (Anderson localization).

---

[41] Already from these simple estimates we can see the special role of space dimensionalities $d = 2$ and $d = 4$, which has the meaning of "lower" and "upper" critical dimensions for Anderson transition.

Let us present explicit results for $d = 3$. For definiteness we just put $x_0 = 1$ and obtain:

$$E_c = \frac{9}{2\pi^4} m^3 (\rho v^2)^2 \qquad (4.263)$$

In terms of dimensionless perturbation theory parameter λ given by (4.256) this corresponds to the critical value $\lambda_c = \frac{d-2}{d} x_0^{2-d} = 1/3$ or

$$\left. \frac{E}{\gamma(E)} \right|_{E=E_c} = \frac{3}{\pi} \qquad p_F l = \frac{3}{\pi} \qquad (4.264)$$

where $l = \frac{v_F}{2\gamma}$ is the mean-free path in Born approximation, which can be used as a measure of disorder. Then (4.264) corresponds to the usual formulation of Ioffe–Regel rule: the mean-free path in a metal can not be shorter than typical interatomic spacing [Mott N.F. (1974)]. With the further growth of disorder the system becomes Anderson insulator. Drude conductivity, corresponding to "critical" mean-free path given by (4.264) is:

$$\sigma_c = \frac{ne^2}{m} \frac{1}{2\gamma(E_c)} = \frac{e^2 p_F}{3\pi^2 \hbar^2} \left. \left(\frac{p_F l}{\hbar} \right) \right|_{E=E_c} = \frac{e^2 p_F}{\pi^3 \hbar^2} \qquad (4.265)$$

where we used $n = \frac{p_F^3}{3\pi^2 \hbar^3}$ and "restored" $\hbar$. Due to $p_F \sim \frac{\hbar}{a}$, where a is interatomic spacing, the value of σ_c (4.265) is of the order of Mott's "minimal metallic conductivity":

$$\sigma_{mm} \approx \frac{1}{\pi^3} \frac{e^2}{\hbar a} \qquad (4.266)$$

which for $a \sim 3\text{Å}$ gives $\sigma_{mm} \sim 2 \cdot 10^2 \text{Ohm}^{-1}\text{cm}^{-1}$.

Writing (4.258) as

$$\tau_E = \frac{1}{2\gamma} \left\{ 1 - \frac{3}{\pi E} \gamma \right\} = \frac{1}{2\gamma} \left\{ 1 - \frac{\gamma(E)}{\gamma(E_c)} \right\} \qquad (4.267)$$

we can rewrite (4.259) in the following form:

$$\sigma = \sigma_0 \left\{ 1 - \frac{\sigma_c}{\sigma_0} \right\} = \sigma_0 - \sigma_c \qquad (4.268)$$

where Drude conductivity $\sigma_0 = \frac{ne^2}{m} \frac{1}{2\gamma}$ enters as the measure of disorder. For weak disorder (large mean-free path) $\sigma_0 \gg \sigma_c$ and from (4.268) follows the usual result: $\sigma \approx \sigma_0$. As disorder grows (mean-free path diminishes) $\sigma \to 0$ as $\sigma_0 \to \sigma_c \approx \sigma_{mm}$. Then we see that σ_{mm} determines characteristic conductivity scale for the *continuous* (in contradiction with early ideas of Mott [Mott N.F. (1974)]) metal–insulator transition induced by disorder. It is rather surprising that these simple expressions are *experimentally* confirmed in many real systems [Sadovskii M.V. (2000)].

From our estimates it is clear that for $d = 3$ the mobility edge E_c belongs to the "strong coupling" region of the width of the order of $E_{sc} = m^3(\rho v^2)^2$ (M.V. Sadovskii, 1977) around the origin on the energy axis, where the perturbation theory parameter $\lambda \sim 1$ and, strictly speaking, we have to take into account *all* diagrams of perturbation theory. On the other hand, for $d \to 2$ it can be seen from (4.260) that $E_c \to \infty$, in accordance with the picture of all states being localized at infinitesimally small disorder for $d = 2$ (P.W. Anderson, E. Abrahams, D.C. Licciardello, D.J. Thouless, 1979).

The fact that mobility edge belongs to the "strong coupling" region is the major difficulty of the theory of electron localization in disordered systems. That is why this theory is still uncompleted and we need the development of some new (non perturbative) approaches.[42]

4.7.2.2   *Anderson insulator*

Consider now the region of localized state $E < E_c$ (Anderson insulator). Let us look for the solution of Eq. (4.246) in the form given by the second expression in (4.252). From the real part of Eq. (4.246), for $\omega \to 0$ we find the following equation determining $\omega_0^2(E)$:

$$1 = d\lambda x_0^{d-2} \int_0^1 dy y^{d-1} \frac{1}{y^2 + \frac{d\omega_0^2(E)}{4x_0^2 E^2}} \qquad (4.269)$$

Similarly, from the imaginary part of (4.246), for $\omega \to 0$, we obtain the following equation for τ_E in localization region:

$$1 - 2\gamma\tau_E = d\lambda x_0^{d-2} \int_0^1 dy y^{d+1} \frac{1}{\left[y^2 + \frac{d\omega_0^2(E)}{4x_0^2 E^2}\right]^2} \qquad (4.270)$$

Using (for $\omega_0^2 \to 0$) the simple estimate of the integral[43] in (4.269) we

---

[42]More details on these problems can be found in original reviews: M.V. Sadovskii. Physics Uspekhi **24**, 96 (1981), I.M. Suslov. Physics Uspekhi **41**, 441 (1998).

[43]Of course, this integral can be calculated exactly, but this will only lead to rather insignificant change of some constants in the final results.

get:

$$
1 \approx d\lambda x_0^{d-2} \int\limits_{\left(\frac{d\omega_0^2(E)}{4x_0^2 E^2}\right)^{1/2}}^{1} dy\, y^{d-3}
$$

$$
= \begin{cases} d\lambda x_0^{d-2} \frac{1}{d-2} \left\{ 1 - \left(\frac{d\omega_0^2(E)}{4x_0^2 E^2}\right)^{\frac{d-2}{2}} \right\} & (2 < d < 4) \\ \lambda \ln \frac{2x_0^2 E^2}{\omega_0^2(E)} & (d = 2) \end{cases} \tag{4.271}
$$

and from (4.260) it follows that:

$$
\omega_0^2(E) = \begin{cases} \frac{4}{d} x_0^2 E^2 \left\{ 1 - \left(\frac{E}{E_c}\right)^{\frac{4-d}{2}} \right\}^{\frac{2}{d-2}} & (2 < d < 4) \\ 2x_0^2 E^2 \exp\left(-\frac{1}{\lambda}\right) & (d = 2) \end{cases} \tag{4.272}
$$

The position of mobility edge on the energy axis is naturally determined by the condition $\omega_0^2(E_c) = 0$. For $d = 2$ we have $\omega_0^2(E) > 0$ for arbitrary E, which corresponds to the picture of localization of all electronic states for any, even weakest possible, disorder (P.W. Anderson, E. Abrahams, D.C. Licciardello, D.J. Thouless, 1979). Note, however, that for weak disorder ($\lambda \ll 1$) $\omega_0^2(E)$ is exponentially small, which really corresponds to *weak* localization.

Using "representation of unity" given by Eq. (4.269) in (4.270), we express τ_E via $\omega_0^2(E)$ as:

$$
2\gamma\tau_E = d\lambda x_0^{d-2} \frac{d\omega_0^2(E)}{4x_0^2 E^2} \int_0^1 dy\, \frac{y^{d-1}}{\left[y^2 + \frac{d\omega_0^2(E)}{4x_0^2 E^2}\right]^2}
$$

$$
\approx d\lambda x_0^{d-2} \frac{1}{d-4} \left\{ \frac{d\omega_0^2(E)}{4x_0^2 E^2} - \left(\frac{d\omega_0^2(E)}{4x_0^2 E^2}\right)^{\frac{d-2}{2}} \right\} \quad (d < 4) \tag{4.273}
$$

so that for $\omega_0^2(E) \to 0$ (close to the mobility edge) we have:

$$
2\gamma\tau_E = \begin{cases} \frac{d}{4-d} \lambda x_0^{d-2} \left(\frac{d\omega_0^2(E)}{4x_0^2 E^2}\right)^{\frac{d-2}{2}} & (2 < d < 4) \\ \lambda \left[1 - \frac{\omega_0^2(E)}{2x_0^2 E^2}\right] & (d = 2) \end{cases} \tag{4.274}
$$

From (4.251) and (4.272) we find localization radius:[44]

$$R_{loc}(E) = \frac{1}{x_0\sqrt{2mE}}\left\{1-\left(\frac{E}{E_c}\right)^{\frac{4-d}{2}}\right\}^{-\frac{1}{d-2}} \sim \frac{1}{p_F}\left|\frac{E-E_c}{E_c}\right|^{-\nu};$$
$$E \leq E_c \qquad\qquad (4.276)$$

where the *critical exponent* (index) ν:

$$\nu = \frac{1}{d-2} \qquad\qquad (4.277)$$

Introduce now characteristic *correlation (localization) length* (coinciding with R_{loc} for $E \leq E_c$, i.e. in insulating phase), defining it as:

$$\xi_{loc}(E) = \frac{1}{x_0\sqrt{2mE}} = \begin{cases} \left[1-\left(\frac{E}{E_c}\right)^{\frac{4-d}{2}}\right]^{-\frac{1}{d-2}}; & E \leq E_c \\ \left[1-\left(\frac{E_c}{E}\right)^{\frac{4-d}{2}}\right]^{-\frac{1}{d-2}}; & E > E_c \end{cases} \qquad (4.278)$$

$$\xi_{loc}(E) \sim \frac{1}{p_F}\left|\frac{E-E_c}{E_c}\right|^{-\nu}; \qquad E \sim E_c \quad (4.279)$$

Now this length is also defined for $E > E_c$, i.e. in metallic phase. Then we can rewrite conductivity (4.259) in metallic phase as:

$$\sigma = \frac{ne^2}{m}\frac{1}{2\gamma}(x_0 p_F\xi_{loc})^{2-d} = \frac{\sigma_0}{(x_0 p_F\xi_{loc})^{d-2}} \sim (E-E_c)^{(d-2)\nu} \quad (4.280)$$

obtaining the so-called Wegner scaling law for conductivity (F. Wegner, 1976) with critical index:

$$s = (d-2)\nu \qquad\qquad (4.281)$$

In particular, for $d = 3$ (assuming $x_0 = 1$) we have:

$$\sigma = \frac{\sigma_0}{p_F\xi_{loc}(E)} \sim (E-E_c); \qquad E > E_c \qquad (4.282)$$

and the critical index of conductivity $s = 1$. Precisely this type of behavior is observed experimentally in the vicinity of metal–insulator

---

[44]Similarly, for $d = 2$ we get

$$R_{loc}(E) = \frac{1}{x_0\sqrt{2mE}}\exp\left\{\frac{\pi E}{m\rho v^2}\right\} \qquad (4.275)$$

so that localization radius is exponentially large in the case of weak disorder.

transition induced by disorder in a number of real systems [Sadovskii M.V. (2000)]. However, in some systems another behavior is observed, corresponding to the critical exponent $s \approx 1/2$. Usually, this discrepancy is attributed to the role of electron–electron interactions, which were neglected in the above analysis. Thus the question of critical behavior of conductivity close to disorder induced metal–insulator transition is, in fact, still open.[45]

Note that all the previous expressions are written in analogy to scaling relations of modern theory of (thermodynamic) critical phenomena for the second order phase transitions [Sadovskii M.V. (2019a)] and correspond to the concept of *scaling at the mobility edge* (P.W. Anderson, E. Abrahams, D.C. Licciardello, D.J. Thouless, 1979). Let us stress (to avoid possible misunderstanding) that Anderson (metal–insulator) transition is in no sense (thermodynamic) phase transition of either order and its description is much more complicated (and, in fact, incomplete). In particular, up to now there is no commonly accepted definition of any *order parameter*, characterizing this transition. As we already mentioned, the difficulties of theoretical description of this transition (even in one-particle approximation, neglecting interactions) are connected with the fact that the mobility edge position at the energy axis belongs to the "strong coupling" region (which is, by the way, quite analogous to the Ginzburg critical region of the usual theory of critical phenomena (M.V. Sadovskii, 1977)), where (in contrast to the theory of critical phenomena!) we have to take into account *all* diagrams of Feynman perturbation series or use essentially non-perturbative methods.[46] It is important to stress, that in the standard theory of critical phenomena interaction of order-parameter fluctuations in the critical region becomes weak, at least in the space of $d = 4 - \varepsilon$ dimensions. But for the Anderson transition we meet quite opposite situation!

Another example of the behavior of physical characteristics at the Anderson transition is dielectric permeability (in the phase of Anderson insulator). From general definition (2.110) we have:

$$\epsilon(\mathbf{q}\omega) = 1 + \frac{4\pi e^2}{q^2}\chi(\mathbf{q}\omega) \qquad (4.283)$$

where $\chi(\mathbf{q}\omega)$ is the retarded density–density response function. Then,

---

[45]This problem is complicated also by the fact that most of the modern numerical simulations of the Anderson transition for non-interacting electrons give the value of conductivity exponent $s \sim 1.5$.

[46]Due to this, we should not be very serious with respect to the explicit relations for critical exponents for the Anderson transition obtained above. At the same time, the continuous nature of this transition for $d = 3$ is rather well established, both theoretically and experimentally (in contradiction with early ideas of Mott [Mott N.F. (1974)]).

using (4.204), (4.205) and (4.252), (4.276), we immediately obtain:

$$\epsilon(0\omega \to 0) = 1 + \frac{\omega_p^2}{\omega_0^2(E)} = 1 + \kappa_D^2 R_{loc}^2(E) \sim \left| \frac{E - E_c}{E_c} \right|^{-2\nu} \qquad (4.284)$$

where $\omega_p^2 = \frac{4\pi n e^2}{m}$ is the square of the plasma frequency and $\kappa_D^2 = 4\pi e^2 N(E)$ is the inverse square of the (Debye) screening length in a metal. From Eq. (4.284) we can see that the static dielectric permeability diverges at the mobility edge as the system approaches the insulator–metal transition (from within the Anderson insulator). This is also the observable effect, which can be used to determine the critical exponent ν of localization length.

4.7.2.3 *Frequency dispersion of the generalized diffusion coefficient*

Results obtained above are valid for $\omega \to 0$. It is, of course, possible to perform a complete analysis of frequency dependence of the generalized diffusion coefficient (D. Vollhardt, P. Wölfle, 1982). We shall give here only a short summary of appropriate results. Eq. (4.247) for the generalized diffusion coefficient can be written in the following form, similar to (4.255):

$$\frac{D_E(\omega)}{D_0} = 1 - \frac{d\lambda}{d-2} x_0^{d-2} + d\lambda x_0^{d-2} \left[-\frac{i\omega D_0}{2\gamma D_E(\omega)} \right] \int_0^1 dy \frac{y^{d-3}}{y^3 - \frac{i\omega D_0}{2\gamma D_E(\omega)}} \qquad (4.285)$$

Under the condition of $\left| \frac{\omega}{2\gamma} \frac{D_0}{D_E(\omega)} \right| \ll 1$, the upper limit of integration in (4.285) can be replaced by infinity, so that we obtain the following algebraic equation for $D_E(\omega)$:

$$\frac{D_E(\omega)}{D_0} = 1 - \frac{\lambda}{\lambda_c} + p_d \frac{\lambda}{\lambda_c} \left\{ -\frac{i\omega}{2\gamma} \frac{D_0}{D_E(\omega)} \right\}^{\frac{d-2}{2}} \qquad (4.286)$$

where $\lambda_c = \frac{d-2}{d} x_0^{2-d}$ is the "critical" value of dimensionless parameter λ at the transition point and $p_d = \Gamma\left(\frac{d}{2}\right) \Gamma\left(2 - \frac{d}{2}\right)$. From (4.286) it follows that $D_E(\omega)$ (and conductivity $\sigma(\omega)$) satisfy the following scaling relation:

$$\frac{D_E(\omega)}{D_0} = \frac{\sigma(\omega)}{\sigma_0} = \left(-\frac{i\omega}{2\gamma} \right)^{\frac{d-2}{d}} F_d\left(\frac{\omega}{\omega_c} \right) \qquad (4.287)$$

where for $\omega \ll \omega_c$ we have:

$$F_d\left(\frac{\omega}{\omega_c} \right) \sim \left(-\frac{i\omega}{\omega_c} \right)^{\frac{2}{d}}; \qquad \lambda > \lambda_c \quad \text{(insulator)} \qquad (4.288)$$

$$F_d\left(\frac{\omega}{\omega_c} \right) \sim \left(-\frac{i\omega}{\omega_c} \right)^{\frac{2-d}{d}}; \qquad \lambda < \lambda_c \quad \text{(metal)} \qquad (4.289)$$

while for $\omega \gg \omega_c$:

$$F_d\left(\frac{\omega}{\omega_c}\right) \sim const \qquad (4.290)$$

Characteristic frequency ω_c is defined as:

$$\omega_c \approx 2\gamma \left|1 - \frac{\lambda}{\lambda_c}\right|^{\frac{d}{d-2}} \sim 2\gamma[p_F\xi_{loc}(E)]^{-d} \qquad (4.291)$$

Inequality used to reduce Eq. (4.285) to (4.286) is satisfied for $\omega_c \ll 2\gamma$, which is definitely valid close to the mobility edge.

For $d = 3$ Eq. (4.286) becomes:

$$\frac{D_E(\omega)}{D_0} = 1 - \frac{\lambda}{\lambda_c} + \frac{\pi}{2}\frac{\lambda}{\lambda_c}\left\{-\frac{i\omega}{2\gamma}\frac{D_0}{D_E(\omega)}\right\}^{1/2}; \qquad \lambda_c = \frac{1}{3x_0} \qquad (4.292)$$

and can be solved explicitly. With sufficient (for many applications) accuracy the generalized diffusion coefficient can be written as [Sadovskii M.V. (2000)]:[47]

$$D_E(\omega) = \begin{cases} D_E; & (\omega \ll \omega_c,\ E \geq E_c) \quad \text{(metal)} \\ D_0\left(-\frac{i\omega}{2\gamma}\right)^{1/3}; & (\omega \gg \omega_c) \quad \text{(both metal and insulator)} \\ D_E\dfrac{-i\omega}{-i\omega + \frac{3D_E}{v_F^2}\omega_0^2(E)}; & (\omega \ll \omega_c,\ E < E_c) \quad \text{(insulator)} \end{cases}$$

$$(4.293)$$

where $D_E = \frac{D_0}{p_F\xi_{loc}(E)}$. At the mobility edge itself we have $\xi_{loc}(E = E_c) = \infty$, so that $\omega_c = 0$, and we obtain the so-called $\omega^{1/3}$-law (W. Götze, 1981):

$$D_E(\omega) = D_0\left(-\frac{i\omega}{2\gamma}\right)^{1/3} \qquad (4.294)$$

The frequency ω_c is in fact determined from $D_E(\omega_c) \sim D_E \sim D_0\left(\frac{\omega_c}{2\gamma}\right)^{1/3}$. The limit of $\omega \to 0$ often used above should be understood in the sense of $\omega \ll \omega_c$. Finally, note that for $\omega \gg 2\gamma$ equations of self-consistent theory of localization describe the transition to the usual Drude-like behavior: $D_E(\omega) \approx D_0\left[1 - \frac{i\omega}{2\gamma}\right]^{-1}$.

4.8 "Triangular" vertex

Existence of diffusion pole (4.140) in "four-leg" vertex $\Gamma_{\mathbf{pp'}}(\mathbf{q}\omega)$ and in two-particle Green's functions in general (cf. e.g. (4.204)) leads to the appearance of similar contributions in other "blocks" of our diagram

---

[47]For $d = 2$ and very small frequencies $\omega \ll \lambda^{-1}e^{-\frac{1}{\lambda}}\gamma$ ($\lambda \ll 1$), self-consistent theory of localization gives $\sigma(\omega) = \frac{ne^2}{m}\frac{\gamma}{\lambda}e^{\frac{2}{\lambda}}\frac{\omega^2}{2(x_0E)^4} \to 0$, for $\omega \to 0$. For $\lambda^{-2}e^{-\frac{1}{\lambda}}\gamma \ll \omega \ll \lambda^2\gamma$ we obtain for $D(\omega)$ the dependence of the type of (4.228).

technique.[48] In particular, it is useful to analyze the "triangular" vertex, defined by diagrams shown in Fig. 4.12(c). In general case, it is defined by the integral equation shown diagrammatically in Fig. 4.22(a). Though in the following we shall mainly consider the case of weak disorder, when we can restrict ourselves by the use of $U_0(\mathbf{p} - \mathbf{p}') = \rho v^2$ ("ladder" approximation), at first we shall present rather general analysis, allowing the generalization in the spirit of self-consistent theory of localization. Returning to the general definition of the two-particle Green's function, shown diagrammatically in Fig. 4.14(a), or analytically in (4.125), we can immediately write down the following expression for "triangular" vertex $\mathcal{T}^{RA}_{\mathbf{p}_+\mathbf{p}_-}(\mathbf{q}\omega)$ via $\Phi^{RA}_{\mathbf{p}\mathbf{p}'}(\mathbf{q}\omega)$:[49]

$$\mathcal{T}^{RA}_{\mathbf{p}_+\mathbf{p}_-}(\mathbf{q}\omega)G^R(E + \omega\mathbf{p}_+)G^A(E\mathbf{p}_-) = -2\pi i \int \frac{d^d p'}{(2\pi)^d} \Phi^{RA}_{\mathbf{p}\mathbf{p}'}(\mathbf{q}\omega)$$

(4.295)

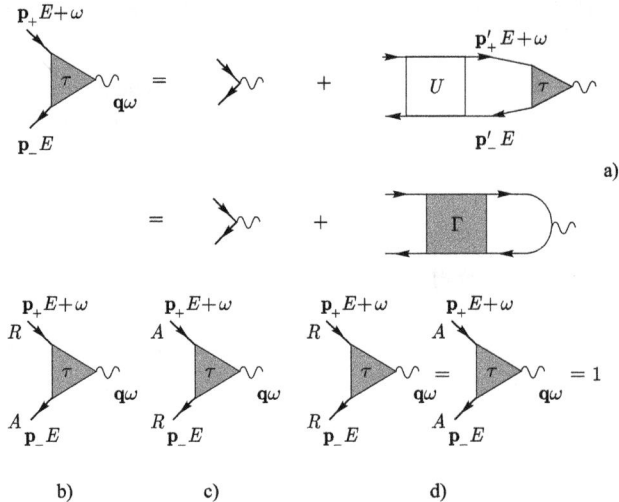

Fig. 4.22 Integral equation for "triangular" vertex (a) and different combinations of (R) and (A) electronic lines (b), (c), (d).

---

[48]The same statement is valid concerning the appearance of "Cooperon"-type contributions related to (4.147).

[49]For definiteness we assume the upper electron line in Fig. 4.22(a) to be retarded.

or

$$\mathcal{T}^{RA}_{\mathbf{p}_+\mathbf{p}_-}(\mathbf{q}\omega) = -\frac{2\pi i}{G^R(E+\omega\mathbf{p}_+)G^A(E\mathbf{p}_-)}\int\frac{d^d p'}{(2\pi)^d}\Phi^{RA}_{\mathbf{p}\mathbf{p}'}(\mathbf{q}\omega) \quad (4.296)$$

Using now (4.192), we can rewrite (4.296) as:

$$\mathcal{T}^{RA}_{\mathbf{p}_+\mathbf{p}_-}(\mathbf{q}\omega) = \frac{1}{\nu(E)}\frac{\Delta G_{\mathbf{p}}}{G^R(E+\omega\mathbf{p}_+)G^A(E\mathbf{p}_-)}$$

$$\times\int\frac{d^d p'}{(2\pi)^d}\int\frac{d^d p''}{(2\pi)^d}\left\{1+\frac{d}{p_F^2}(\mathbf{p}\hat{\mathbf{q}})(\mathbf{p}''\hat{\mathbf{q}})\right\}\Phi^{RA}_{\mathbf{p}''\mathbf{p}'}(\mathbf{q}\omega)$$

$$= \frac{1}{\nu(E)}\frac{\Delta G_{\mathbf{p}}}{G^R(E+\omega\mathbf{p}_+)G^A(E\mathbf{p}_-)}\left\{\Phi^{RA}(\mathbf{q}\omega)+d(\hat{\mathbf{p}}\hat{\mathbf{q}})\Phi^{RA}_1(\mathbf{q}\omega)\right\}$$

$$(4.297)$$

where we have used also (4.88) and (4.185). Using Eqs. (4.184) and (4.193) for $\Phi^{RA}(\mathbf{q}\omega)$ and $\Phi^{RA}_1(\mathbf{q}\omega)$ we have obtained (4.202) and (4.203), so that, in particular, we have:

$$\Phi^{RA}_1(\mathbf{q}\omega) = \frac{1}{v_F q}\chi(\mathbf{q}\omega) \quad (4.298)$$

Then:

$$d(\hat{\mathbf{p}}\hat{\mathbf{q}})\Phi^{RA}_1(\mathbf{q}\omega) = (\mathbf{p}\mathbf{q})\frac{d}{p_F v_F q^2}\chi(\mathbf{q}\omega) \quad (4.299)$$

so that after the use of (4.149):

$$\frac{\Delta G_{\mathbf{p}}}{G^R(E+\omega\mathbf{p}_+)G^A(E\mathbf{p}_-)}$$

$$= -\left\{\omega-\frac{1}{m}(\mathbf{p}\mathbf{q})-\Sigma^R(\mathbf{p}_+E+\omega)+\Sigma^A(\mathbf{p}_-E)\right\} \quad (4.300)$$

and (4.297) is rewritten as:

$$\mathcal{T}^{RA}_{\mathbf{p}_+\mathbf{p}_-}(\mathbf{q}\omega) = \frac{1}{\nu(E)}\left\{-\omega+\frac{1}{m}(\mathbf{p}\mathbf{q})+\Sigma^R(\mathbf{p}_+E+\omega)-\Sigma^A(\mathbf{p}_-E)\right\}$$

$$\times\left\{\Phi^{RA}(\mathbf{q}\omega)+\frac{1}{m}(\mathbf{p}\mathbf{q})\frac{d}{v_F^2 q^2}\chi(\mathbf{q}\omega)\right\}$$

$$= -\left\{-\omega+\frac{1}{m}(\mathbf{p}\mathbf{q})+\Sigma^R(\mathbf{p}_+E+\omega)-\Sigma^A(\mathbf{p}_-E)\right\}\frac{\omega+M(\mathbf{q}\omega)+\frac{1}{m}(\mathbf{p}\mathbf{q})}{\omega^2+\omega M(\mathbf{q}\omega)-\frac{1}{d}v_F^2 q^2}$$

$$= \left\{\omega-\frac{1}{m}(\mathbf{p}\mathbf{q})+2i\gamma\right\}\frac{\omega+M(\mathbf{q}\omega)+\frac{1}{m}(\mathbf{p}\mathbf{q})}{\omega^2+\omega M(\mathbf{q}\omega)-\frac{1}{d}v_F^2 q^2} \quad (4.301)$$

where we have used also (4.202) and (4.203), as well as, at the end, the simplest approximation for self-energies. As a result, in the limit of $\omega \to 0$ and $q \to 0$, introducing again the generalized diffusion coefficient (4.206), we obtain the following simple form of "triangular" vertex in RA-channel and shown in Fig. 4.22(b):

$$\mathcal{T}^{RA}(\mathbf{q}\omega) \approx \frac{2\gamma}{-i\omega + D_E(\mathbf{q}\omega)q^2} \tag{4.302}$$

Of course, in the "ladder" approximation (weak disorder!), when we use (4.127), expression of the type of (4.302) (with the replacement $D_E(\mathbf{q}\omega) \to D_0$) is directly obtained from the equation shown in Fig. 4.22(a) after elementary calculations. More complicated derivation given above allows generalization to the case of strong disorder $(p_F l \sim 1)$, when for $D_E(\mathbf{q}\omega)$ we may use expressions derived within self-consistent theory of localization.

Similarly, we can show that the "triangular" vertex in AR-channel, shown in Fig. 4.22(a), is given by:

$$\mathcal{T}^{AR}(\mathbf{q}\omega) \approx \frac{2\gamma}{i\omega + D_E(\mathbf{q}\omega)q^2} \tag{4.303}$$

while for vertices in RR and AA channels, shown in Fig. 4.22(d) we have:

$$\mathcal{T}^{RR}(\mathbf{q}\omega) = \mathcal{T}^{AA}(\mathbf{q}\omega) = 1 \tag{4.304}$$

so that diffusion pole here is absent.

In Matsubara formalism the general form of "triangular" vertex is now also quite clear:[50]

$$\mathcal{T}(\mathbf{q}\omega_m \varepsilon_n) = \theta(\varepsilon_n)\theta(\varepsilon_n + \omega_m) + \theta(-\varepsilon_n)\theta(-\varepsilon_n - \omega_m)$$
$$+ 2\gamma \left\{ \frac{\theta(\varepsilon_n)\theta(-\varepsilon_n - \omega_m)}{-\omega_m + D_0 q^2} + \frac{\theta(-\varepsilon_n)\theta(\varepsilon_n + \omega_m)}{\omega_m + D_0 q^2} \right\} \tag{4.305}$$

where we have only written expressions with $D_0 = \frac{1}{d}v_F\tau = \frac{v_F^2}{2d\gamma}$, as in the rest of this chapter we shall be interested only in the case of weak disorder $(p_F l \gg 1)$.

---

[50]We have taken into account that $G(\varepsilon_n > 0\mathbf{p})$ is continued to $G^R(E\mathbf{p})$, $G(\varepsilon_n < 0\mathbf{p})$ to $G^A(E\mathbf{p})$, $i\omega_m \to \omega \pm i\delta$ for $m > 0$ and $m < 0$ etc.

4.9 The role of electron–electron interaction

In real disordered metals we deal, of course, with electrons interacting via Coulomb repulsion. The task of joint account of both disorder (impurity scattering) and interaction effects is very difficult problem, which is not finally solved up to now [Altshuler B.L., Aronov A.G. (1985); Lee P.A., Ramakrishnan T.V. (1985)]. Below we limit ourselves only to the analysis of some simple examples of interactions effects in disordered systems, mainly concerning the density of states close to the Fermi level.

Consider the simplest interaction correction to single-electron Green's function shown diagrammatically in Fig. 4.23(a). Here the wave-like line corresponds to electron–electron interaction, while "triangular" vertices describe renormalization of this interaction due to the multiple scattering of electrons by impurities.

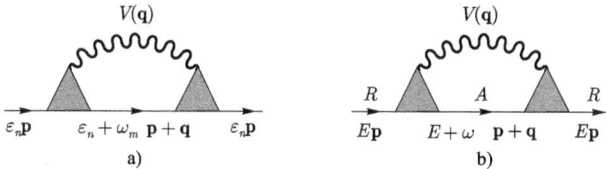

Fig. 4.23 Simplest correction to the single-particle Green's function due to electron–electron interactions in disordered metal.

First we shall make calculations for $T = 0$. Analytic expression for the Green's function correction in this case has the following form:[51]

$$\delta G(E\mathbf{p}) = iG^2(E\mathbf{p}) \int \frac{d^d q}{(2\pi)^d} \int_{-\infty}^{\infty} \frac{d\omega}{2\pi} V(\mathbf{q})\mathcal{T}^2(\mathbf{q}\omega)G(E + \omega\mathbf{p} + \mathbf{q})$$

(4.306)

where $V(\mathbf{q})$ is the Fourier transform of interaction potential. Appropriate correction to the density states is given by:

$$\delta N(E) = -\frac{N(E_F)}{\pi} Im \int_{-\infty}^{\infty} d\xi \int \frac{d^d q}{(2\pi)^d} \int \frac{d\omega}{2\pi} V(\mathbf{q})\mathcal{T}^2(\mathbf{q}\omega)$$

$$\times \frac{i}{E - \omega - \xi(\mathbf{p} + \mathbf{q}) + i\gamma sign(E - \omega)} \left(\frac{1}{E - \xi + i\gamma signE} \right)^2 \quad (4.307)$$

---

[51]Note that in the following E denotes not the Fermi energy (as above), but the energy calculated *with respect to* the Fermi energy, which will be denoted now as E_F.

For $E > 0$ the integral over ξ is different from zero if $sign(E - \omega) < 0$, i.e. for $\omega > E$. Accordingly we have:

$$\frac{\delta N(E)}{N(E_F)} \approx -\frac{1}{\pi} Im\, i \int_{-\infty}^{\infty} d\xi(p) \int \frac{d^d q}{(2\pi)^d} \int_E^{\infty} \frac{d\omega}{2\pi} V(\mathbf{q}) [\mathcal{T}^{RA}(\mathbf{q}\omega)]^2$$
$$\times G^A(E - \omega \mathbf{p} + \mathbf{q}) [G^R(E\mathbf{p})]^2 \qquad (4.308)$$

which corresponds to diagram shown in Fig. 4.23(b). Here we already can use explicit expression for the vertex given by (4.302). Substituting to (4.308) standard expressions for $G^A(E\mathbf{p})$ and $G^R(E\mathbf{p})$ (dependence of G^A on ω and q can be neglected in the limit of $\omega \ll \gamma$ and $v_F q \ll \gamma$), we obtain:

$$\frac{\delta N(E)}{N(E_F)} = -\frac{1}{\pi} Im\, i \int \frac{d^d q}{(2\pi)^d} \int_E^{\infty} \frac{d\omega}{2\pi} V(\mathbf{q}) \times [\mathcal{T}^{RA}(\mathbf{q}\omega)]^2$$
$$\times \int_{-\infty}^{\infty} d\xi \frac{1}{E - \xi - i\gamma} \frac{1}{(E - \xi + i\gamma)^2}$$
$$= -\frac{1}{2\gamma^2} Im \int \frac{d^d q}{(2\pi)^d} \int_E^{\infty} \frac{d\omega}{2\pi} [\mathcal{T}^{RA}(\mathbf{q}\omega)]^2 V(\mathbf{q}) \quad (4.309)$$

which, after the use of (4.302), finally gives for $T = 0$:

$$\frac{\delta N(E)}{N(E_F)} = -\frac{1}{\pi} Im \int \frac{d^d q}{(2\pi)^d} \int_E^{\infty} d\omega \frac{1}{(-i\omega + D_0 q^2)^2} V(\mathbf{q}) \qquad (4.310)$$

Strictly speaking we also have to introduce here the cut-off for ω-integration at $\omega \sim 1/\tau$ to guarantee the validity of diffusion approximation, but this is unnecessary, due to fast convergence of the integral at infinity.

Surely, the same result can be obtained also from Matsubara formalism in the limit of $T \to 0$. Let us see how it can be done. Since we are seeking the density of states correction, define:

$$\delta N(\varepsilon_n) = -\frac{2}{\pi} \int \frac{d^d p}{(2\pi)^d} \delta G(\varepsilon_n \mathbf{p}) \qquad (4.311)$$

which in accordance with Fig. 4.23(a), reduces to:

$$\delta N(\varepsilon_n) = \frac{2}{\pi} \int \frac{d^d q}{(2\pi)^d} T \sum_n V(\mathbf{q}) \mathcal{T}^2(\mathbf{q}\omega_m)$$
$$\times \int \frac{d^d p}{(2\pi)^d} G^2(\varepsilon_n \mathbf{p}) G(\varepsilon_n + \omega_m \mathbf{p} + \mathbf{q}) \qquad (4.312)$$

Let us first calculate the integral over $\mathbf{p}$, where for $v_F q \ll \gamma$ we can completely neglect dependence on q:

$$\int \frac{d^d p}{(2\pi)^d} G^2(\varepsilon_n \mathbf{p}) G(\varepsilon_n + \omega_m \mathbf{p} + \mathbf{q}) \approx \int \frac{d^d p}{(2\pi)^d} G^2(\varepsilon_n p) G(\varepsilon_n + \omega_m p)$$

$$\approx \nu(E_F) \int_{-\infty}^{\infty} d\xi \frac{1}{(i\varepsilon_n - \xi + i\gamma \operatorname{sign}\varepsilon_n)^2} \frac{1}{i\varepsilon_n + i\omega_m - \xi + i\gamma \operatorname{sign}(\varepsilon_n + \omega_m)}$$

$$= 2\pi i\nu(E_F) \left\{ \frac{\theta(\varepsilon_n)\theta(-\varepsilon_n - \omega_m)}{(2i\gamma - i\omega_m)^2} - \frac{\theta(\varepsilon_n + \omega_m)\theta(-\varepsilon_n)}{(2i\gamma + i\omega_m)^2} \right\}$$

$$\approx -i\nu(E_F) 2\pi\tau^2 \{\theta(\varepsilon_n)\theta(-\varepsilon_n - \omega_m) - \theta(\varepsilon_n + \omega_m)\theta(-\varepsilon_n)\} \quad (4.313)$$

Here the integral over ξ was calculated in a standard way via residues (it is different from zero on for different signs of ε_n and $\varepsilon_n + \omega_m$), and in the final expression we have used $|\omega_m \tau| \ll 1$.

Consider for definiteness $\varepsilon_n > 0$. Then only first pair of θ-functions in (4.313) contributes: $\varepsilon_n > 0$, $\varepsilon_n + \omega_m < 0$, which gives $\omega_m < -\varepsilon_n$. Then (4.312) reduces to:

$$\delta N(\varepsilon_n) = -2iTN(E_F)\tau^2 \sum_{\omega_m=-\infty}^{-\varepsilon_n} \int \frac{d^d q}{(2\pi)^d} V(q) \mathcal{T}^2(q\omega_m)$$

$$= -2iTN(E_F) \sum_{\omega_m=-\infty}^{-\varepsilon_n} \int \frac{d^d q}{(2\pi)^d} \frac{V(q)}{(-\omega_m + D_0 q^2)^2} \quad (4.314)$$

For $T \to 0$ we replace $i\omega_m \to \omega + i\delta$, i.e. $\omega_m \to -i(\omega + i\delta)$, $i\varepsilon_n \to E + i\delta$ and, accordingly, $iT\sum_m \cdots \to \int \frac{d\omega}{2\pi} \cdots$, so that in this limit we get:

$$\frac{\delta N(E)}{N(E_F)} = -2Im \int_{-\infty}^{-E} \frac{d\omega}{2\pi} \int \frac{d^d q}{(2\pi)^d} V(q) \frac{1}{(i\omega + D_0 q^2)^2}$$

$$= -2Im \int_{E}^{\infty} \frac{d\omega}{2\pi} \int \frac{d^d q}{(2\pi)^d} V(q) \frac{1}{(-i\omega + D_0 q^2)^2} \quad (4.315)$$

which obviously coincides with Eq. (4.310), obtained via diagram technique for $T = 0$.

The rest is elementary. We have:

$$\frac{\delta N(E)}{N(E_F)} = -2Im \int_{E}^{\infty} \frac{d\omega}{2\pi} \int \frac{d^d q}{(2\pi)^d} V(q) \frac{(i\omega + D_0 q^2)^2}{[\omega^2 + (D_0 q^2)^2]^2}$$

$$= -2 \int_{E}^{\infty} \frac{d\omega}{2\pi} \int \frac{d^d q}{(2\pi)^d} V(q) \frac{2D_0 q^2 \omega}{[\omega^2 + (D_0 q^2)^2]^2}$$

$$= 2 \int_{E}^{\infty} \frac{d\omega}{2\pi} \int \frac{d^d q}{(2\pi)^d} V(q) \frac{d}{d\omega} \frac{D_0 q^2}{\omega^2 + (D_0 q^2)^2} \quad (4.316)$$

so that finally we get:

$$\frac{\delta N(E)}{N(E_F)} = -\frac{1}{\pi} \int \frac{d^d q}{(2\pi)^d} V(q) \frac{D_0 q^2}{E^2 + (D_0 q^2)^2} \qquad (4.317)$$

Let us estimate this correction to the density of states for the case of point-like electron–electron interaction $V(q) = V_0$. Then we have:

$$\frac{\delta N(E)}{N(E_F)} = -\frac{V_0}{\pi} S_d \int_0^{p_0} dq\, q^{d-1} \frac{D_0 q^2}{E^2 + (D_0 q^2)^2}$$

$$= -\frac{V_0}{\pi} S_d \frac{1}{D_0^{d/2}} \int_0^{\tilde{E}^{1/2}} dx \frac{x^{d+1}}{E^2 + x^4} \approx -\frac{V_0}{\pi} S_d \frac{1}{D_0^{d/2}} \int_{E^{1/2}}^{\tilde{E}^{1/2}} dx\, x^{d-3}$$

$$(4.318)$$

where we have introduced the upper limit cut-off $p_0 \sim l^{-1}$, which corresponds to $\tilde{E} = D_0 p_0^2$, and $S_d = \Omega_d/(2\pi)^d = 2^{-(d-1)} \pi^{-\frac{d}{2}}/\Gamma\left(\frac{d}{2}\right)$. The last equality gives in (4.318) gives the simple estimate of the integral. Of course, it can be calculated exactly, but this estimate is valid up to insignificant constants of the order of unity. Finally we obtain the following correction to the density of states close to the Fermi level (A.G. Aronov, B.L. Altshuler, 1979):

$$\frac{\delta N(E)}{N(E_F)} = \frac{V_0}{\pi} \begin{cases} \frac{1}{D_0^{d/2}} S_d \frac{1}{d-2} \left\{ |E|^{\frac{d-2}{2}} - \tilde{E}^{\frac{d-2}{2}} \right\} & (d > 2) \\ \frac{1}{D_0} S_2 \ln \frac{|E|}{\tilde{E}} & (d = 2) \\ \frac{1}{D_0^{1/2}} \left\{ \frac{1}{\tilde{E}^{1/2}} - \frac{1}{|E|^{1/2}} \right\} & (d = 1) \end{cases} \qquad (4.319)$$

In particular, for $d = 3$ this gives the famous Aronov–Altshuler "sea gull" form of the density of states around the Fermi level:

$$\frac{\delta N(E)}{N(E_F)} \sim \frac{\sqrt{|E|}}{D_0^{3/2}} \qquad (4.320)$$

which is shown in Fig. 4.24. It is remarkable, that precisely this behavior obtained for the density of states in many tunneling experiments in disordered metals [Altshuler B.L., Aronov A.G. (1985)]. With the growth of disorder diffusion coefficient D_0 is suppressed, so that this anomaly in the density of states grows. What happens close to the metal–insulator transition (when $p_F l \sim 1$) is at present unclear, the complete theory of this transition with the account of electron–electron interactions is still absent [Altshuler B.L., Aronov A.G. (1985);

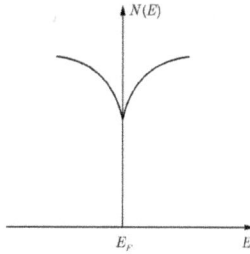

Fig. 4.24 Typical form of interaction correction to the density of states in a disordered metal.

Lee P.A., Ramakrishnan T.V. (1985)]. The majority opinion is that this Altshuler–Aronov anomaly smoothly transforms to the so-called "Coulomb gap" of Efros and Shklovskii, which forms in the density of states deeply in localized phase [Shklovskii B.I., Efros A.L. (1984)]. In principle, the behavior of this type is derived in simple generalizations of the above theory in the spirit of the self-consistent theory of localization [Sadovskii M.V. (2000)], but we shall not discuss these complicated (unsolved) problems here.

Instead we shall generalize our analysis for the more realistic case of Coulomb (long-range) interaction, when we have to take into account the effects of dynamic screening. Also we shall consider the finite temperature effects, making calculations in Matsubara technique. Now the wave-like line of the diagram in Fig. 4.23(a) represents the screened interaction. Returning to (4.312) we rewrite this expression similarly to (4.314):

$$\delta N(\varepsilon_n) = -2iN(E_F)\tau^2 T \sum_m \int \frac{d^d q}{(2\pi)^d} \mathcal{V}(q\omega_m)\mathcal{T}^2(q\omega_m)$$

$$\times \{\theta(\varepsilon_n)\theta(-\varepsilon_n - \omega_m) - \theta(-\varepsilon_n)\theta(\varepsilon_n + \omega_m)\}$$

$$= -2iN(E_F)T \sum_m \int \frac{d^d q}{(2\pi)^d} \mathcal{V}(q\omega_m) \left\{ \frac{\theta(\varepsilon_n)\theta(-\varepsilon_n - \omega_m)}{[-\omega_m + D_0 q^2]^2} - \frac{\theta(-\varepsilon_n)\theta(\varepsilon_n + \omega_m)}{[\omega_m + D_0 q^2]^2} \right\}$$

$$(4.321)$$

which for $\varepsilon_n > 0$ reduces to an expression similar to (4.314):

$$\delta N(\varepsilon_n) = -2iTN(E_F) \sum_{\omega_m=-\infty}^{-\varepsilon_n} \int \frac{d^d q}{(2\pi)^d} \frac{\mathcal{V}(q\omega_m)}{(-\omega_m + D_0 q^2)^2} \qquad (4.322)$$

Here (cf. (2.7), (2.8)):

$$\mathcal{V}(q\omega_m) = \frac{4\pi e^2}{q^2 \epsilon(q\omega_m)} \qquad (4.323)$$

$$\epsilon(q\omega_m) = 1 - \frac{4\pi e^2}{q^2}\Pi(q\omega_m) \qquad (4.324)$$

where the polarization operator, with the account of impurity scattering, is determined by diagrams of Fig. 4.12(a), and analytically by (4.89):

$$\Pi(\mathbf{q}\omega_m) = 2T \sum_n \int \frac{d^d p}{(2\pi)^d} G(\mathbf{p}\varepsilon_n)G(\mathbf{p} + \mathbf{q}\varepsilon_n + \omega_m)\mathcal{T}(\mathbf{q}\omega_m) \quad (4.325)$$

Using again (4.313), (4.305) and assuming, for example, $\omega_m < 0$, we obtain:

$$\Pi^{RA}(q\omega_m) = T \sum_{0<\varepsilon_n<-\omega_m} \frac{2\pi N(E_F)}{-\omega_m + D_0 q^2} = N(E_F)\frac{-\omega_m}{-\omega_m + D_0 q^2}$$
$$(4.326)$$

The case of $\omega_m > 0$ is calculated in a similar way, so that:

$$\Pi^{RA}(q\omega_m) = N(E_F)\left\{\frac{\omega_m \theta(\omega_m)}{\omega_m + D_0 q^2} + \frac{-\omega_m \theta(-\omega_m)}{-\omega_m + D_0 q^2}\right\} \qquad (4.327)$$

Now we have to find also the contribution from RR and AA-channels, when, according to (4.305):

$$\mathcal{T}(\mathbf{q}\omega_m \varepsilon_n) = \theta(\varepsilon_n)\theta(\varepsilon_n + \omega_m) + \theta(-\varepsilon_n)\theta(-\varepsilon_n - \omega_m) \qquad (4.328)$$

As diffusion pole is absent here, we may just put $\omega_m = 0$, $q = 0$ and write:

$$\Pi^{RR}(00) + \Pi^{AA}(00) = 2T \sum_n \sum_{\mathbf{p}} G^2(\varepsilon_n) = -N(E_F) \qquad (4.329)$$

In slightly more details:

$$\Pi^{RR}(00) + \Pi^{AA}(00) = 2T \sum_n \int \frac{d^d p}{(2\pi)^d} \frac{1}{(i\varepsilon_n - \xi(p) + i\gamma sign\varepsilon_n)^2}$$

$$= T \sum_n N(E_F) \int_{-\infty}^{\infty} d\xi \frac{\partial}{\partial \xi} \frac{1}{i\varepsilon_n - \xi + i\gamma sign\varepsilon_n}$$

$$\approx N(E_F) \sum_n \int_{-\infty}^{\infty} d\xi \frac{\partial}{\partial \xi} \frac{1}{i\varepsilon_n - \xi} = N(E_F) \int_{-\infty}^{\infty} d\xi \left(-\frac{\partial n(\xi)}{\partial \xi}\right) = -N(E_F)$$
$$(4.330)$$

where $n(\xi)$ is Fermi distribution.

As a result, to (4.327) we have to add $-N(E_F)\{\theta(\omega_m) + \theta(-\omega_m)\}$, so that the final expression for polarization operator of a metal with impurities is:

$$\Pi(q\omega_m) = -N(E_F)D_0 q^2 \left\{ \frac{\theta(\omega_m)}{\omega_m + D_0 q^2} + \frac{\theta(-\omega_m)}{-\omega_m + D_0 q^2} \right\} \quad (4.331)$$

From here, using (4.324), we immediately get:

$$\epsilon(q\omega_m) = 1 + \frac{D_0 \kappa_D^2}{\omega_m + D_0 q^2}\theta(\omega_m) + \frac{D_0 \kappa_D^2}{-\omega_m + D_0 q^2}\theta(-\omega_m) \quad (4.332)$$

where $\kappa_D^2 = 4\pi e^2 N(E_F)$ is the inverse square of screening length. For $\omega_m = 0$ from (4.332) we obtain the usual result:[52]

$$\epsilon(q0) = 1 + \frac{\kappa_D^2}{q^2} \quad (4.333)$$

for the static screening of Coulomb interaction. In general case Eq. (4.332) determines the dielectric function (permeability) of a metal with impurities in the limit of $|\omega_m \tau| \ll 1$ and $ql \ll 1$.

As a result, for small ω_m and q we obtain the effective interelectron interaction (4.323) in the following form:

$$\begin{aligned}
\mathcal{V}(q\omega_m) &= \frac{4\pi e^2}{q^2 - 4\pi e^2 \Pi(q\omega_m)} \\
&\approx \frac{4\pi e^2(-\omega_m + D_0 q^2)}{4\pi e^2 N(E_F)D_0 q^2} = \frac{-\omega_m + D_0 q^2}{N(E_F)D_0 q^2} \quad (4.334)
\end{aligned}$$

where, for definiteness, we have assumed $\omega_m < 0$. It is interesting to note, that in this approximation dependence on electric charge e^2 just canceled out.[53]

As a result, dynamically screened Coulomb interaction (4.322) reduces to:

$$\delta N(\varepsilon_n > 0) \approx -2iT \sum_{\omega_m = -\infty}^{-\varepsilon_n} \frac{1}{D_0} \int \frac{d^d q}{(2\pi)^d} \frac{q^{-2}}{-\omega_m + D_0 q^2} \quad (4.335)$$

For point-like interaction the similar expression, following from (4.314), has the form:

$$\delta N(\varepsilon_n > 0) \approx -2iN(E_F)T \sum_{\omega_m = -\infty}^{-\varepsilon_n} \int \frac{d^d q}{(2\pi)^d} \frac{V_0}{[-\omega_m + D_0 q^2]^2} \quad (4.336)$$

---

[52] Take into account the definition: $\theta(\omega_m) = \begin{cases} 1 & m \geq 0 \\ 0 & m < 0 \end{cases}$.

[53] This result is not universal and e^2 dependence reappears in more refined approximations.

Let us make explicit calculations for the case of $d = 3$. We have:

$$\delta N(\varepsilon_n > 0) = -\frac{iT}{\pi^2} \sum_{\omega_m=-\infty}^{-\varepsilon_n} \frac{1}{D_0} \int_0^{p_0} \frac{dq}{D_0 q^2 - \omega_m} \qquad \text{(Coulomb)}$$

(4.337)

$$\delta N(\varepsilon_n > 0) = -\frac{iT}{\pi^2} g \sum_{\omega_m=-\infty}^{-\varepsilon_n} \int_0^{p_0} \frac{dq\, q^2}{[D_0 q^2 - \omega_m]^2} \qquad \text{(point-like)}$$

(4.338)

where $g = N(E_F)V_0$ is dimensionless coupling constant for the case of point-like potential.

After the variable change $x^2 = D_0 q^2$, these expressions are rewritten as:

$$\delta N(\varepsilon_n > 0) = -\frac{iT}{\pi^2 D_0^{3/2}} \sum_{\omega_m=-\infty}^{-\varepsilon_n} \Phi(\omega_m) \qquad (4.339)$$

where

$$\Phi(\omega_m) = \begin{cases} \int_0^{x_0} \frac{dx}{x^2 - \omega_m} & \text{(Coulomb)} \\ g \int_0^{x_0} \frac{x^2\, dx}{(x^2 - \omega_m)^2} & \text{(point-like)} \end{cases} \qquad (4.340)$$

Now we have to calculate the sum over Matsubara (Bose) frequencies: $\sum_{\omega_m=-\infty}^{-\varepsilon_n} \Phi(\omega_m) = \sum_{m=-\infty}^{-n} \Phi(\omega_m)$. Using $\omega_m = 2\pi m T$ and $\varepsilon_n = (2n+1)\pi T$, we write this sum as the following integral over the contour, shown in Fig. 4.25:

$$T \sum_{m=-\infty}^{-n} \Phi(i\omega_m) = \oint \frac{dz}{2\pi i} n_B(z) \Phi(z) = \int_{-i\varepsilon_n+\infty}^{-i\varepsilon_n-\infty} \frac{dz}{2\pi i} n_B(z) \Phi(z)$$

$$= -\int_{-\infty}^{\infty} \frac{dz}{2\pi i} n_B(z - i\varepsilon_n) \Phi(z - i\varepsilon_n) = \int_{-\infty}^{\infty} \frac{dz}{2\pi i} n(z) \Phi(z - i\varepsilon_n)$$

(4.341)

where $n_B(z) = \frac{1}{e^{\frac{z}{T}} - 1}$ is Bose and $n(z) = \frac{1}{e^{\frac{z}{T}} + 1}$ is Fermi distribution. All this "works" if $\Phi(z)$ does not possess singularities for $Im\, z < -\varepsilon_n$ and vanishes (sufficiently fast!) for $|z| \to \infty$. Then, for Coulomb case we can write:

$$\Phi(z) = \int_0^{\infty} \frac{dx}{x^2 + iz} \qquad (4.342)$$

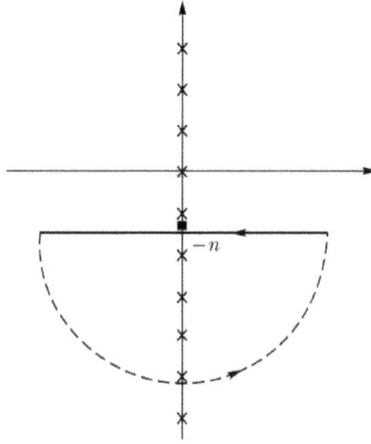

Fig. 4.25   Integration contour used in the calculations of the sum over Matsubara frequencies.

while for point-like interaction:

$$\Phi(z) = g \int_0^\infty \frac{dx\, x^2}{(x^2 + iz)^2} \qquad (4.343)$$

This function vanishes at $|z| \to \infty$ too slow, but this gives (as we shall see in a moment) only some infinite constant irrelevant for us (in difference with (4.340) here we write integrals without the upper limit cut-off!). Making in (4.341) continuation $i\varepsilon_n \to E + i\delta$ (for $\varepsilon_n > 0$) and calculating the imaginary part $Im\Phi(z) = -Im\Phi(-z)$, we get:

$$\int_{-\infty}^\infty dz\, n(z) Im\Phi(z) = \frac{1}{2} \int_{-\infty}^\infty dz [n(z + E) - n(-z + E)] Im\Phi(z)$$

$$= \int_0^\infty dz [n(z + E) - n(-z + E)] Im\Phi(z) \qquad (4.344)$$

The the density of states correction is given by:

$$\delta N(E) = -\frac{1}{2\pi^3 D_0^{3/2}} \int_0^\infty dz [n(z+E) + n(z-E) - 1] Im\Phi(z) \quad (4.345)$$

where we have used $-n(-z) = 1 + n(z)$, which allowed us to separate the last term in square brackets, which gives divergent contribution (to be cut-off!) independent of E, and thus irrelevant to us.

186   *Diagrammatics*

For the Coulomb case we have:

$$Im\Phi(z) = -Im \int_0^\infty dx \frac{x^2 - iz}{(x^2 + iz)(x^2 - iz)} = z \int_0^\infty \frac{dx}{x^4 + z^2}$$
$$= \frac{1}{\sqrt{z}} \int_0^\infty \frac{dx}{x^4 + 1} = \frac{\pi}{2\sqrt{2z}} \qquad (4.346)$$

Then using (4.346) in (4.345) (dropping the irrelevant contribution of the last term, depending on the cut-off) we get:

$$\delta N(E) = -\frac{1}{2^{5/2}\pi^2 D_0^{3/2}} \int_0^\infty d\omega \frac{1}{\sqrt{\omega}} [n(\omega + E) + n(\omega - E)] \quad (4.347)$$

or, after the partial integration:

$$\delta N(E) = \frac{T^{1/2}}{2^{3/2}\pi^2 D_0^{3/2}} \varphi\left(\frac{E}{2T}\right) \qquad (4.348)$$

where

$$\varphi(x) = \frac{1}{\sqrt{2}} \int_0^\infty dy\, y^{1/2} \left\{ \frac{1}{ch^2(x-y)} + \frac{1}{ch^2(x+y)} \right\} \qquad (4.349)$$

These expressions describe the "sea gull" behavior of the density of states at finite temperatures. Note the asymptotic dependencies:

$$\varphi(x) = \begin{cases} \sqrt{\pi}(1 - \sqrt{2})\zeta(1/2) \approx 1.07 & x \to 0 \\ \sqrt{2x} & x \gg 1 \end{cases} \qquad (4.350)$$

Thus, for Coulomb case (for $T \to 0$) we also have:

$$\delta N(E) \sim \frac{\sqrt{|E|}}{D_0^{3/2}} \qquad (4.351)$$

which is similar to (4.320) obtained for the case of short-range interaction.

These corrections to the density of states can be illustrated by the following heuristic estimates. Consider interactions of an electron in some quantum state n with energy E with some other electron from the Fermi surface. The relative correction to the wave function of our electron can be estimated in the first order of perturbation theory as:

$$\frac{\delta\varphi_n}{\varphi_n} \sim \int_0^\infty dt\, H_{int}(t) \qquad (4.352)$$

where $t = 0$ is the moment, when interaction is switched on, and $H_{int}(t)$ is interaction Hamiltonian (in interaction representation). During the time interval t our electron diffuses in disordered metal within the volume of the order of $\sim (D_0 t)^{d/2}$.

Then the matrix element of the interaction due to short-range interaction can be estimated as $V_0(D_0 t)^{-d/2}$. Accordingly we have:

$$\frac{\delta\varphi_n}{\varphi_n} \sim V_0 \int_{t_{min}}^{t_{max}} dt (D_0 t)^{-d/2} \sim \frac{V_0}{D_0^{d/2}} \left\{ t_{min}^{1-\frac{d}{2}} - t_{max}^{1-\frac{d}{2}} \right\} \qquad (4.353)$$

Here t_{min} is naturally defined by the limit of validity of diffusion approximation: $(D_0 t_{min})^{1/2} \sim l$, which gives $t_{min} \sim (D_0 l^{-2})^{-1} \sim \tilde{E}^{-1}$. The time t_{max} is determined as some $t \geq |E|^{-1}$, as at these times the matrix element is effectively suppressed by fast (time) oscillations of wave functions. Then, assuming $\frac{\delta N(E)}{N(E_F)} \sim \frac{\delta\varphi_n}{\varphi_n}$ from (4.353) we immediately obtain (4.319).

All this is not the end, but only the start of an unfinished story about the role of electron–electron interactions in disordered systems. Above we limited our analysis to the study of only single lowest order (so-called "Fock") process, described by the diagram of Fig. 4.23. There are many other contributions (diagrams), which are to be accounted for even in the limit of weak disorder $p_F l \gg 1$. Important corrections (qualitatively similar to quantum corrections considered above) appear not only in the density of states, but also in conductivity. The interested reader can find the detailed discussion of these problems in original reviews [Altshuler B.L., Aronov A.G. (1985); Alshuler B.L., Aronov A.G., Khmelnitskii D.E., Larkin A.I. (1982); Lee P.A., Ramakrishnan T.V. (1985)]. Even more complicated is the problem of the role of interaction effects in the region of strong disorder, when $p_F l \sim 1$, especially in the vicinity of metal–insulator transition induced by disorder. Despite the significant progress achieved in many theoretical studies, we are still rather far from complete understanding of this region [Sadovskii M.V. (2000)].

Chapter 5

Superconductivity

5.1 Cooper instability

Consider scattering of two-electrons due to phonon exchange, shown in Fig. 5.1. Dashed line here corresponds to:

$$g^2 D(\varepsilon_3 - \varepsilon_1; \mathbf{p}_3 - \mathbf{p}_1) = g^2 \frac{\omega^2_{\mathbf{p}_3 - \mathbf{p}_1}}{(\varepsilon_3 - \varepsilon_1)^2 - \omega^2_{\mathbf{p}_3 - \mathbf{p}_1}} \qquad (5.1)$$

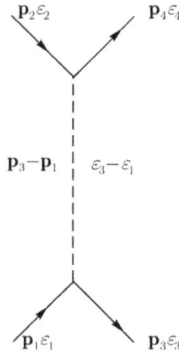

Fig. 5.1 Elementary process of electron–electron interaction due to phonon exchange.

If we consider interacting electrons with small *sum* of momenta, so that $\mathbf{p}_3 + \mathbf{p}_1 \sim 0$ (nearly opposite momenta!), the transferred momentum $\mathbf{p}_3 - \mathbf{p}_1$ is not small and its absolute value $\sim 2p_F$. At the same time, for electrons, which are close to the Fermi surface, we have $\varepsilon_3 \sim \varepsilon_1 \sim 0$. Then (5.1) in fact reduces to:

$$g^2 D(\varepsilon_3 - \varepsilon_1; \mathbf{p}_3 - \mathbf{p}_1) = -g^2 < 0 \qquad (5.2)$$

which corresponds to the appearance of electron–electron attraction. This leads to the general idea that electrons in metals with opposite momenta and spins (Pauli principle!) attract each other due to phonon exchange, which is of basic importance to BCS approach to super-conductivity [Abrikosov A.A., Gorkov L.P., Dzyaloshinskii I.E. (1963); De Gennes P.G. (1966); Schrieffer J.R. (1964)]. In the simplest approach using BCS model Hamiltonian the real interaction due to phonon exchange is replaced by an effective point-like attraction, which is different from zero only for electrons from the layer of the width of $\sim 2\omega_D$ around the Fermi surface [Abrikosov A.A., Gorkov L.P., Dzyaloshinskii I.E. (1963); De Gennes P.G. (1966); Schrieffer J.R. (1964)].

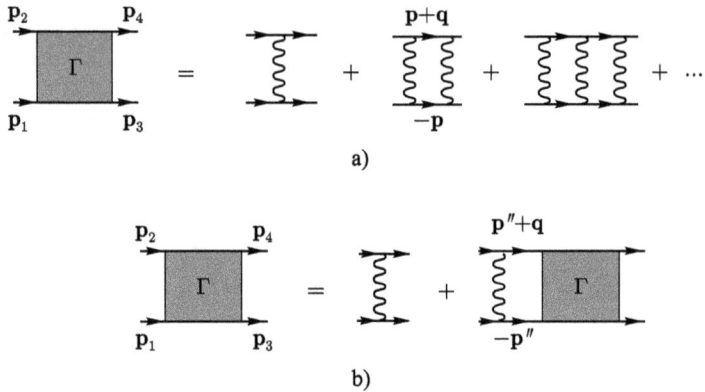

Fig. 5.2 "Ladder" in Cooper channel (a) and integral equation for the appropriate vertex-part (b).

Cooper instability of the normal metallic phase due to this attraction can be analyzed if we consider "ladder" diagrams, describing interaction of two quasiparticles (electrons) close to the Fermi surface, shown in Fig. 5.2(a), where the wave-like line denotes this attractive interaction.[1] The sum of this series (without external "tails") is given by the vertex Γ, which is determined by the integral equation shown in Fig. 5.2(b)

---

[1] We assume the following choice of external 4-momenta: $p_1 = p+q$; $p_2 = -p$; $p_3 = p'+q$; $p_4 = -p'$, so that q is the small sum of (incoming) 4-momenta.

and has the following (analytic) form:

$$\Gamma(p_3p_4;p_1p_2)=<p_3p_4|\Gamma|p_1p_2>=<p'+q,-p'|\Gamma|p+q,-p>=V(p-p')$$
$$+i\int\frac{d^4p''}{(2\pi)^4}V(p'-p'')G_0(p''+q)G_0(-p'')<p''+q,-p''|\Gamma|p+q,-p> \tag{5.3}$$

We can check the validity of Eq. (5.3) by iterations — as a result we just have the "ladder" series. In BCS model interaction "potential" $V(p-p')$ is taken in the following form:

$$V(p-p')\to V(p,p')=\lambda w_{\mathbf{p}}w_{\mathbf{p}'} \tag{5.4}$$

where

$$w_{\mathbf{p}}=\begin{cases}1 & |\xi_{\mathbf{p}}|<\omega_D\\0 & |\xi_{\mathbf{p}}|>\omega_D\end{cases} \tag{5.5}$$

From the simplest estimate given above for electron–phonon exchange we can take $\lambda=-g^2$. Now Eq. (5.3) is easily solved:

$$\Gamma(p_3p_4;p_1p_2)=<p'+q,-p'|\Gamma|p+q,-p>=\frac{\lambda w_{\mathbf{p}'+\mathbf{q}}w_{\mathbf{p}+\mathbf{q}}}{1-i\lambda\int\frac{d^4p}{(2\pi)^4}w^2_{\mathbf{p}+\mathbf{q}}G_0(p+q)G_0(-p)} \tag{5.6}$$

which is checked by direct substitution into Eq. (5.3). Consider the integral entering this expression:

$$i\lambda\int\frac{d^4p}{(2\pi)^4}w^2_{\mathbf{p}+\mathbf{q}}G_0(p+q)G_0(-p)=i\lambda\int\frac{d^4p}{(2\pi)^4}w^2_{\mathbf{p}}G_0(p)G_0(q-p)$$
$$=i\lambda\int\frac{d^3p}{(2\pi)^3}\int\frac{d\varepsilon}{2\pi}w^2_{\mathbf{p}}\frac{1}{\varepsilon-\xi(\mathbf{p})+i\delta\,sign\xi(\mathbf{p})}\frac{1}{\omega_0-\varepsilon-\xi(\mathbf{q}-\mathbf{p})+i\delta\,sign\xi(\mathbf{q}-\mathbf{p})} \tag{5.7}$$

where $q=[\omega_0,\mathbf{q}]=[\varepsilon_1+\varepsilon_2;\mathbf{p}_1+\mathbf{p}_2]$. Performing elementary contour integration, we get:

$$\int\frac{d\varepsilon}{2\pi}\frac{1}{\varepsilon-\xi(\mathbf{p})+i\delta\,sign\xi(\mathbf{p})}\frac{1}{\omega_0-\varepsilon-\xi(\mathbf{q}-\mathbf{p})+i\delta\,sign\xi(\mathbf{q}-\mathbf{p})}$$
$$=\begin{cases}-i\frac{1}{\omega_0-\xi(\mathbf{p})-\xi(\mathbf{q}-\mathbf{p})+i\delta} & \text{for}\quad \xi(\mathbf{p})>0;\ \xi(\mathbf{q}-\mathbf{p})>0\\i\frac{1}{\omega_0-\xi(\mathbf{p})-\xi(\mathbf{q}-\mathbf{p})-i\delta} & \text{for}\quad \xi(\mathbf{p})<0;\ \xi(\mathbf{q}-\mathbf{p})<0\end{cases} \tag{5.8}$$

Substituting this into (5.7), changing integration variable to $\xi = \xi(p)$, with the account of factor w_p^2 cutting-off this integration at Debye frequency ω_D, we obtain:

$$i\lambda \int \frac{d^4p}{(2\pi)^4} w_p^2 G_0(p)G_0(q-p)$$

$$\approx -\lambda \frac{mp_F}{2\pi^2} \int_0^{\omega_D} d\xi \int_0^1 dx \left\{ \frac{1}{\omega_0 + 2\xi + v_F qx - i\delta} + \frac{1}{2\xi + v_F qx - \omega_0 - i\delta} \right\}$$

$$(5.9)$$

where we have used $\xi(\mathbf{q}-\mathbf{p}) \approx \xi(\mathbf{p}) - v_F q \cos\theta$ and introduced $x = \cos\theta$. The remaining integrations are elementary and we get:

$$i\lambda \int \frac{d^4p}{(2\pi)^4} w_p^2 G_0(p)G_0(q-p)$$

$$\approx -\lambda \frac{mp_F}{2\pi^2} \left\{ 1 + \frac{1}{2} \ln \frac{2\omega_D - i\delta}{\omega_0 + v_F q - i\delta} + \frac{1}{2} \ln \frac{2\omega_D - i\delta}{-\omega_0 + v_F q - i\delta} \right.$$

$$\left. + \frac{\omega_0}{2v_F q} \left(\ln \frac{\omega_0 - i\delta}{\omega_0 + v_F q - i\delta} + \ln \frac{v_F q - \omega_0 - i\delta}{-\omega_0 - i\delta} \right) \right\}$$

$$(5.10)$$

The main (dominating) contribution to this expression at small (in comparison to ω_D) ω_0 and $v_F q$ is of the following form:

$$-\lambda \frac{mp_F}{2\pi^2} \ln \frac{\omega_D}{Max[2\omega_0; \, v_F q]}$$

$$(5.11)$$

i.e. we obtain large logarithmic factor (logarithmic divergence as limiting behavior). Finally, for the vertex part (5.6) of interest to us we get:

$$\Gamma(p_3 p_4; p_1 p_2) = < p' + q, -p' |\Gamma| p + q, -p > \equiv \Gamma(q) w_{\mathbf{p'}+\mathbf{q}} w_{\mathbf{p}+\mathbf{q}} \quad (5.12)$$

where (for $\omega_0 > v_F q)^2$

$$\Gamma(q) = \lambda \left\{ 1 + \lambda \frac{mp_F}{2\pi^2} \left[\ln e \left| \frac{2\omega_D}{\omega_0} \right| + \frac{i\pi}{2} + \frac{1}{2} \ln \left| \frac{\omega_0^2}{\omega_0^2 - v_F q^2} \right| + \frac{\omega_0}{2v_F q} \ln \left| \frac{\omega_0 - v_F q}{\omega_0 + v_F q} \right| \right] \right\}^{-1}$$

$$(5.15)$$

---

[2] Here we are using the well known relation:

$$\ln z = \ln |z| + i \, arg \, z \qquad (5.13)$$

where

$$arg \, z = arg(x + iy) = \begin{cases} arctg \frac{y}{x} & \text{for} & x > 0 \\ \pi + arctg \frac{y}{x} & \text{for} & x < 0; \; y > 0 \\ -\pi + arctg \frac{y}{x} & \text{for} & x < 0; \; y < 0 \end{cases} \qquad (5.14)$$

Let us analyze now the properties of the vertex-part (5.15). For simplicity, consider first the case of $q = 0$. For real and positive ω_0 we have:

$$\Gamma(\omega_0) = \frac{\lambda}{1 + \lambda \frac{mp_F}{2\pi^2} \left[\ln \left| \frac{2\omega_D}{\omega_0} \right| + \frac{i\pi}{2} \right]} \qquad (5.16)$$

Consider now $\Gamma(\omega_0)$ as the function of a complex variable ω_0, defining it as an analytic continuation of (5.16) to the upper half-plane, where $Im\omega_0 > 0$. Then, putting in Eq. (5.16) $\omega_0 = |\omega_0| e^{i\varphi}$, we get:

$$\Gamma(\omega_0) = \frac{\lambda}{1 + \lambda \frac{mp_F}{2\pi^2} \left[\ln \left| \frac{2\omega_D}{\omega_0} \right| + \frac{i\pi}{2} - i\varphi \right]} \qquad (5.17)$$

If interaction of electrons is attractive, i.e. $\lambda < 0$ (5.17) has a pole, defined by the equation:[3]

$$1 + \lambda \frac{mp_F}{2\pi^2} \left[\ln \left| \frac{2\omega_D}{\omega_0} \right| - i \left(\frac{\pi}{2} - \varphi \right) \right] = 0 \qquad (5.18)$$

giving $\varphi = \frac{\pi}{2}$ and $1 + \lambda \frac{mp_F}{2\pi^2} \left[\ln \left| \frac{2\omega_D}{\omega_0} \right| \right] = 0$. In other words, the pole appears at *imaginary* frequency, $\omega_0 = i\tilde{\omega}$, where:

$$\tilde{\omega} = 2\omega_D \exp \left(-\frac{2\pi^2}{mp_F |\lambda|} \right) \qquad (5.19)$$

Close to the pole $\Gamma(\omega_0)$ has the following form:

$$\Gamma(\omega_0) \approx -\frac{2\pi^2}{mp_F} \frac{i\tilde{\omega}}{\omega_0 - i\tilde{\omega}} \qquad (5.20)$$

This corresponds to *Cooper instability* — the pole in the vertex part in the upper half-plane of frequency formally signifies the appearance of an unstable collective mode with exponentially growing (in time) amplitude: $e^{-i\omega_0 t} \sim e^{-ii\tilde{\omega}t} \sim e^{\tilde{\omega}t}$! This leads to instability of the system and reconstruction of its ground state and spectra of excitations.

---

[3]In case of repulsion $\lambda > 0$ and there is nothing interesting. In this case Eq. (5.17) just gives the sum of all "ladder" corrections to the "bare" interaction λ. The large logarithm leads only to the effective suppression of this repulsion and there is no "pathology" at all.

For nonzero values of $v_F q$ Eq. (5.15) may be rewritten (for $\omega_0 > v_F q$) as:

$$\Gamma(\mathbf{q}, \omega_0) = \lambda \left\{ 1 + \lambda \frac{m p_F}{2\pi^2} \left[\ln e \left| \frac{2\omega_D}{\omega_0} \right| + \frac{i\pi}{2} - i\varphi \right. \right.$$
$$\left. \left. - \frac{1}{2} \ln \left(1 - \frac{v_F^2 q^2}{\omega_0^2} \right) + \frac{\omega_0}{2 v_F q} \ln \left(\frac{\omega_0 - v_F q}{\omega_0 + v_F q} \right) \right] \right\}^{-1} \qquad (5.21)$$

so that after the continuation to the half-plane of $Im\,\omega_0 > 0$ and the use of definition of $\tilde{\omega}$ given in (5.19), we find:

$$\Gamma(\mathbf{q}, \omega_0) = -\frac{2\pi^2}{m p_F} \left\{ \ln \frac{\omega_0}{i\tilde{\omega}} - 1 + \frac{1}{2} \ln \left(1 - \frac{v_F^2 q^2}{\omega_0^2} \right) \right.$$
$$\left. - \frac{\omega_0}{2 v_F q} \ln \left(\frac{\omega_0 - v_F q}{\omega_0 + v_F q} \right) \right\}^{-1} \qquad (5.22)$$

For small $v_F q \ll \tilde{\omega}$ we have:

$$\Gamma(\mathbf{q}, \omega_0) \approx -\frac{2\pi^2}{m p_F} \frac{i\tilde{\omega}}{\omega_0 - i\tilde{\omega} + i\frac{v_F^2 q^2}{6\tilde{\omega}}} \qquad (5.23)$$

Then we find the pole position as a function of q:

$$\omega_0 = i\tilde{\omega} \left(1 - \frac{v_F^2 q^2}{6\tilde{\omega}^2} \right) \qquad (5.24)$$

so that the absolute value of ω_0 diminishes with the growth of q. For some $v_F q_{max}$ the pole position ω_0 goes to zero, and for larger values of $v_F q$ the pole in Γ is just absent. As $\mathbf{q}$ gives the sum of the momenta of two electrons, this result means that the tendency to pairing is stronger for electrons with nearly opposite momenta.

Cooper "ladder" contributes to electron self-energy via diagrams shown in Fig. 5.3(a). Naturally, the existence of the pole in the "ladder" leads to singularity in $\Sigma(p)$ and in the vertex part, shown in Fig. 5.3(b).

Let us stress once again that these results signify the instability of the usual (normal) ground state $(T = 0)$ of Fermi-gas due to attractive interaction. The physical meaning of this instability reduces to the ability of particles (with almost zero momentum of their center of inertia) to form bound pairs, i.e. some kind of Bose particles, which may "condense" in the ground state. The temperature, corresponding to the appearance of this instability, defines the temperature of superconducting transition. To understand this more deeply, let us analyze the same problem within Matsubara formalism.

a)

b)

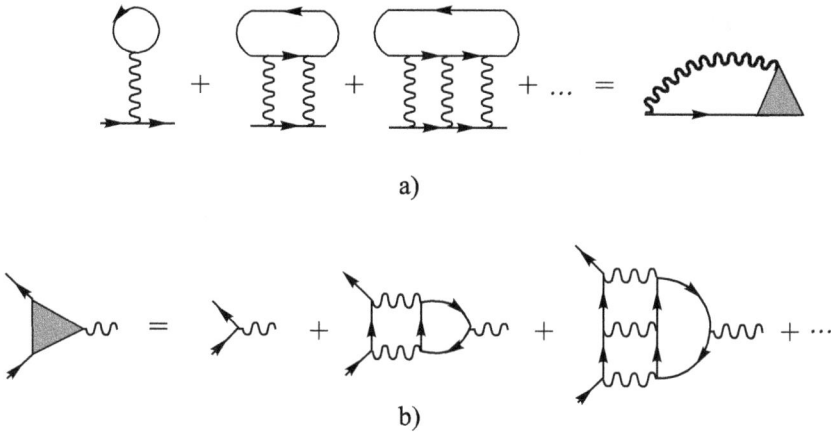

Fig. 5.3 Corrections to electron self-energy due to scattering in Cooper channel (a) and diagrams for appropriate vertex-part (b).

If we neglect scattering of bound pairs on each other, the ideal Bose-gas of Cooper pairs is formed and Matsubara Green's function of this gas can be written as:

$$G(\mathbf{q}, \omega_m) = \frac{1}{i\omega_m - \frac{q^2}{2m^*} + \mu} \qquad (5.25)$$

where $\omega_m = 2\pi mT$, $\mathbf{q}$ is the momentum of the bound pair, m^* — its mass, which is equal to two masses of an electron. For $\omega_m = 0$ (5.25) reduces to $[\mu - q^2/2m^*]^{-1}$. At the temperature of Bose condensation $T = T_0$ this function diverges for $q = 0$, so that T_0 is determined from the equation for $\mu = 0$, in accordance with the standard analysis of Bose-condensation [Sadovskii M.V. (2019a)].

If we take into account the internal structure of Cooper pair, the analogue of (5.25) is the two-particle Fermion Green's function. At the transition point its analytic behavior have to be similar to that of Bose-gas Green's function, in the sense of its dependence on $\omega_{0m} = (\varepsilon_1 + \varepsilon_2)_n$ and $\mathbf{q} = \mathbf{p}_1 + \mathbf{p}_2$ (corresponding to the center of inertia of a pair). Single-Fermion Green's functions do not have these singularities and we have to consider the appropriate vertex-part $\Gamma(\mathbf{q}\omega_0)$, which is given by the same "ladder" diagrams as above, but written in Matsubara formalism.

The difference is that instead of Eq. (5.7) we have to consider:

$$I = -\lambda T \sum_n \int \frac{d^3 p}{(2\pi)^3} w_{\mathbf{p}}^2 \frac{1}{i\varepsilon_n - \xi(\mathbf{p})} \frac{1}{i\omega_{0m} - i\varepsilon_n - \xi(\mathbf{q} - \mathbf{p})} \quad (5.26)$$

where $\omega_{0m} = 2\pi m_0 T$ (where m_0 is an integer). We shall not calculate (5.26) for arbitrary ω_{0m} and $\mathbf{q}$, because it is clear that (as in Bose-gas) the pole in $\Gamma(\mathbf{q}\omega_0)$ appears first for $\omega_0 = q = 0$. Thus it is sufficient to analyze only this case and we have to calculate:

$$I = \frac{\lambda}{(2\pi)^3} T \sum_n \int d^3 p \, w_p^2 \frac{1}{i\varepsilon_n - \xi(p)} \frac{1}{i\varepsilon_n + \xi(p)}$$

$$\approx -\lambda \frac{m p_F}{2\pi^2} \int_{-\omega_D}^{\omega_D} d\xi T \sum_n \frac{1}{\varepsilon_n^2 + \xi^2} = -\lambda \frac{m p_F}{2\pi^2} \int_0^{\omega_D} \frac{d\xi}{\xi} th \frac{\xi}{2T} \quad (5.27)$$

where the sum over frequencies was calculated using (2.100). After partial integration we get:

$$I = -\lambda \frac{m p_F}{2\pi^2} \left(\ln \frac{\omega_D}{2T} - \int_0^\infty dx \frac{\ln x}{ch^2 x} \right) \quad (5.28)$$

where in the remaining integral we replaced the upper limit $x = \frac{\omega_D}{2T}$ by infinity, due to fast convergence and $T \ll \omega_D$ (of interest to us). Now this integral is equal to $\ln \frac{\pi}{4\gamma}$, where $\ln \gamma = C = 0.577...$ (Euler constant), so that $\gamma \approx 1.78$ and $\frac{2\gamma}{\pi} \approx 1.14$. Finally we get:

$$\Gamma(0,0) = \frac{\lambda}{1 + \lambda \frac{m p_F}{2\pi^2} \ln \frac{2\gamma\omega_D}{\pi T}} \quad (5.29)$$

For $\lambda < 0$ we again get a pole, close to which we have:

$$\Gamma(0,0) = -\frac{2\pi^2}{m p_F} \frac{T_c}{T - T_c} \quad (5.30)$$

where *temperature of superconducting transition* T_c is defined by BCS expression (J. Bardeen, L. Cooper, J. Schrieffer, 1957):

$$T_c = \frac{2\gamma}{\pi} \omega_D \exp \left(-\frac{2\pi^2}{|\lambda| m p_F} \right) = \frac{2\gamma}{\pi} \omega_D \exp \left(-\frac{1}{|\lambda| \nu_F} \right) \quad (5.31)$$

Note that here enters the density of states at the Fermi level ν_F for the *single* spin projection. The value of $|\lambda| \nu_F$ determines dimensionless coupling constant of pairing interaction. It is important to stress that dependence on this constant in (5.31) is *nonanalytic* and this expression can not be expanded in powers of λ for $\lambda \to 0$!

The frequency $\tilde{\omega}$ introduced in Eq. (5.19) and characterizing instability of the system at $T = 0$ is directly connected with T_c:

$$\tilde{\omega} = \frac{\pi}{\gamma}T_c \qquad (5.32)$$

5.2 Gorkov equations

From the previous results it is clear that special analysis is required for the temperature region $T < T_c$ [Abrikosov A.A., Gorkov L.P., Dzyaloshinskii I.E. (1963); Lifshits E.M., Pitaevskii L.P. (1980)]. We shall consider the simplified model of the Fermi-gas with point-like attraction described by the Hamiltonian:

$$\begin{aligned}
H = H_0 &+ H_{int} \\
&= \sum_\alpha \int d\mathbf{r}\psi_\alpha^+(\mathbf{r})\left[-\frac{1}{2m}\nabla^2 - \mu\right]\psi_\alpha(\mathbf{r}) \\
&+ \frac{\lambda}{2}\sum_{\alpha\beta}\int d\mathbf{r}\psi_\alpha^+(\mathbf{r})\psi_\beta^+(\mathbf{r})\psi_\beta(\mathbf{r})\psi_\alpha(\mathbf{r}) \qquad (5.33)
\end{aligned}$$

Formally this corresponds to interaction potential $V(\mathbf{r} - \mathbf{r}') = \lambda\delta(\mathbf{r} - \mathbf{r}')$, but during calculations of integrals and sums we shall take into account the limitation of the type given by Eqs. (5.4), (5.5), to mimic electron–phonon nature of this attraction in real metals.

In Heisenberg representation we can write down the standard equations of motion for electronic operators:

$$i\frac{\partial}{\partial t}\psi_\alpha(x) = [\psi_\alpha, H] \qquad (5.34)$$

Commutator in the right-hand side is calculated directly using commutation relations for operators $\psi_\alpha(x)$:

$$\psi_\alpha(\mathbf{r}, t)\psi_\beta^+(\mathbf{r}', t) + \psi_\beta^+(\mathbf{r}', t)\psi_\alpha(\mathbf{r}, t) = \delta_{\alpha\beta}\delta(\mathbf{r} - \mathbf{r}') \qquad (5.35)$$

$$\psi_\alpha(\mathbf{r}, t)\psi_\beta(\mathbf{r}', t) + \psi_\beta(\mathbf{r}', t)\psi_\alpha(\mathbf{r}, t) = 0$$
$$\psi_\alpha^+(\mathbf{r}, t)\psi_\beta^+(\mathbf{r}', t) + \psi_\beta^+(\mathbf{r}', t)\psi_\alpha^+(\mathbf{r}, t) = 0 \qquad (5.36)$$

Calculating commutators of operators with separate terms of the Hamiltonian (5.33) we get:

$$[\psi_\alpha(x), H_0] = -\left(\frac{\nabla^2}{2m} + \mu\right)\psi_\alpha(x) \qquad (5.37)$$

$$[\psi_\alpha(x), H_{int}] = -\lambda \sum_\beta \psi_\beta^+(x)\psi_\beta(x)\psi_\alpha(x) \tag{5.38}$$

$$[\psi_\alpha^+(x), H_0] = \left(\frac{\nabla^2}{2m} + \mu\right)\psi_\alpha^+(x) \tag{5.39}$$

$$[\psi_\alpha^+(x), H_{int}] = \lambda \sum_\beta \psi_\alpha^+(x)\psi_\beta^+(x)\psi_\beta(x) \tag{5.40}$$

so that explicitly equations of motion are:

$$i\frac{\partial\psi_\alpha}{\partial t} = -\left(\frac{\nabla^2}{2m} + \mu\right)\psi_\alpha - \lambda\psi_\gamma^+\psi_\gamma\psi_\alpha \tag{5.41}$$

$$i\frac{\partial\psi_\alpha^+}{\partial t} = \left(\frac{\nabla^2}{2m} + \mu\right)\psi_\alpha^+ + \lambda\psi_\alpha^+\psi_\gamma^+\psi_\gamma \tag{5.42}$$

where we implicitly assume summation over repeating Greek (spin) indices.

Our qualitative picture of the ground state of a superconductor assumes that at $T = 0$ we are dealing with *condensate* of Cooper pairs with enormous (macroscopic) number of particles. Physically, it is obvious that this state does not change at all if we change the number of pairs in the condensate by one.[4] Mathematically this is expressed by the appearance of nonzero (in the limit of number of particles $N \to \infty$) values of matrix elements of the following form:

$$\lim_{N\to\infty} <m, N|\psi_\beta(x_2)\psi_\alpha(x_1)|m, N+2> =$$
$$\lim_{N\to\infty} <m, N+2|\psi_\alpha^+(x_1)\psi_\beta^+(x_2)|m, N>^* \neq 0 \tag{5.43}$$

Here $\psi_\beta(x_2)\psi_\alpha(x_1)$ is the operator of annihilation of two electrons, while similarly $\psi_\alpha^+(x_1)\psi_\beta^+(x_2)$ is the pair creation operator. In the following, for shortness, we drop the symbol of limit and diagonal matrix index, enumerating "the same" states of the system with different number of particles.

Thus, superconducting transition is characterized by *spontaneous breaking of gauge symmetry*,[5] corresponding to particle number (or

---

[4]Note the obvious analogy of this assumption with similar hypothesis in Bogoliubov's theory of weakly interacting Bose-gas [Sadovskii M.V. (2019a)].

[5]The concept of spontaneous symmetry breaking plays the central role in the modern theory of second order phase transitions [Mattuck R.D. (1968)]. The ground state of "condensed" phase, appearing at temperatures below the transition point T_c, always possess the symmetry, which is *lower* than the symmetry of the Hamiltonian, describing this phase transition [Sadovskii M.V. (2019a)].

charge!) conservation — electron pairs may "disappear" in condensate, or "appear" from condensate, without change of the macroscopic
state of the system (for $N \to \infty$!).

Thus, besides the usual (normal) Green's function:

$$iG_{\alpha\beta}(x_1, x_2) =< N|T\psi_\alpha(x_1)\psi_\beta^+(x_2)|N > \tag{5.44}$$

it is necessary to introduce the so-called *anomalous* Green's functions:[6]

$$iF_{\alpha\beta}(x_1, x_2) =< N|T\psi_\alpha(x_1)\psi_\beta(x_2)|N+2 > \tag{5.45}$$

$$iF_{\alpha\beta}^+(x_1, x_2) =< N+2|T\psi_\alpha^+(x_1)\psi_\beta^+(x_2)|N > \tag{5.46}$$

Anomalous Green's functions $F_{\alpha\beta}$ and $F_{\alpha\beta}^+$ satisfy the following general
symmetry properties, which follow directly from commutation relations
for electron operators:

$$F_{\alpha\beta}(x_1, x_2) = -F_{\beta\alpha}(x_2, x_1) \qquad F_{\alpha\beta}^+(x_1, x_2) = -F_{\beta\alpha}^+(x_2, x_1) \tag{5.47}$$

In the following we consider only (spin) singlet Cooper pairing, which is realized in majority of known metallic superconductors.[7] Let us separate spin
dependence using the following representation:

$$F_{\alpha\beta} = A_{\alpha\beta}F \qquad F_{\alpha\beta}^+ = B_{\alpha\beta}F^+ \tag{5.48}$$

Due to Pauli principle $< \psi_\alpha(x)\psi_\alpha(x) >= 0$, so that $F_{\alpha\alpha} = F_{\alpha\alpha}^+ = 0$. Accordingly
$A_{\alpha\alpha} = B_{\alpha\alpha} = 0$ and matrices A and B can be written as:

$$A = \begin{pmatrix} 0 & a_1 \\ a_2 & 0 \end{pmatrix} \qquad B = \begin{pmatrix} 0 & b_1 \\ b_2 & 0 \end{pmatrix} \tag{5.49}$$

For singlet pairing $F(\mathbf{r} - \mathbf{r}') = F(\mathbf{r}' - \mathbf{r})$ and from (5.47) it follows that $A_{\alpha\beta} = -A_{\beta\alpha}$ and $B_{\alpha\beta} = -B_{\beta\alpha}$, so that:[8]

$$A = \begin{pmatrix} 0 & a \\ -a & 0 \end{pmatrix} \qquad B = \begin{pmatrix} 0 & b \\ -b & 0 \end{pmatrix} \tag{5.50}$$

and from $(F_{\alpha\beta}^+)^* = -F_{\alpha\beta}$ we get $B^* = -A$, so that also $b^* = -a$. Accordingly:

$$A = a \begin{pmatrix} 0 & 1 \\ -1 & 0 \end{pmatrix} \qquad B = -a^* \begin{pmatrix} 0 & 1 \\ -1 & 0 \end{pmatrix} \tag{5.51}$$

---

[6]Note the close connection of the appearance of anomalous Green's functions with
Bogoliubov's ideology of quasi-averages [Sadovskii M.V. (2019a); Mattuck R.D. (1968);
Bogoliubov N.N. (1991b)].

[7]Triplet pairing is observed in superfluid phases of He^3 and in some metallic compounds,
e.g. in Sr_2RuO_4.

[8]For triplet pairing the coordinate part of anomalous Green's function is antisymmetric,
while spin part is symmetric.

Then the spin dependence of anomalous Green's functions reduces to unit antisymmetric spinor of the second rank:

$$g_{\alpha\beta} = \begin{pmatrix} 0 & 1 \\ -1 & 0 \end{pmatrix} = i\sigma_{\alpha\beta}^y \qquad (\hat{g}^2)_{\alpha\beta} = -\delta_{\alpha\beta} \qquad (5.52)$$

where $\sigma^y = \begin{pmatrix} 0 & -i \\ i & 0 \end{pmatrix}$. Thus we can write anomalous Green's functions as:

$$F_{\alpha\beta}(x_1, x_2) = g_{\alpha\beta}F(x_1, x_2) \qquad F_{\alpha\beta}^+(x_1, x_2) = g_{\alpha\beta}F^+(x_1, x_2) \qquad (5.53)$$

where in the r.h.s. we have functions symmetric over x_1 and x_2. Spin dependence of the normal Green's function $G_{\alpha\beta}$ (for nonmagnetic system) reduces to $G_{\alpha\beta} = \delta_{\alpha\beta}G$. In homogeneous and stationary system Green's functions G, F, F^+ depend only on the differences of coordinates and moments of time.

Let us introduce anomalous functions at coinciding points:

$$\Xi(x) = iF(x, x) \qquad \Xi^*(x) = -iF^+(x, x) \qquad (5.54)$$

which sometimes are called condensate wave-functions of Cooper pairs. In stationary and spatially homogeneous system $\Xi(x)$ reduces to a constant and with the appropriate choice of phases of ψ-operators this constant may be made real.

Let us now find these Green's functions in our model of Fermi-gas with attraction, defined by the Hamiltonian (5.33). Equations of motion for operators ψ and ψ^+ were given above in Eqs. (5.41) and (5.42). Now we have:[9]

$$\frac{\partial}{\partial t_1}G_{\alpha\beta} = -i\left\langle T\frac{\partial\psi_\alpha(x_1)}{\partial t_1}\psi_\beta^+(x_2)\right\rangle - i\delta_{\alpha\beta}\delta(\mathbf{r}_1 - \mathbf{r}_2)\delta(t_1 - t_2) \quad (5.56)$$

so that substituting here (5.41) we obtain the equation of motion for the normal Green's function:

$$\left(i\frac{\partial}{\partial t} + \frac{\nabla^2}{2m} + \mu\right)G_{\alpha\beta}(x - x') - i\lambda <N|T\psi_\gamma^+(x)\psi_\gamma(x)\psi_\alpha(x)\psi_\beta^+(x')|N>$$
$$= \delta_{\alpha\beta}\delta(x - x') \qquad (5.57)$$

[9]To obtain the second term in (5.56) we have to remember that $G_{\alpha\beta}$ is discontinuous at $t_1 = t_2$:

$$G_{\alpha\beta}|_{t_1=t_2+0} - G_{\alpha\beta}|_{t_1=t_2-0} = -i < \psi_\alpha(t_1, \mathbf{r}_1)\psi^+(t_1, \mathbf{r}_2) + \psi_\beta^+(t_1, \mathbf{r}_2)\psi_\alpha(t_1, \mathbf{r}_1) >$$
$$= -i\delta_{\alpha\beta}\delta(\mathbf{r}_1 - \mathbf{r}_2) \qquad (5.55)$$

Here the diagonal matrix element of the product of four ψ-operators can be (approximately) expanded a'la Wick theorem via all "pairings", i.e. into the sum of matrix elements of products of pairs of operators:

$$< N|T\psi_\gamma^+(x)\psi_\gamma(x)\psi_\alpha(x)\psi_\beta^+(x')|N >$$
$$\approx G_{\gamma\gamma}(0)G_{\alpha\beta}(x - x') - G_{\alpha\gamma}(0)G_{\gamma\beta}(x - x')$$
$$+ < N|T\psi_\gamma(x)\psi_\alpha(x)|N + 2 >< N + 2|T\psi_\gamma^+(x)\psi_\beta^+(x')|N > \quad (5.58)$$

The terms of the type of GG are not very interesting to us, as they just lead to some irrelevant renormalization of the energy spectrum of the normal state. Thus, in the following we consider only the contributions of the last terms in (5.58), containing the matrix elements of transitions changing the number of particles $N \leftrightarrow N + 2$, so that:[10]

$$< N|T\psi_\gamma^+(x)\psi_\gamma(x)\psi_\alpha(x)\psi_\beta^+(x')|N >$$
$$\to < N|T\psi_\gamma(x)\psi_\alpha(x)|N + 2 >< N + 2|T\psi_\gamma^+(x)\psi_\beta^+(x')|N >$$
$$= -F_{\gamma\alpha}(x, x)F_{\gamma\beta}^+(x, x') = -\delta_{\alpha\beta}F(0)F^+(x - x') \quad (5.59)$$

where we have taken into account (5.45), (5.46) and (5.52), (5.53). Using (5.59) Eq. (5.57) can be reduced to:

$$\left(i\frac{\partial}{\partial t} + \frac{\nabla^2}{2m} + \mu\right)G(x) + \lambda\Xi F^+(x) = \delta(x) \quad (5.60)$$

where we have changed $x - x'$ by x, and denoted a constant $iF(0)$ as Ξ.

But now we have to write down an equation of motion also for the anomalous Green's function $F^+(x)$! To do this, calculate first the derivative:

$$i\frac{\partial}{\partial t_1}F_{\alpha\beta}^+(x - x') = \left\langle N + 2 \left| T\frac{\partial\psi_\alpha^+(x)}{\partial t_1}\psi_\beta^+(x') \right| N \right\rangle \quad (5.61)$$

Note that term with δ-function, of the type of the second term in (5.56), does not appear here as $F_{\alpha\beta}(x - x')$ (opposite to $G_{\alpha\beta}(x - x')$) is continuous at $t = t'$ (due to anticommutation of $\psi_\alpha^+(t, \mathbf{r})$ and $\psi_\beta^+(t, \mathbf{r}')$). Substituting (5.41) into (5.61) and separating condensate contribution, similarly to (5.59), we obtain the equation:

$$\left(i\frac{\partial}{\partial t} - \frac{\nabla^2}{2m} - \mu\right)F^+(x) + \lambda\Xi^*G(x) = 0 \quad (5.62)$$

---

[10]In the normal system of fermions (without condensate these matrix elements are, of course, zero). But here, these anomalous contributions lead to qualitatively new results.

Thus we obtain the closed system of Eqs. (5.60) and (5.62), determining the Green's functions of a superconductor (L.P. Gorkov, 1958).

In the momentum representation Gorkov equations can be written as $\left(p = (\varepsilon, \mathbf{p}), \xi(\mathbf{p}) = \frac{\mathbf{p}^2}{2m} - \mu\right)$:

$$(\varepsilon - \xi(\mathbf{p}))G(p) + \lambda \Xi F^+(p) = 1 \qquad (5.63)$$

$$(\varepsilon + \xi(\mathbf{p}))F^+(p) + \lambda \Xi^* G(p) = 0 \qquad (5.64)$$

Substituting F^+ from the second equation into the first we immediately find:

$$(\varepsilon^2 - \xi^2(\mathbf{p}) - |\Delta|^2)G(p) = \varepsilon + \xi(\mathbf{p}) \qquad (5.65)$$

where we have introduced notation:

$$\Delta = \lambda \Xi \qquad (5.66)$$

Below we shall see that this quantity will play the role of the *energy gap* in the spectrum of elementary excitations of a superconductor and simultaneously that of the *order parameter* for superconducting transition.

Formal solution of (5.65) is:

$$G(p) = \frac{\varepsilon + \xi(\mathbf{p})}{\varepsilon^2 - \varepsilon^2(\mathbf{p})} = \frac{u_p^2}{\varepsilon - \varepsilon(\mathbf{p})} + \frac{v_p^2}{\varepsilon + \varepsilon(\mathbf{p})} \qquad (5.67)$$

where

$$\varepsilon(\mathbf{p}) = \sqrt{\xi^2(\mathbf{p}) + |\Delta|^2} \qquad (5.68)$$

is the spectrum of elementary excitations of BCS theory (with energy gap $2|\Delta|$), while u_p and v_p are the well known Bogoliubov's coefficients [Sadovskii M.V. (2019a)]:

$$\left.\begin{array}{c} u_p^2 \\ v_p^2 \end{array}\right\} = \frac{1}{2}\left(1 \pm \frac{\xi(\mathbf{p})}{\sqrt{\xi^2(\mathbf{p}) + |\Delta|^2}}\right) \qquad (5.69)$$

Note that in Eq. (5.67) the imaginary part of G remains undetermined. It obviously contains contribution of the type of $\delta(\varepsilon \pm \varepsilon(\mathbf{p}))$ which vanishes after multiplication by $(\varepsilon^2 - \varepsilon^2(\mathbf{p}))$ in (5.65). According to general properties of analyticity [Abrikosov A.A., Gorkov L.P., Dzyaloshinskii I.E. (1963)] the sign of the imaginary part of the Green's function should be opposite to the sign of ε. Thus, in terms with positive and negative frequencies this variable should be changed to $\varepsilon \pm i\delta$, so that

$$G(\varepsilon \mathbf{p}) = \frac{u_p^2}{\varepsilon - \varepsilon(\mathbf{p}) + i\delta} + \frac{v_p^2}{\varepsilon + \varepsilon(\mathbf{p}) - i\delta}$$

$$= \frac{\varepsilon + \xi(\mathbf{p})}{(\varepsilon - \varepsilon(\mathbf{p}) + i\delta)(\varepsilon + \varepsilon(\mathbf{p}) - i\delta)} \tag{5.70}$$

Now from (5.64) we find also the anomalous function F^+:

$$F^+(\varepsilon \mathbf{p}) = -\frac{\lambda \Xi^*}{(\varepsilon - \varepsilon(\mathbf{p}) + i\delta)(\varepsilon + \varepsilon(\mathbf{p}) - i\delta)}$$

$$= -\frac{\Delta^*}{(\varepsilon - \varepsilon(\mathbf{p}) + i\delta)(\varepsilon + \varepsilon(\mathbf{p}) - i\delta)} \tag{5.71}$$

At the same time, by definition, we have:

$$i\Xi^* = F^+(x = 0) = \int_{-\infty}^{\infty} \frac{d\varepsilon}{2\pi} \int \frac{d^3 p}{(2\pi)^3} F^+(\varepsilon \mathbf{p}) \tag{5.72}$$

Substituting here (5.71) we perform integration over ε closing the contour in the upper half-plane, so that integral is expressed via the residue at the pole at $\varepsilon = \varepsilon(\mathbf{p})$. As a result, after division by Ξ^*, we get:

$$1 = -\frac{\lambda}{2(2\pi)^3} \int d^3 p \frac{1}{\sqrt{\xi^2(\mathbf{p}) + \Delta^2}} \tag{5.73}$$

— the gap equation of BCS theory. Remember that we consider $\lambda < 0$, also we now write Δ already without the sign of modulus as now we can consider the gap to be real. The divergence of integral in (5.73) is cut-off, as usual in BCS approach, due to the fact that only electrons from the layer of the width of $2\omega_D$ around the Fermi level attract each other (cf. (5.4), (5.5)). Then we get:

$$\frac{|\lambda|}{2(2\pi)^3} \int dp \frac{4\pi p^2}{\sqrt{\xi^2(\mathbf{p}) + \Delta^2}} \rightarrow \frac{|\lambda| p_F^2}{4\pi^2 v_F} \int_{-\omega_D}^{\omega_D} \frac{d\xi}{\sqrt{\xi^2 + \Delta^2}} = \frac{|\lambda| m p_F}{2\pi^2} \ln \frac{2\omega_D}{\Delta} \tag{5.74}$$

so that (5.73) reduces to:

$$1 = |\lambda| \frac{m p_F}{2\pi^2} \ln \frac{2\omega_D}{\Delta} \tag{5.75}$$

which gives the standard result of BCS theory:

$$\Delta_0 = 2\omega_D e^{-\frac{2\pi^2}{|\lambda| m p_F}} = 2\omega_D e^{-\frac{1}{|\lambda| \nu_F}} \tag{5.76}$$

Comparing with (5.31), we have:

$$\Delta_0 = \frac{\pi}{\gamma} T_c \tag{5.77}$$

It is instructive to calculate the density of electronic states of a super-conductor. Using the general definition of the density of states given in (4.99) and (5.67), (5.70), we obtain:

$$N(\varepsilon) = -\frac{2}{\pi}\nu_F \int_{-\infty}^{\infty} d\xi ImG^R(\varepsilon,\xi) = -\frac{2}{\pi}\nu_F sign\varepsilon Im \int_{-\infty}^{\infty} d\xi \frac{\varepsilon+\xi}{\varepsilon^2 - \xi^2 - \Delta^2}$$

$$= \frac{2}{\pi}|\varepsilon| Im \int_{-\infty}^{\infty} d\xi \frac{1}{\xi^2 + \Delta^2 - \varepsilon^2} = 2\nu_F|\varepsilon| Im \frac{1}{\sqrt{\Delta^2 - \varepsilon^2}} \tag{5.78}$$

where we have taken into account that for $T = 0$ $ImG^R(\varepsilon\mathbf{p}) = sign\varepsilon ImG(\varepsilon\mathbf{p})$ [Abrikosov A.A., Gorkov L.P., Dzyaloshinskii I.E. (1963)]. Introducing the density of states of free electrons at the Fermi level (for both spin projections) $N(E_F) = 2\nu_F$, we have:

$$\frac{N(\varepsilon)}{N(E_F)} = \begin{cases} \frac{|\varepsilon|}{\sqrt{\varepsilon^2 - \Delta^2}} & \text{for} \quad |\varepsilon| > \Delta \\ 0 & \text{for} \quad |\varepsilon| < \Delta \end{cases} \tag{5.79}$$

This is the famous result of BCS theory. Density of states is zero within the gap, i.e. in the energy region of the width of 2Δ around the Fermi level. For $\varepsilon = \pm\Delta$ there are square root divergences, while for $|\varepsilon| \to \infty$ the density of states tends asymptotically to its free value $N(E_F)$.

Gorkov equations may be expressed diagrammatically as shown in Fig. 5.4, where zig-zag line denotes the "coherent" (condensate!) field of the order parameter Δ, while for the anomalous Green's function we use the standard notation with opposite arrows[11] In analytic form these diagrams correspond to the following system of equations:[12]

$$G(p) = \frac{1}{\varepsilon - \xi(\mathbf{p})} - \frac{1}{\varepsilon - \xi(\mathbf{p})}\Delta F^+(p) \tag{5.80}$$

$$F^+(p) = -\frac{1}{\varepsilon + \xi(\mathbf{p})}\Delta^* G(p) \tag{5.81}$$

which is equivalent to (5.63), (5.64).

---

[11]Interaction with Δ, shown in this diagram, can be interpreted as "annihilation" of electron pairs into condensate or their "creation" from the condensate.
[12]For brevity we drop here infinitesimally small imaginary terms in the denominators of Green's functions of free electrons.

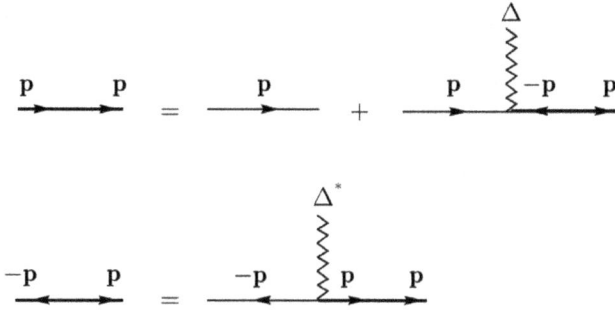

Fig. 5.4   Diagrammatic representation of Gorkov equations.

Using (5.66) and equation conjugate to (5.72), we can write:

$$\Delta = \lambda\Xi = i\lambda \int_{-\infty}^{\infty} \frac{d\varepsilon}{2\pi} \int \frac{d^3p}{(2\pi)^3} F(\varepsilon\mathbf{p}) \qquad (5.82)$$

which can be diagrammatically expressed as shown in Fig. 5.5(a). Then the diagrams of Fig. 5.4 lead to graphical representation of Gorkov equations shown in Fig. 5.5(b).[13]

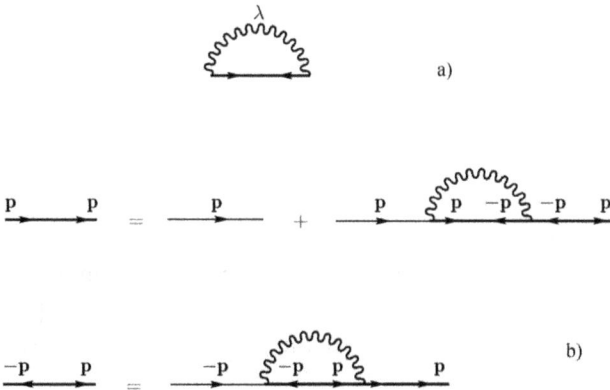

Fig. 5.5   Self-energy part due to pairing interaction built on the anomalous Green's function (a) and another form of diagrammatic representation of Gorkov equations (b).

Often it is convenient to use matrix form of Gorkov equations, introduced first by Nambu. Let us define, along with G, F^+ and F (i.e. (5.44), (5.45) and

[13]Note that here as above we just neglect the pairing interaction contribution to the self-energy built on the normal Green's function.

Green's function conjugated to (5.45)), an additional anomalous function:

$$i\tilde{G}_{\alpha\beta}(x_1, x_2) = < N|T\psi_\alpha^+(x_1)\psi_\beta(x_2)|N > \tag{5.83}$$

which enters the system of equations, similar to (5.80), (5.81):

$$\tilde{G}(p) = \frac{1}{\varepsilon + \xi(\mathbf{p})} - \frac{1}{\varepsilon + \xi(\mathbf{p})}\Delta^* F(p) \tag{5.84}$$

$$F(p) = -\frac{1}{\varepsilon - \xi(\mathbf{p})}\Delta\tilde{G}(p) \tag{5.85}$$

Then we can define the matrix Green's function as:

$$\hat{G}(p) = \begin{pmatrix} G(p) & F(p) \\ F^+(p) & \tilde{G}(p) \end{pmatrix} \tag{5.86}$$

and write both pairs of Gorkov equations ((5.80), (5.81) and (5.84), (5.85)) as a single matrix equation:

$$\left\{ \varepsilon\hat{1} - \hat{\sigma}_z\xi(\mathbf{p}) + \frac{1}{2}\left(\Delta\hat{\sigma}^+ + \Delta^*\hat{\sigma}^-\right) \right\}\hat{G}(p) = \hat{1} \tag{5.87}$$

where we have introduced the standard Pauli matrices $\hat{\sigma}_x$, $\hat{\sigma}_y$, $\hat{\sigma}_z$ and their linear combinations $\hat{\sigma}^\pm = \hat{\sigma}_x \pm i\hat{\sigma}_y$.

It is easily checked that

$$i\hat{G}(x_1, x_2) = < T\hat{\psi}(x_1)\hat{\psi}^+(x_2) > \tag{5.88}$$

where

$$\hat{\psi}(x_1) = \begin{pmatrix} \psi_\alpha(x_1) \\ \psi_\beta^+(x_1) \end{pmatrix} \qquad \hat{\psi}^+(x_2) = (\psi_\alpha^+(x_2), \psi_\beta(x_2)) \tag{5.89}$$

are the so-called Nambu spinors.

In the presence of an external electromagnetic field Green's functions depend not only on difference of coordinates. Electromagnetic field can be easily introduced into Gorkov equations written as (5.60), (5.62). This is done by the usual substitution of covariant derivatives:

$$\nabla\psi \to (\nabla - ie\mathbf{A})\psi \qquad \nabla\psi^+ \to (\nabla + ie\mathbf{A})\psi^+ \tag{5.90}$$

where $\mathbf{A}$ is an external vector potential.[14] Then equations for G and F^+ take the form:

$$\left\{ i\frac{\partial}{\partial t} + \frac{1}{2m}(\nabla - ie\mathbf{A})^2 + \mu \right\}G(x, x') + i\lambda F(x, x)F^+(x, x') = \delta(x - x') \tag{5.91}$$

---

[14]We assume the use of the gauge with zero scalar potential $\phi = 0$.

$$\left\{ i\frac{\partial}{\partial t} - \frac{1}{2m}(\nabla + ie\mathbf{A})^2 - \mu \right\} F^+(x, x') + i\lambda F^-(x, x)G(x, x') = 0$$

(5.92)

Under the gauge transformation:

$$\mathbf{A} \rightarrow \mathbf{A} + \nabla\varphi$$

(5.93)

Green's functions G, F and F^+ are transformed as:

$$G(x, x') \rightarrow G(x, x')e^{ie[\varphi(\mathbf{r}) - \varphi(\mathbf{r}')]}$$

(5.94)

$$F(x, x') \rightarrow F(x, x')e^{+ie[\varphi(\mathbf{r}) + \varphi(\mathbf{r}')]}$$

(5.95)

$$F^+(x, x') \rightarrow F^+(x, x')e^{-ie[\varphi(\mathbf{r}) + \varphi(\mathbf{r}')]}$$

(5.96)

as *charged* electron operators (fields) are transformed according to:

$$\psi(x) \rightarrow \psi e^{ie\varphi(\mathbf{r})} \qquad \psi^+(x) \rightarrow \psi^+ e^{-ie\varphi(\mathbf{r})}$$

(5.97)

Then the gap functions $\Delta(x) \sim |\lambda|F(x, x)$ or $\Delta^*(x) \sim |\lambda|F^+(x, x)$, which in external field are, in general, functions of x, are transformed as:

$$F(x, x) \rightarrow F(x, x)e^{2ie\varphi(\mathbf{r})} \qquad F^+(x, x) \rightarrow F^+(x, x)e^{-2ie\varphi(\mathbf{r})}$$

(5.98)

which, in accordance with general ideology of gauge theories [Sadovskii M.V. (2019b)], means that the order parameter of a superconductor ("gap") Δ is a *charged* field with electric charge $2e$, i.e. the double charge of an electron (Cooper pair condensate)!

Let us consider now the case of finite temperatures. In Matsubara formalism, along with "normal" Green's function of an electron, in superconducting state we have to introduce also the "anomalous" function:

$$F_{\alpha\beta}(\tau_1, \mathbf{r}_1; \tau_2, \mathbf{r}_2) = Sp\left\{ e^{\frac{\Omega + \mu N - H}{T}} T_\tau(\psi_\alpha(\tau_1\mathbf{r}_1)\psi_\beta(\tau_2\mathbf{r}_2)) \right\}$$

(5.99)

$$F^+_{\alpha\beta}(\tau_1, \mathbf{r}_1; \tau_2, \mathbf{r}_2) = Sp\left\{ e^{\frac{\Omega + \mu N - H}{T}} T_\tau(\bar{\psi}_\alpha(\tau_1\mathbf{r}_1)\bar{\psi}_\beta(\tau_2\mathbf{r}_2)) \right\}$$

(5.100)

where

$$\psi_\alpha(\tau\mathbf{r}) = e^{\tau(H - \mu N)}\psi_\alpha(\mathbf{r})e^{-\tau(H - \mu N)}$$
$$\bar{\psi}_\beta(\tau\mathbf{r}) = e^{\tau(H - \mu N)}\psi^+_\beta(\mathbf{r})e^{-\tau(H - \mu N)}$$

(5.101)

where averaging in (5.99) and (5.100) is performed over the grand canonical ensemble (Ω is the appropriate thermodynamic potential), T_τ is "time ordering" operator over imaginary (Matsubara) time.[15]

Spin dependence of these functions is separated (similarly to (5.53))[16] as:

$$F_{\alpha\beta} = g_{\alpha\beta}F \qquad F_{\alpha\beta}^+ = -g_{\alpha\beta}F^+ \tag{5.103}$$

Similarly to G, both functions F and F^+ depend on the difference $\tau = \tau_1 - \tau_2$ and satisfy "antiperiodicity" condition:

$$F(\tau) = -F\left(\tau + \frac{1}{T}\right) \qquad F^+(\tau) = -F^+\left(\tau + \frac{1}{T}\right) \tag{5.104}$$

so that Fourier series expansions of these functions over τ contain only odd Matsubara frequencies $\varepsilon_n = \pi T(2n + 1)$. Matsubara ψ-operators at $\tau = 0$ coincide with Heisenberg operators at $t = 0$, and comparing the definitions (5.99), (5.100) with (5.45), (5.46) and (5.54), we find:

$$F(0, \mathbf{r}; 0, \mathbf{r}) = \Xi(\mathbf{r}) \qquad F^+(0, \mathbf{r}; 0, \mathbf{r}) = \Xi^*(\mathbf{r}) \tag{5.105}$$

and thus defined Ξ can be considered as the condensate "wave function", averaged over the Gibbs ensemble.

Equations of "motion" for Matsubara Green's functions of a superconductor G, F, F^+ are derived similarly to the derivation of (5.60), (5.62), only instead of differentiating by time t we have to calculate derivatives over τ and substituting in (5.34) $it \to \tau$. Then we obtain:

$$\left(-\frac{\partial}{\partial\tau} + \frac{\nabla^2}{2m} + \mu\right)G(\tau, \mathbf{r}; \tau', \mathbf{r}') + \lambda\Xi F^+(\tau, \mathbf{r}; \tau', \mathbf{r}') = \delta(\tau - \tau')\delta(\mathbf{r} - \mathbf{r}') \tag{5.106}$$

$$\left(\frac{\partial}{\partial\tau} + \frac{\nabla^2}{2m} + \mu\right)F^+(\tau, \mathbf{r}; \tau', \mathbf{r}') - \lambda\Xi^*G(\tau, \mathbf{r}; \tau', \mathbf{r}') = 0 \tag{5.107}$$

After Fourier transformation these equations take the form:

$$(i\varepsilon_n - \xi(\mathbf{p}))G(\varepsilon_n\mathbf{p}) + \Delta F^+(\varepsilon_n\mathbf{p}) = 1 \tag{5.108}$$

---

[15] For normal Green's function we use the standard definition [Abrikosov A.A., Gorkov L.P., Dzyaloshinskii I.E. (1963)]:

$$G_{\alpha\beta}(\tau_1, \mathbf{r}_1; \tau_2, \mathbf{r}_2) = -Sp\left\{e^{\frac{\Omega+\mu N - H}{T}} T_\tau(\psi_\alpha(\tau_1\mathbf{r}_1)\bar\psi_\beta(\tau_2\mathbf{r}_2))\right\} \tag{5.102}$$

[16] The sign difference in comparison to (5.53) is connected here with the absence of the factor of i in definitions of (5.99), (5.100), which is present in $T = 0$ formalism.

$$- (i\varepsilon_n + \xi(\mathbf{p}))F^+(\varepsilon_n\mathbf{p}) - \Delta^* G(\varepsilon_n\mathbf{p}) = 0 \qquad (5.109)$$

where we have introduced:

$$\Delta = \lambda\Xi = \lambda F(0\mathbf{r};0\mathbf{r}) \qquad \Delta^* = \lambda\Xi^* = \lambda F^+(0\mathbf{r};0\mathbf{r}) \qquad (5.110)$$

Solution of Eqs. (5.108), (5.109) is:

$$G(\varepsilon_n\mathbf{p}) = -\frac{i\varepsilon_n + \xi(\mathbf{p})}{\varepsilon_n^2 + \xi^2(\mathbf{p}) + |\Delta|^2} = -\frac{i\varepsilon_n + \xi(\mathbf{p})}{\varepsilon_n^2 + \varepsilon^2(\mathbf{p})} \qquad (5.111)$$

$$F^+(\varepsilon_n\mathbf{p}) = \frac{\Delta^*}{\varepsilon_n^2 + \xi^2(\mathbf{p}) + |\Delta|^2} = \frac{\Delta^*}{\varepsilon_n^2 + \varepsilon^2(\mathbf{p})} \qquad (5.112)$$

where $\varepsilon(\mathbf{p})$ is again given by (5.68):

$$\varepsilon(\mathbf{p}) = \sqrt{\xi^2(\mathbf{p}) + |\Delta|^2} \qquad (5.113)$$

In contrast to the case of $T = 0$ everything here is well defined and we do not need any additional clarifications, based on analytic properties.

The gap equation can be found from:

$$\Xi^* = F^+(\tau = 0, \mathbf{r} = 0) = T \sum_{n=-\infty}^{\infty} \int \frac{d^3p}{(2\pi)^3} F^+(\varepsilon_n\mathbf{p}) \qquad (5.114)$$

which, after the substitution of (5.112), gives:

$$1 = \frac{|\lambda|T}{(2\pi)^3} \sum_{n=-\infty}^{\infty} \int \frac{d^3p}{\varepsilon_n^2 + \xi^2(\mathbf{p}) + |\Delta|^2} = \frac{|\lambda|T}{(2\pi)^3} \sum_{n=-\infty}^{\infty} \int \frac{d^3p}{\varepsilon_n^2 + \varepsilon^2(\mathbf{p})} \qquad (5.115)$$

Summation over frequencies is again performed using (2.100), which gives gap equation of BCS theory for the case of finite temperatures:

$$1 = \frac{|\lambda|}{2} \int \frac{d^3p}{(2\pi)^3} \frac{1}{\sqrt{\xi^2(p) + \Delta^2(T)}} th \frac{\sqrt{\xi^2(p) + \Delta^2(T)}}{2T} \qquad (5.116)$$

The properties of this equation are well known [Lifshits E.M., Pitaevskii L.P. (1980); Schrieffer J.R. (1964); Sadovskii M.V. (2019a); De Gennes P.G. (1966)] and we shall not analyze (5.116) in details, noting only that it gives the famous temperature dependence of the gap $\Delta(T)$ of BCS theory. In particular, the gap becomes zero for $T \geq T_c$, which is defined by (5.31). This can be seen directly, as for $\Delta = 0$ Eq. (5.116) reduces to:

$$1 = \frac{|\lambda|mp_F}{2\pi^2} \int_0^{\omega_D} \frac{d\xi}{\xi} th \frac{\xi}{2T} \qquad (5.117)$$

which, in fact, coincides with equation, determining the pole position
in Cooper "ladder" (5.30). For $T = 0$ (5.116) reduces to (5.73), so that
we again obtain (5.76), (5.77) and $\Delta_0 = \frac{\pi}{\gamma} T_c$.

Gorkov equations (5.108) and (5.109) can be rewritten as:

$$G(\varepsilon_n \mathbf{p}) = \frac{1}{i\varepsilon_n - \xi(\mathbf{p})} - \frac{1}{i\varepsilon_n - \xi(\mathbf{p})} \Delta F^+(\varepsilon_n \mathbf{p}) \qquad (5.118)$$

$$F^+(\varepsilon_n \mathbf{p}) = \frac{1}{-i\varepsilon_n - \xi(-\mathbf{p})} \Delta^* G(\varepsilon_n \mathbf{p}) = -\frac{1}{i\varepsilon_n + \xi(\mathbf{p})} \Delta^* G(\varepsilon_n \mathbf{p})$$
$$(5.119)$$

which is similar to (5.80), (5.81) and can be represented diagrammati-
cally as shown in Fig. 5.4, with $p = (\varepsilon_n, \mathbf{p})$. Iterating (i.e. substituting
many times (5.119) into (5.118)), we obtain for the normal Green's
function perturbation theory expansion in powers of Δ:

$$\begin{aligned}
G(\varepsilon_n \mathbf{p}) &= \frac{1}{i\varepsilon_n - \xi(\mathbf{p})} + \frac{1}{i\varepsilon_n - \xi(\mathbf{p})} \frac{|\Delta|^2}{(i\varepsilon + \xi(\mathbf{p}))(i\varepsilon - \xi(\mathbf{p}))} \\
&+ \frac{1}{i\varepsilon_n - \xi(\mathbf{p})} \frac{|\Delta|^4}{(i\varepsilon_n + \xi(\mathbf{p}))^2 (i\varepsilon_n - \xi(\mathbf{p}))^2} + \cdots \\
&= \frac{i\varepsilon_n + \xi(\mathbf{p})}{(i\varepsilon_n)^2 - \xi^2(\mathbf{p}) - |\Delta|^2} \qquad (5.120)
\end{aligned}$$

shown diagrammatically in Fig. 5.6(a). Summation of this series obviously gives (5.111). Similarly, we can substitute Eq. (5.118) into (5.119)
and "generate" diagrammatic series for $F^+(\varepsilon_n \mathbf{p})$, shown in Fig. 5.6(b),
where in every term we have an odd number of interaction lines with
"field" Δ, and which gives (5.112). Note, that in the series for anomalous Green's function the zeroth-order term is just absent — there is no
"free" anomalous Green's function at all.[17] Note also that the anomalous Green's function F^+ explicitly depends on the *phase* of the "field"
(order parameter) Δ.

Similarly to (5.82), expressions (5.110) and (5.114) for the gap can
be written as:

$$\Delta = \lambda \Xi = \lambda T \sum_{n=-\infty}^{\infty} \int \frac{d^3 p}{(2\pi)^3} F(\varepsilon_n \mathbf{p}) \qquad (5.121)$$

---

[17]It appears if we introduce an infinitesimally weak Bogoliubov's "source" of Cooper
pairs, which is introduced in the concept of quasi-averages [Sadovskii M.V. (2019a);
Bogoliubov N.N. (1991b)], and is equivalent to infinitesimal field Δ.

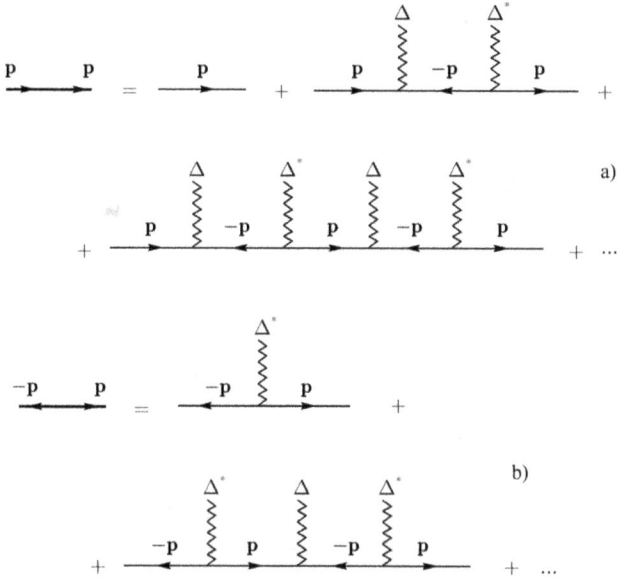

Fig. 5.6 Diagrammatic expansion of the normal (a) and anomalous (b) Green's functions in powers of Δ.

which can be represented by the diagram shown in Fig. 5.5(a), so that Gorkov equations take the form shown graphically in Fig. 5.5(b).

In Matsubara formalism we also can use Nambu matrices, when in addition to (5.99), (5.100) and (5.102) we introduce:

$$\tilde{G}_{\alpha\beta}(\tau_1, \mathbf{r}_1; \tau_2, \mathbf{r}_2) = -Sp\left\{ e^{\frac{\Omega + \mu N - H}{T}} T_\tau(\bar{\psi}_\alpha(\tau_1 \mathbf{r}_1)\psi_\beta(\tau_2 \mathbf{r}_2)) \right\} \qquad (5.122)$$

and define the matrix Green's function:

$$\hat{G}(\varepsilon_n \mathbf{p}) = \begin{pmatrix} G(\varepsilon_n \mathbf{p}) & -F(\varepsilon_n \mathbf{p}) \\ -F^+(\varepsilon_n \mathbf{p}) & \tilde{G}(\varepsilon_n \mathbf{p}) \end{pmatrix} \qquad (5.123)$$

which satisfies the following equation:

$$\left\{ i\varepsilon_n \hat{1} - \hat{\sigma}_z \xi(\mathbf{p}) - \frac{1}{2}\left(\Delta\hat{\sigma}^+ + \Delta^*\hat{\sigma}^- \right) \right\} \hat{G}(p) = \hat{1} \qquad (5.124)$$

which contains (5.108), (5.109) and similar equations for $\tilde{G}$ and F:

$$(i\varepsilon_n - \xi(\mathbf{p}))F(\varepsilon_n \mathbf{p}) + \Delta\tilde{G}(\varepsilon_n \mathbf{p}) = 0 \qquad (5.125)$$

$$(i\varepsilon_n + \xi(\mathbf{p}))\tilde{G}(\varepsilon_n \mathbf{p}) + \Delta^* F(\varepsilon_n \mathbf{p}) = 1 \qquad (5.126)$$

Matrix Green's function (5.123) can be written as ($x = (\tau, \mathbf{r})$):

$$\hat{G}(x_1, x_2) = \begin{pmatrix} - < T\psi_\alpha(x_1)\bar{\psi}_\beta(x_2) > & - < T\psi_\alpha(x_1)\psi_\beta(x_2) > \\ - < T\bar{\psi}_\alpha(x_1)\bar{\psi}_\beta(x_2) > & - < T\bar{\psi}_\alpha(x_1)\psi_\beta(x_2) > \end{pmatrix}$$

$$= - < T\hat{\psi}(x_1)\hat{\psi}^+(x_2) > \tag{5.127}$$

where angular brackets denote averaging over the Gibbs ensemble, while Nambu spinors are defined as:

$$\hat{\psi}(x_1) = \begin{pmatrix} \psi_\alpha(x_1) \\ \bar{\psi}_\beta(x_1) \end{pmatrix} \qquad \hat{\psi}^+(x_2) = (\bar{\psi}_\alpha(x_2), \psi_\beta(x_2)) \tag{5.128}$$

The appearance of matrix Green's functions of the Nambu type, which contain "anomalous" functions, is typical in systems with spontaneous symmetry breaking (phase transitions of second order) [Mattuck R.D. (1968)]. This formalism is convenient as now we can draw Feynman diagrams as in the "normal" state, with particle lines corresponding to matrix Green's functions, so that in fact we obtain systems of equations for "normal" and "anomalous" Green's functions.

5.3 Basics of Eliashberg–McMillan theory

We can construct more refined microscopic theory of superconductivity, based not BCS–Gorkov model of pairing interaction (5.33), but on more "realistic" model of electron–phonon interaction (Fröhlih Hamiltonian). The theory is based on the system of equations for normal and anomalous Green's functions of a superconductor, shown diagrammatically on Fig. 5.7. The structure of these equations (G.M. Eliashberg, 1960) is clear without further explanations. It is cleat, that solving these *integral* equations, with the account of peculiarities of the real phonon spectrum, poses rather difficult problem. However, a significant progress here was achieved (W.L. McMillan, 1968) and the theory of traditional superconductors, based on the pairing due to electron–phonon interaction is an example of quite successful application of Green's functions formalism. Detailed presentation of the analysis of Eliashberg equations and the results of its applications

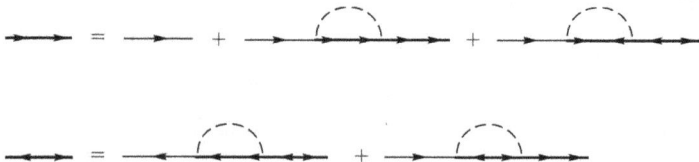

Fig. 5.7 Diagrammatic representation of Eliashberg equations. Dashed line denotes phonon Green's function.

can be found in [Vonsovsky S.V., Izyumov Yu.A., Kurmaev E.Z. (1977)]. Below we present somehow simplified derivation of these equations, dropping some technical details. In particular, we shall not consider the role of direct Coulomb repulsion of electrons within Cooper pair, which is taken into account in the full Eliashberg–McMillan theory [Vonsovsky S.V., Izyumov Yu.A., Kurmaev E.Z. (1977)], limiting ourselves only to electron–phonon interaction, as it is actually already shown in Fig. 5.7.

Taking into account that in adiabatic approximation, according to Migdal's theorem, vertex corrections are irrelevant, Eliashberg equations can be derived by calculating the diagram, shown in Fig. 3.1, where electronic Green's function in superconducting state is taken in Nambu matrix representation. All calculations are, in principle, similar to derivation of Eqs. (3.60)–(3.68), (3.75). We again perform calculations in Matsubara technique ($T \neq 0$) using the notations of Eqs. (3.9)–(3.15) [Schrieffer J.R. (1964)]. In Nambu formalism, the electronic Green's function of a superconductor is written in the standard form as ($\hat{\sigma}_i$-Pauli matrices) [Vonsovsky S.V., Izyumov Yu.A., Kurmaev E.Z. (1977)]:

$$\hat{G}^{-1}(i\omega_n, \mathbf{p}) = i\omega_n \hat{1} - \varepsilon_{\mathbf{p}} \hat{\sigma}_z - \hat{\Sigma}(i\omega_n, \mathbf{p}) \tag{5.129}$$

where the matrix self-energy is represented as:[18]

$$\hat{\Sigma}(i\omega_n, \mathbf{p}) = (1 - Z(i\omega_n))i\omega_n \hat{1} + Z(i\omega_n)\Delta(i\omega_n)\hat{\sigma}_x \tag{5.130}$$

Here we introduced a number of simplifications, like independence of renormalization factor $Z(i\omega_n)$ and gap function $\Delta(i\omega_n)$ on the momentum [Vonsovsky S.V., Izyumov Yu.A., Kurmaev E.Z. (1977)]. Then we have:

$$\hat{G}(i\omega_n, \mathbf{p}) = \frac{Z(i\omega_n)i\omega_n \hat{1} + \varepsilon_{\mathbf{p}} \hat{\sigma}_z + Z(i\omega_n)\Delta(i\omega_n)\hat{\sigma}_x}{Z^2(i\omega_n)(i\omega_n)^2 - Z^2(i\omega_n)\Delta^2(i\omega_n) - \varepsilon_{\mathbf{p}}^2} \tag{5.131}$$

Self-energy part, corresponding to diagram of Fig. 3.1 with matrix Green's function of electron (5.131), can be written as:

$$\hat{\Sigma}(\omega_n, \mathbf{p}) = -T \sum_m \sum_{\mathbf{p}'} |\bar{g}_{\mathbf{p}\mathbf{p}'}|^2 D(i\omega_n - i\omega_m, \mathbf{p} - \mathbf{p}')\hat{\sigma}_z \hat{G}(i\omega_m, \mathbf{p}')\hat{\sigma}_z \tag{5.132}$$

where phonon Green's function $D(i\omega_n - i\omega_m, \mathbf{p} - \mathbf{p}')$ can be taken as in (3.10), denoting the phonon frequency by $\Omega_{\mathbf{p}-\mathbf{p}'}$, as in Eqs. (3.60)–(3.64).

Calculations can be conveniently done using the standard spectral representation of electronic Green's function [Abrikosov A.A., Gorkov L.P., Dzyaloshinskii I.E. (1963)]:

$$\hat{G}(i\omega_n, \mathbf{p}) = -\frac{1}{\pi} \int_\infty^\infty d\varepsilon' \frac{Im\hat{G}(\varepsilon' + i\delta)}{i\omega_n - \varepsilon'} \tag{5.133}$$

---

[18]Possible contribution proportional to $\hat{\sigma}_y$ here can be dropped after appropriate choice of the phase of the order parameter, while contribution proportional to $\hat{\sigma}_z$ reduces to the renormalization of the chemical potential [Vonsovsky S.V., Izyumov Yu.A., Kurmaev E.Z. (1977)].

As a result in Eq. (5.132) appears the following sum over Matsubara frequencies:

$$-T\sum_m D(i\omega_n - i\omega_m, \mathbf{p} - \mathbf{p}')\hat{G}(i\omega_m, \mathbf{p}')$$

$$= -T\frac{1}{\pi}\int_\infty^\infty d\varepsilon' Im\hat{G}(\varepsilon' + i\delta)\sum_m \frac{1}{i\omega_n - \varepsilon'}\frac{2\Omega_{\mathbf{p}-\mathbf{p}'}}{(i\omega_n - i\omega_m)^2 - \Omega_{\mathbf{p}-\mathbf{p}'}^2} \quad (5.134)$$

which is calculated similarly to (3.60):

$$-T\sum_m \frac{1}{i\omega_n - \varepsilon'}\frac{2\Omega_{\mathbf{p}-\mathbf{p}'}}{(i\omega_n - i\omega_m)^2 - \Omega_{\mathbf{p}-\mathbf{p}'}^2} = \int_C \frac{dz}{2\pi i}f(z)\frac{2\Omega_{\mathbf{p}-\mathbf{p}'}}{(i\omega_n - z)^2 - \Omega_{\mathbf{p}-\mathbf{p}'}^2}$$

$$(5.135)$$

where integration contour C goes counterclockwise around the imaginary axis where discrete points $i\omega_n$ are placed. Calculating the residues and obtain this sum as:

$$[1 - f(\varepsilon') + n(\Omega_q)]\frac{1}{\varepsilon' + \Omega_q - i\omega_n} + [f(\varepsilon') + n(\Omega_q)]\frac{1}{\varepsilon' - \Omega_q - i\omega_n} \quad (5.136)$$

where again appeared Fermi $f(\varepsilon')$ and Bose (Planck) $n(\Omega_q)$ distributions and we introduced $\mathbf{q} = \mathbf{p} - \mathbf{p}'$ just to shorten notations. In obtaining this expression we have taken into account that $f(i\omega_n - \Omega) = -n(\Omega)$ (cf. Eq. (3.62)).

Collecting now all these expressions and making the standard analytic continuation $i\omega_n \to \varepsilon + i\delta$ (for $\omega_n > 0$), we write down the self-energy part (5.132) as:

$$\hat{\Sigma}(\varepsilon, \mathbf{p}) = \frac{1}{\pi}\int_{-\infty}^\infty d\varepsilon'\sum_{\mathbf{p}'}|\bar{g}_{\mathbf{p}\mathbf{p}'}|^2\int_0^\infty d\omega\delta(\omega - \Omega_{\mathbf{p}-\mathbf{p}'})\hat{\sigma}_z Im\hat{G}(\varepsilon' + i\delta, \mathbf{p}')\hat{\sigma}_z$$

$$\times\left[\frac{1 - f(\varepsilon') + n(\omega)}{\varepsilon' + \omega - \varepsilon - i\delta} + \frac{f(\varepsilon') + n(\omega)}{\varepsilon' - \omega - \varepsilon - i\delta}\right] \quad (5.137)$$

where we have introduced the additional integration with $\delta(\omega - \Omega_{\mathbf{p}-\mathbf{p}'})$ similar to used in writing Eq. (3.66) above.

Note that for real frequencies, after analytic continuation $i\omega_n \to \varepsilon + i\delta$ Green's function (5.129) itself is rewritten as:

$$G(\varepsilon, \mathbf{p}) = \frac{Z(\varepsilon)\varepsilon\hat{1} + \varepsilon_{\mathbf{p}}\hat{\sigma}_z + Z(\varepsilon)\Delta(\varepsilon)\hat{\sigma}_x}{Z^2(\varepsilon)(\varepsilon + i\delta)^2 - Z^2(\varepsilon)\Delta^2(\varepsilon) - \varepsilon_{\mathbf{p}}^2} \quad (5.138)$$

which corresponds to matrix self-energy of the following form:

$$\Sigma(\varepsilon, \mathbf{p}) = [1 - Z(\varepsilon)]\varepsilon\hat{1} + Z(\varepsilon)\Delta(\varepsilon)\hat{\sigma}_x \quad (5.139)$$

As we know, all physics related to superconductivity, is confined in a narrow layer of the width of the order of $2\omega_D \ll E_F$ near the Fermi surface. Then we can make here the substitution (3.67) and obtain for (5.137) the following expression for self-energy part, averaged over momenta on the Fermi surface (similarly to (3.68)):

$$\hat{\Sigma}(\varepsilon) = \frac{1}{\pi}\int_{-\infty}^\infty d\varepsilon'\int_0^\infty d\omega\alpha^2(\omega)F(\omega)\int_{-\infty}^\infty d\varepsilon_{\mathbf{p}'}\hat{\sigma}_z Im\hat{G}(\varepsilon' + i\delta, \mathbf{p}') \times \hat{\sigma}_z$$

$$\times\left[\frac{1 - f(\varepsilon') + n(\omega)}{\varepsilon' + \omega - \varepsilon - i\delta} + \frac{f(\varepsilon') + n(\omega)}{\varepsilon' - \omega - \varepsilon - i\delta}\right] \quad (5.140)$$

Calculating here the integral over $\varepsilon_{p'}$ in infinite (due to rapid convergence) limits, using the explicit expression for the Green's function (5.138), we obtain:

$$\int_{-\infty}^{\infty} d\varepsilon_{p'} \hat{G}(\varepsilon' + i\delta, \mathbf{p}') = -i\pi \frac{Z(\varepsilon')\varepsilon'\hat{1} + Z(\varepsilon')\Delta(\varepsilon')\hat{\sigma}_x}{[Z^2(\varepsilon')\varepsilon'^2 - Z^2(\varepsilon')\Delta^2(\varepsilon')]^{1/2}} \, sign \, \varepsilon' \quad (5.141)$$

Then Eq. (5.140) is rewritten as:

$$\hat{\Sigma}(\varepsilon) = \frac{1}{\pi} \int_{-\infty}^{\infty} d\varepsilon' \int_0^{\infty} d\omega \alpha^2(\omega) F(\omega) Re \left\{ \frac{\hat{\sigma}_z[Z(\varepsilon')\varepsilon'\hat{1} + Z(\varepsilon')\Delta(\varepsilon')\hat{\sigma}_x]\hat{\sigma}_z}{\sqrt{Z^2(\varepsilon')\varepsilon'^2 - Z^2(\varepsilon')\Delta^2(\varepsilon')}} \right\} sign \, \varepsilon'$$

$$\times \left[\frac{1 - f(\varepsilon') + n(\omega)}{\varepsilon' + \omega - \varepsilon - i\delta} + \frac{f(\varepsilon') + n(\omega)}{\varepsilon' - \omega - \varepsilon - i\delta} \right] \quad (5.142)$$

Now we can write down the l.h.s. of this expression as (5.139) and make equal the coefficients at $\hat{1}$ and $\hat{\sigma}_x$ (taking into account $\hat{\sigma}_z\hat{\sigma}_x\hat{\sigma}_z = -\hat{\sigma}_x$) and immediately obtain the following Eliashberg (integral) equations for (complex in general!) functions of mass renormalization $Z(\varepsilon)$ and energy gap $\Delta(\varepsilon)$:

$$[1 - Z(\varepsilon)]\varepsilon = -\int_{-\infty}^{\infty} d\varepsilon' K(\varepsilon', \varepsilon) Re \frac{\varepsilon'}{\sqrt{\varepsilon'^2 - \Delta^2(\varepsilon')}} sign \, \varepsilon' \quad (5.143)$$

$$Z(\varepsilon)\Delta(\varepsilon) = \int_{-\infty}^{\infty} K(\varepsilon', \varepsilon) Re \frac{\Delta(\varepsilon')}{\sqrt{\varepsilon'^2 - \Delta^2(\varepsilon')}} sign \, \varepsilon' \quad (5.144)$$

where the integral kernel has the form:

$$K(\varepsilon', \varepsilon) = \int_0^{\infty} d\omega \alpha^2(\omega) F(\omega) \left\{ \frac{1 - f(\varepsilon') + n(\omega)}{\varepsilon' + \omega - \varepsilon - i\delta} - \frac{f(\varepsilon') + n(\omega)}{\varepsilon' - \omega - \varepsilon - i\delta} \right\}$$

$$= \frac{1}{2} \int_0^{\infty} d\omega \alpha^2(\omega) F(\omega) \left\{ \frac{th\frac{\varepsilon'}{2T} + cth\frac{\omega}{2T}}{\varepsilon' + \omega - \varepsilon - i\delta} - \frac{th\frac{\varepsilon'}{2T} - cth\frac{\omega}{2T}}{\varepsilon' - \omega - \varepsilon - i\delta} \right\} \quad (5.145)$$

Here we used the identities $th\frac{z}{2T} = 1 - 2f(z)$ and $cth\frac{z}{2T} = 1 + 2n(z)$. After the substitution of (5.145) into (5.143) and (5.144), after obvious transformations, we can rewrite Eliashberg equations as:

$$[1 - Z(\varepsilon)]\varepsilon = \frac{1}{2} \int_0^{\infty} d\omega \alpha^2(\omega) F(\omega) \int_0^{\infty} d\varepsilon' Re \frac{\varepsilon'}{\sqrt{\varepsilon'^2 - \Delta^2(\varepsilon')}}$$

$$\times \left\{ \left(th\frac{\varepsilon'}{2T} + cth\frac{\omega}{2T} \right) \left(\frac{1}{\varepsilon' + \omega + \varepsilon + i\delta} - \frac{1}{\varepsilon' + \omega - \varepsilon - i\delta} \right) \right.$$

$$\left. - \left(th\frac{\varepsilon'}{2T} - cth\frac{\omega}{2T} \right) \left(\frac{1}{\varepsilon' - \omega + \varepsilon + i\delta} - \frac{1}{\varepsilon' - \omega - \varepsilon - i\delta} \right) \right\} \quad (5.146)$$

$$Z(\varepsilon)\Delta(\varepsilon) = \frac{1}{2} \int_0^{\infty} d\omega \alpha^2(\omega) F(\omega) \int_0^{\infty} d\varepsilon' Re \frac{\Delta(\varepsilon')}{\sqrt{\varepsilon'^2 - \Delta^2(\varepsilon')}}$$

$$\times \left\{ \left(th\frac{\varepsilon'}{2T} + cth\frac{\omega}{2T} \right) \left(\frac{1}{\varepsilon' + \omega + \varepsilon + i\delta} + \frac{1}{\varepsilon' + \omega - \varepsilon - i\delta} \right) \right.$$

$$\left. - \left(th\frac{\varepsilon'}{2T} - cth\frac{\omega}{2T} \right) \left(\frac{1}{\varepsilon' - \omega + \varepsilon + i\delta} + \frac{1}{\varepsilon' - \omega - \varepsilon - i\delta} \right) \right\} \quad (5.147)$$

In fact, these equations replace Eqs. (3.68) and (3.75) for normal metal after its transition to superconducting phase.

To determine the temperature of superconducting transition it is sufficient to consider the linearized (over $\Delta(\varepsilon)$) Eliashberg equations, which for the case of characteristic phonon frequencies much exceeding T_c has the following form [Vonsovsky S.V., Izyumov Yu.A., Kurmaev E.Z. (1977)]:

$$[1 - Z(\varepsilon)]\varepsilon = \int_0^\infty d\varepsilon' \int_0^\infty d\omega \alpha^2(\omega) F(\omega) f(-\varepsilon')$$
$$\times \left(\frac{1}{\varepsilon' + \varepsilon + \omega + i\delta} - \frac{1}{\varepsilon' - \varepsilon + \omega - i\delta} \right) \tag{5.148}$$

$$Z(\varepsilon)\Delta(\varepsilon) = \int_0^\infty \frac{d\varepsilon'}{\varepsilon'} th\frac{\varepsilon'}{2T_c} Re\Delta(\varepsilon') \int_0^\infty d\omega \alpha^2(\omega) F(\omega)$$
$$\times \left(\frac{1}{\varepsilon' + \varepsilon + \omega + i\delta} + \frac{1}{\varepsilon' - \varepsilon + \omega - i\delta} \right) \tag{5.149}$$

which is the system of homogeneous integral equations determining T_c to be solved, e.g. numerically.

For simple estimates it is sufficient to consider in these equations the limit of $\varepsilon \to 0$ and look for solutions $Z(0) = Z$ and $\Delta(0) = \Delta$. Then, from (5.148) we obtain:

$$[1 - Z]\varepsilon = -2\varepsilon \int_0^\infty d\omega \alpha^2(\omega) F(\omega) \int_0^\infty \frac{d\varepsilon'}{(\varepsilon' + \omega)^2} = -2\varepsilon \int_0^\infty \frac{d\omega}{\omega} \alpha^2(\omega) F(\omega) \tag{5.150}$$

or

$$Z = 1 + \lambda \tag{5.151}$$

where constant λ was defined above in Eqs. (3.72), (3.74). Thus precisely this effective constant determines mass renormalization both in normal and superconducting phases. Note that factor Z introduced in (5.151) is, in fact, inverse to renormalization factor, defined according to (3.69) for the normal state. This simply reflects the difference in standard notations used in the literature.

In the limit of $\varepsilon \to 0$, using (5.151) from (5.149) we immediately obtain the following equation for T_c:

$$1 + \tilde{\lambda} = 2 \int_0^\infty d\omega \alpha^2(\omega) F(\omega) \int_0^\infty \frac{d\varepsilon'}{\varepsilon'(\varepsilon' + \omega)} th\frac{\varepsilon'}{2T_c} \tag{5.152}$$

In the model with one Einstein phonon of frequency Ω_0 we have $F(\omega) = \delta(\omega - \Omega_0)$, so that Eq. (5.152) takes the form:

$$1 + \tilde{\lambda} = 2\alpha^2(\Omega_0) \int_0^\infty \frac{d\varepsilon'}{\varepsilon'(\varepsilon' + \Omega_0)} th\frac{\varepsilon'}{2T_c} \tag{5.153}$$

where pairing interaction constant λ is defined as:

$$\lambda = 2 \int_0^\infty \frac{d\omega}{\omega} \alpha^2(\omega) F(\omega) = \alpha^2(\Omega_0)\frac{2}{\Omega_0} \tag{5.154}$$

Then we immediately obtain an estimate for T_c in Eliashberg theory for the case of intermediate coupling:

$$T_c \sim \Omega_0 \exp\left(-\frac{1+\lambda}{\lambda}\right) \qquad (5.155)$$

which for the values of $\lambda \sim 1$ rather significantly changes the standard result of BCS theory, which follows from here only in the weak-coupling limit $\lambda \ll 1$.

Consider now the more general model with discrete set of Einstein phonons, when the phonon density of states can be written as:

$$F(\omega) = \sum_i \delta(\omega - \Omega_i) \qquad (5.156)$$

where discrete frequencies Ω_i just model optical phonons. Then, from (3.72) we get:

$$\lambda = 2\sum_i \frac{\alpha^2(\Omega_i)}{\Omega_i} \equiv \sum_i \lambda_i \qquad (5.157)$$

Correspondingly, for this model:

$$\alpha^2(\omega)F(\omega) = \sum_i \alpha^2(\Omega_i)\delta(\omega - \Omega_i)$$

$$= \sum_i \frac{\lambda_i}{2}\Omega_i \delta(\omega - \Omega_i) \qquad (5.158)$$

so that Eq. (5.152) is written as:

$$1 + \lambda = 2\sum_i \alpha^2(\Omega_i) \int_0^D \frac{d\varepsilon'}{\varepsilon'(\varepsilon' + \Omega_i)} th\frac{\varepsilon'}{2T_c} \qquad (5.159)$$

This equation is easily solved and gives:

$$T_c \sim \prod_i \Omega_i^{\frac{\lambda_i}{\lambda}} \exp\left(-\frac{1+\lambda}{\lambda}\right) \qquad (5.160)$$

In particular, for the case of two Einstein phonons with frequencies Ω_1 and Ω_2 we have:

$$T_c \sim \Omega_1^{\frac{\lambda_1}{\lambda}} \Omega_2^{\frac{\lambda_2}{\lambda}} \exp\left(-\frac{1+\lambda}{\lambda}\right) \qquad (5.161)$$

where $\lambda = \lambda_1 + \lambda_2$. Thus, the preexponential factor in the expression for the critical temperature of superconducting transition is determined by partial contributions of different phonons, with appropriate partial coupling constants.

Eq. (5.160) is easily rewritten as:

$$T_c \sim \omega_{log} \exp\left(-\frac{1+\lambda}{\lambda}\right) \qquad (5.162)$$

where we have introduced the logarithmic average of the frequency $\langle\Omega\rangle$ as:

$$\omega_{log} = \ln\prod_i \Omega_i^{\frac{\lambda_i}{\lambda}} = \sum_i \frac{\lambda_i}{\lambda}\ln\Omega_i \qquad (5.163)$$

In the limit of continuous distribution of phonon frequencies the last expression reduces to:

$$\ln \omega_{log} = \frac{2}{\lambda} \int_0^\infty \frac{d\omega}{\omega} \alpha^2(\omega) F(\omega) \ln \omega = \frac{\int_0^\infty \frac{d\omega}{\omega} \ln \omega \alpha^2(\omega) F(\omega)}{\int_0^\infty \frac{d\omega}{\omega} \alpha^2(\omega) F(\omega)} \qquad (5.164)$$

where the total coupling constant λ is determined by the usual expression (3.72). Thus, in general case, preexponential factor in the expression for T_c is determined by average logarithm of the phonon frequency (5.164), where averaging is performed over the whole phonon spectrum.

There exists a vast literature on numerical solutions of Eliashberg equations [Vonsovsky S.V., Izyumov Yu.A., Kurmaev E.Z. (1977); Allen P.B., Mitrović B. (1982)]. Using these solutions, a number of authors proposed the appropriate analytic expressions for T_c, approximating the results of numerical calculations. As an example, we shall quote the popular interpolation formula, proposed by Allen and Dynes [Allen P.B., Mitrović B. (1982)], which is formally valid for the wide range of dimensionless coupling constant of electron–phonon interaction λ, including the region of strong coupling $\lambda > 1$:

$$T_c = \frac{f_1 f_2}{1.20} \omega_{log} \exp\left\{-\frac{1.04(1+\lambda)}{\lambda - \mu^*(1+0.62\lambda)}\right\} \qquad (5.165)$$

where

$$f_1 = [1 + (\lambda/\Lambda_1)^{3/2}]^{1/3}; \qquad f_2 = 1 + \frac{[<\omega^2>^{1/2}/\omega_{log} - 1]\lambda^2}{\lambda^2 + \Lambda_2^2}$$

$$\Lambda_1 = 2.46(1 + 3.8\mu^*); \qquad \Lambda_2 = 1.82(1 + 6.3\mu^*)\frac{<\omega^2>^{1/2}}{\omega_{log}} \qquad (5.166)$$

Here ω_{log} is average logarithmic frequency of phonons introduced above in (5.164), while $<\omega^2>$ is the average (over the phonon spectrum) square of phonon frequency, defined as:

$$<\omega^2> = \frac{2}{\lambda} \int_0^\infty \frac{d\omega}{\omega} \alpha^2(\omega) F(\omega)\omega = \frac{\int_0^\infty d\omega \omega \alpha^2(\omega) F(\omega)}{\int_0^\infty \frac{d\omega}{\omega} \alpha^2(\omega) F(\omega)} \qquad (5.167)$$

Coulomb pseudopotential μ^* determines electron repulsion within the Cooper pair. According to most of calculations [Vonsovsky S.V., Izyumov Yu.A., Kurmaev E.Z. (1977); Allen P.B., Mitrović B. (1982)] its values are small and vary in the interval 0.1–0.15. It is interesting, that in the limit of very strong coupling with $\lambda > 10$ these formulas give the following expression for T_c:

$$T_c \approx 0.15\sqrt{\lambda <\omega^2>} \qquad (5.168)$$

where the BCS-like exponential factor is just absent, so that in principle, pretty large values of T_c can be achieved even for electron–phonon mechanism of Cooper pairing.

5.4 Superconductivity in disordered metals

Consider a superconductor with impurities with potential (or, more precisely its Fourier transform) $v(\mathbf{p})$, randomly distributed in space (with density ρ). In principle, this problem can be analyzed similarly to the case of impurities in a normal metal. However, superconductor is different as here we have both "normal" and "anomalous" Green's functions and we have to write down the system of equations for both functions, averaged over random configurations of impurities. "Impurity" diagram technique has the usual form, only diagrams are now built on Green's functions G and F^+. However, there is one delicate point — introduction of impurities leads, in general, to impurity dependence of the energy gap (order parameter) $\Delta(\mathbf{r})$, $\Delta^*(\mathbf{r})$. This may much complicate the diagram technique, as corrections to Δ will be determined by an integral equation (as $\Delta(\mathbf{r}) = \lambda F(x,x)$). Thus, it is usually assumed that superconducting order parameter (gap) is *self-averaging* (non random): $< \Delta(\mathbf{r}) > = \Delta^{(0)}$ (where angular brackets denote averaging over impurities), $< \Delta^2(\mathbf{r}) > - < \Delta(\mathbf{r}) >^2 = 0$, so that all corrections due to impurity scattering vanish. This assumption[19] will be confirmed by the final result, when we shall see that all quantities of the type of $F(x,x)$ are not changed by (nonmagnetic) impurities. Thus, the diagram technique for impurity scattering is usual — dashed line with a cross denotes $\rho v^2(\mathbf{q})$, and the frequency of electronic line does not change in the interaction vertex. All estimates allowing us to neglect diagrams with intersecting interaction lines are valid here as in the normal metal. Equations for the averaged Green's functions $G(p)$ and $F^+(p)$ are shown graphically in Fig. 5.8, which is clear without additional justifications.

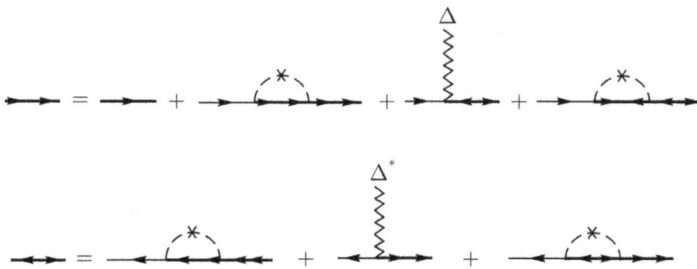

Fig. 5.8   Gorkov equations for superconductor with impurities.

---

[19]In fact it can be justified in the limit of weak disorder when $p_F l \gg 1$, $E_F \tau \gg 1$.

Using the explicit form of Green's function of the "clean" superconductor $G^{(0)}$ and $F^{+(0)}$, which we found above, we can reduce Gorkov
equations for impure superconductor to the following (relatively simple)
form [Abrikosov A.A., Gorkov L.P., Dzyaloshinskii I.E. (1963)]:[20]

$$(i\varepsilon_n - \xi(\mathbf{p}) - \bar{G}_\varepsilon)G(p) - (\Delta + \bar{F}_\varepsilon)F^+(p) = 1 \qquad (5.170)$$

$$-(i\varepsilon_n + \xi(\mathbf{p}) - \bar{G}_{-\varepsilon})F^+(p) + (\Delta^* + \bar{F}_\varepsilon^+)G(p) = 0 \qquad (5.171)$$

where:

$$\bar{G}_\varepsilon = \rho \int \frac{d^3 p'}{(2\pi)^3} |v(\mathbf{p} - \mathbf{p}')|^2 G(p') \qquad (5.172)$$

$$\bar{F}_\varepsilon^+ = \rho \int \frac{d^3 p'}{(2\pi)^3} |v(\mathbf{p} - \mathbf{p}')|^2 F^+(p') \qquad (5.173)$$

where we again use the notation $p = (\mathbf{p}, \varepsilon_n)$.

Solution of this system takes the following form (below we shall see
that $\bar{G}_\varepsilon = -\bar{G}_{-\varepsilon}$):

$$G(p) = -\frac{i\varepsilon_n - \bar{G}_\varepsilon + \xi(\mathbf{p})}{-(i\varepsilon_n - \bar{G}_\varepsilon)^2 + \xi^2(\mathbf{p}) + |\Delta + F_\varepsilon^+|^2} \qquad (5.174)$$

$$F^+(p) = -\frac{\Delta + \bar{F}_\varepsilon^+}{-(i\varepsilon_n - \bar{G}_\varepsilon)^2 + \xi^2(\mathbf{p}) + |\Delta + F_\varepsilon^+|^2} \qquad (5.175)$$

Substitution of these expressions to (5.172) and (5.173) gives two equa-
tions determining $\bar{G}_\varepsilon$ and $\bar{F}_\varepsilon^+$. As before (for impurity scattering in
normal metals), self-energy part of $\bar{G}_\varepsilon$ contains a constant, which may
be considered as an additive contribution to the chemical potential.
This is independent of temperature and originates mainly from integration $d^3 p'$ far from the Fermi surface. Thus, this contribution is, in fact,
the same as in the normal metal:

$$\delta\mu \approx \rho \int \frac{d^3 p'}{(2\pi)^3} |v(\mathbf{p} - \mathbf{p}')|^2 \frac{1}{\xi(\mathbf{p}')} \qquad (5.176)$$

---

[20]For $\rho v(\mathbf{p})^2 \to 0$ Eqs. (5.170), (5.171) reduce to (5.108) and (5.109). We only have to
take into account that, in accordance with notations of [Abrikosov A.A., Gorkov L.P.,
Dzyaloshinskii I.E. (1963)], instead of (5.110) and (5.121), we define here:

$$\Delta = |\lambda| F(x, x) \qquad \Delta^* = |\lambda| F^+(x, x) \qquad (5.169)$$

which leads to some difference of signs in comparison to (5.108) and (5.109).

Then, in Eqs. (5.172) and (5.173) we can limit ourselves to integration over the linearized spectrum (close to the Fermi surface), so that (5.170) and (5.171) are rewritten as:

$$(i\tilde{\varepsilon}_n - \xi(\mathbf{p}))G(p) - \tilde{\Delta}F^+(p) = 1 \qquad (5.177)$$

$$(i\tilde{\varepsilon}_n + \xi(\mathbf{p}))F^+(p) - \tilde{\Delta}^*G(p) = 0 \qquad (5.178)$$

where

$$i\tilde{\varepsilon}_n = i\varepsilon_n - \bar{G}_\varepsilon = i\varepsilon_n + \rho v^2 \nu_F \int_{-\infty}^{\infty} d\xi \frac{i\tilde{\varepsilon}_n + \xi}{-(i\tilde{\varepsilon}_n)^2 + \xi^2 + \tilde{\Delta}^2} \qquad (5.179)$$

$$\tilde{\Delta} = \Delta + \bar{F}_\varepsilon = \Delta + \rho v^2 \nu_F \int_{-\infty}^{\infty} d\xi \frac{\tilde{\Delta}}{-(i\tilde{\varepsilon}_n)^2 + \xi^2 + \tilde{\Delta}^2} \qquad (5.180)$$

where we are already considering point-like impurities. Note that contribution of the second term under the integral in (5.179) is zero (as integrand is odd), so that:

$$-\frac{\bar{G}_\varepsilon}{i\varepsilon_n} = \frac{\bar{F}_\varepsilon}{\Delta} \qquad (5.181)$$

Now we can write:

$$\tilde{\Delta} = \Delta + \bar{F}_\varepsilon = \Delta\eta_\varepsilon \qquad (5.182)$$

$$i\tilde{\varepsilon}_n = i\varepsilon_n - \bar{G}_\varepsilon = i\varepsilon_n\eta_\varepsilon \qquad (5.183)$$

where η_ε is defined (cf. (5.179), (5.180)) by the equation:

$$\eta_\varepsilon = 1 + \frac{\eta_\varepsilon}{2\pi\tau} \int_{-\infty}^{\infty} \frac{d\xi}{\xi^2 + (\varepsilon_n^2 + \Delta^2)\eta_\varepsilon^2} = 1 + \frac{\eta_\varepsilon}{2\pi\tau} \frac{\pi}{\eta_\varepsilon\sqrt{\varepsilon_n^2 + \Delta^2}} \qquad (5.184)$$

where we have taken into account that $\rho v^2 \nu_F = \frac{1}{2\pi\tau}$. Finally, we obtain:

$$\eta_\varepsilon = 1 + \frac{1}{2\tau\sqrt{\varepsilon_n^2 + \Delta^2}} \qquad (5.185)$$

Thus, Green's functions $G(p)$ and $F^+(p)$ of a superconductor averaged over random configurations of impurities are obtained from appropriate Gorkov's functions of the "clean" superconductor (5.111) and (5.112) via the simple substitution:

$$\{\varepsilon_n, \Delta\} \to \{\varepsilon_n\eta_\varepsilon, \Delta\eta_\varepsilon\} \qquad (5.186)$$

Then, repeating calculations leading to Eqs. (5.114), (5.115) and (5.116), we can convince ourselves that η_ε just drops out the equation for superconducting gap $\Delta(T)$ after the change of integration variable $\xi \to \xi/\eta$, which changes nothing. Let us see it in details. The replacement (5.186) in (5.114) and (5.115) gives the gap equation in the following form:

$$1 = \frac{|\lambda|T}{(2\pi)^3} \sum_{n=-\infty}^{\infty} \int d^3p \frac{\eta_\varepsilon}{\varepsilon_n^2 \eta_\varepsilon^2 + \xi^2(\mathbf{p}) + |\Delta|^2 \eta_\varepsilon^2}$$

$$\approx |\lambda|\nu_F T \sum_{n=-\infty}^{\infty} \int_{-\omega_D}^{\omega_D} d\xi \frac{\eta_\varepsilon}{\varepsilon_n^2 \eta_\varepsilon^2 + \xi^2 + |\Delta|^2 \eta_\varepsilon^2} \qquad (5.187)$$

Equation for T_c can be obtained by putting $\Delta = 0$ in (5.187):

$$1 = \frac{|\lambda|T}{(2\pi)^3} \sum_{n=-\infty}^{\infty} \int d^3p \frac{\eta_\varepsilon}{\varepsilon_n^2 \eta_\varepsilon^2 + \xi^2} \approx |\lambda|\nu_F T \sum_{n=-\infty}^{\infty} \int_{-\omega_D}^{\omega_D} d\xi \frac{\eta_\varepsilon}{\varepsilon_n^2 \eta_\varepsilon^2 + \xi^2}$$

$$(5.188)$$

where

$$\eta_\varepsilon = 1 + \frac{1}{2|\varepsilon_n|\tau} \qquad (5.189)$$

Now let us add $\pm \frac{1}{\varepsilon_n^2 + \xi^2}$ to the integrand in (5.188), and again use (2.100) to rewrite equation for T_c as:

$$1 = |\lambda|\nu_F \int_{-\omega_D}^{\omega_D} \frac{d\xi}{2\xi} th \frac{\xi}{2T_c}$$

$$+ |\lambda|\nu_F \sum_{n=-\infty}^{\infty} \int_{-\omega_D}^{\omega_D} d\xi \left\{ \frac{\eta_\varepsilon}{\varepsilon_n^2 \eta_\varepsilon^2 + \xi^2} - \frac{1}{\varepsilon_n^2 + \xi^2} \right\} \qquad (5.190)$$

In the second integral here (due to the fast convergence) we may put $\omega_D \to \infty$. Then, after the change of integration variable $\xi \to \xi/\eta_\varepsilon$ in the first term of the integral it is precisely canceled by the second one! As a result Eq. (5.190) reduces to Eq. (5.117), which defines T_c of the "clean" superconductor (5.31).

Similarly we can analyze also the "full" equation (5.187), determining $\Delta(T)$ of the impure superconductor. It is easily seen that it also reduces to Eq. (5.116) for the "clean" case.

Thus, both T_c and $\Delta(T)$ of a superconductor with "normal" (nonmagnetic) impurities do not depend on the presence of impurities at all

(A.A. Abrikosov, L.P. Gorkov, 1959).[21] We obtained this result in the limit of weak disorder, when $p_F l \gg 1$, $E_F \tau \gg 1$, and we can neglect the contribution of diagrams with intersecting impurity interaction lines. In fact, this statement has even more wide region of validity [Sadovskii M.V. (2000)], if we assume the self-averaging nature of superconducting order parameter (gap). These results are in rather good agreement with experimentally known *relative* stability of superconducting state in many metals to the introduction of more or less small amount of nonmagnetic impurities (disordering).

Of course, due to the oversimplified nature of BCS model, these results are only approximate. In fact, disordering leads e.g. to the growth of effective repulsion (due to the growth of the so-called Coulomb pseudopotential [Sadovskii M.V. (2019a); De Gennes P.G. (1966)]) of electrons forming the Cooper pair and appropriate lowering of T_c [Sadovskii M.V. (2000)]. "Anderson theorem" is also invalid within the BCS model, when we consider superconductors with gap anisotropy at the Fermi surface (e.g. for systems with Cooper pairs with higher orbital moments, like the case of d-wave pairing in copper oxide high-temperature superconductors). In such cases, disordering usually strongly suppresses superconductivity [Sadovskii M.V. (2000)]. Spectacular example of superconductivity suppression by disordering is the case of magnetic (paramagnetic) impurities (i.e. impurities with "free" (uncompensated) spin **S**) (A.A. Abrikosov, L.P. Gorkov, 1960). Let us consider this problem briefly.

If superconductor contains paramagnetic impurities, potential of electron interaction with impurity contains an exchange term:

$$V(\mathbf{r}) = v(\mathbf{r}) + J(\mathbf{r})(\mathbf{S} \cdot \mathbf{s}) \qquad (5.191)$$

where $J(\mathbf{r})$ is appropriate exchange integral, **S** — impurity spin, $\mathbf{s} = \frac{1}{2}\sigma$ — spin of conduction electron. In this case, in diagrams of impurity scattering in Gorkov equations, shown in Fig. 5.8, we have to take into account the spin structure of G and F^+. Above we have seen that $G_{\alpha\beta} = G\delta_{\alpha\beta}$, $F_{\alpha\beta} = g_{\alpha\beta}F$, $F^+_{\alpha\beta} = -g_{\alpha\beta}F^+$, where $g_{\alpha\beta} = i\sigma^y_{\alpha\beta}$ (cf. (5.52)). Then it happens that contributions of the second term in (5.191) into self-energies of Fig. 5.8 due to impurity scattering, built on normal and anomalous Green's functions, are of *different sign*.[22] Then, instead of the *same* renormalization of $i\varepsilon_n$ and Δ of the type given in (5.182), (5.183) and (5.185), we obtain *equations*:

$$i\tilde{\varepsilon}_n = i\varepsilon_n + \frac{i\tilde{\varepsilon}_n}{2\tau_1\sqrt{\tilde{\varepsilon}_n^2 + \tilde{\Delta}^2}} \qquad (5.192)$$

---

[21]Sometimes this is called "Anderson theorem" as he also obtained this result by a different method [De Gennes P.G. (1966)].

[22]This follows from $\sigma_{\alpha\gamma}\delta_{\gamma\delta}\sigma_{\delta\beta} = \sigma_{\alpha\delta}\sigma_{\delta\beta} = \sigma^2\delta_{\alpha\beta}$ and $\sigma_{\alpha\gamma}g_{\gamma\delta}\sigma_{\delta\beta} = -\sigma^2 g_{\alpha\beta}$, where $\sigma^2 = 3$.

$$\tilde{\Delta} = \Delta + \frac{\tilde{\Delta}}{2\tau_2\sqrt{\tilde{\varepsilon}_n^2 + \tilde{\Delta}^2}} \tag{5.193}$$

where

$$\frac{1}{\tau_1} = 2\pi\rho\nu_F\left(v^2 + \frac{1}{4}S(S+1)J^2\right) \tag{5.194}$$

$$\frac{1}{\tau_2} = 2\pi\rho\nu_F\left(v^2 - \frac{1}{4}S(S+1)J^2\right) \tag{5.195}$$

where we have assumed the point-like nature for both impurity potential and exchange integral, and also averaged over the orientations of impurity spins, writing $< \mathbf{S}^2 > = \frac{1}{3}S(S+1)$. Besides, we have taken into account that for electron spin $\mathbf{s}^2 = \frac{1}{4}\sigma^2 = \frac{3}{4}$. The difference of (5.194) and (5.195) reduces to:

$$\frac{1}{\tau_1} - \frac{1}{\tau_2} = \frac{2}{\tau_s} \quad \text{i.e.} \quad \frac{1}{\tau_s} = \frac{\pi}{2}\pi\rho J^2\nu_F S(S+1) \tag{5.196}$$

and defines the scattering rate (including spin–flip processes) due to exchange potential. Though, in fact, we always have inequality $\frac{1}{\tau_s} \ll \frac{1}{\tau}$, where $\frac{1}{\tau}$ is scattering rate due to potential scattering, it is $\frac{1}{\tau_s}$ which is the relevant parameter for superconductivity. Complete analysis[23] of the gap equation, derived from (5.174), (5.175) or (5.177), (5.178) with the use of (5.192), (5.193), leads to the following conclusions. Scattering by magnetic impurities leads to the strong suppression of superconducting critical temperature, described by the notorious Abrikosov–Gorkov equation:

$$\ln\frac{T_{c0}}{T_c} = \psi\left(\frac{1}{2} + \frac{1}{2\pi T_c\tau_s}\right) - \psi\left(\frac{1}{2}\right) \tag{5.197}$$

where T_{c0} is transition temperature in the absence of impurities, while

$$\psi(z) = \frac{\Gamma'(z)}{\Gamma(z)} = -\ln\gamma - \frac{1}{z} + \sum_{n=1}^{\infty}\left(\frac{1}{n} - \frac{1}{n+z}\right) \tag{5.198}$$

is the logarithmic derivative of the Γ-function (digamma function), $\ln\gamma = C = 0.577...$ is Euler constant. Characteristic form of dependence of T_c on the rate of exchange scattering, following from (5.197), is shown in Fig. 5.9. There exists the critical scattering rate (determining the critical concentration of magnetic impurities):

$$\frac{1}{\tau_s^c} = \frac{\pi T_{c0}}{2\gamma} = \frac{\Delta_0}{2} \qquad \Delta_0 = \frac{\pi}{\gamma}T_{c0} \tag{5.199}$$

When this critical value is reached (e.g. with the growth of the concentration of magnetic impurities) superconductivity vanishes (T_c becomes zero). For weak exchange scattering (small impurity concentration) $\tau_s \to \infty$, and from (5.197) we get:

$$T_c \approx T_{c0} - \frac{\pi}{4\tau_s} \tag{5.200}$$

---

[23]Details of calculations for a similar problem will be given in the next Chapter.

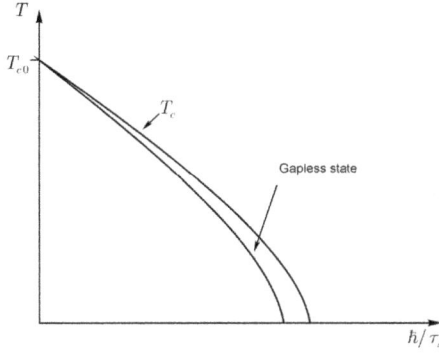

Fig. 5.9   Dependence of superconducting critical temperature on the scattering rate due to magnetic impurities and the region of "gapless" superconductivity.

so that we have a small suppression of T_c. Note that dependence of T_c on concentration of magnetic impurities, determined by (5.197), is directly confirmed experimentally.

Another remarkable property of the model with magnetic impurities is the existence on the "phase diagram" shown in Fig. 5.9 of a narrow region of the so-called "gapless" superconductivity. It follows from the detailed calculations based on Eqs. (5.177), (5.178) and (5.192), (5.193), that in this region both T_c and the order parameter Δ remain finite (Δ becomes zero at $T = T_c$), while the energy gap in the spectrum of elementary excitations (or, more precisely, in the density of states) is absent. The thing is that in the presence of (magnetic) impurities the order parameter Δ does not coincide with the gap in the spectrum and the scattering by impurities "smears" this gap (leads to the overlapping "tail" formation within the gap), leading to characteristic form of the density of states with "pseudogap", (different from that of BCS theory, which was given in (5.79)), and shown qualitatively in Fig. 5.10. Superconducting response of the system in this unusual state persists.

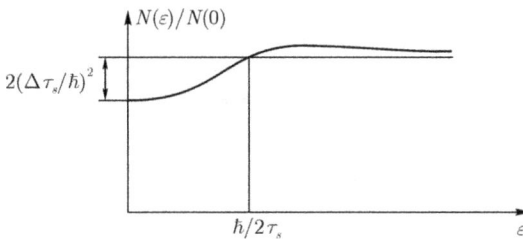

Fig. 5.10   Density of electronic states in "gapless" superconductor.

Why superconducting state is stable towards introduction of normal impurities, but unstable towards magnetic impurities? The reason is very simple — the usual (potential) scattering acts on both electrons of the Cooper pair in the "same way" and the pair survives, while magnetic scattering acts on the opposite spins of electrons in the pair differently. For singlet pairing (the only pairing we consider here) this scattering leads to depairing of electrons or destruction of Cooper pairing.

5.5   Ginzburg–Landau expansion

It is well known how important is the phenomenological approach to superconductivity, proposed by Ginzburg and Landau (1950), and based on the expansion of free energy in powers of the order parameter, allowing to describe main properties of superconductors close to superconducting transition temperature [Lifshits E.M., Pitaevskii L.P. (1980); Sadovskii M.V. (2019a); De Gennes P.G. (1966)]. Let us show, how this expansion can be *derived* from microscopic BCS theory. It was first done by Gorkov (1959), but below we shall use slightly different approach [Sadovskii M.V. (2000)].

In fact, it is sufficient to analyze the case of electrons in a normal metal $(T > T_c)$, propagating in a random "field" of thermodynamic fluctuations[24] of superconducting order parameter, which we describe by a single Fourier component, $\Delta_{\mathbf{q}}$, characterized by some fixed (small) wave vector $\mathbf{q}$. Then we can write down the following Hamiltonian for electron interaction with these fluctuations:[25]

$$H_{int} = \sum_{\mathbf{p}} \left\{ \Delta_{\mathbf{q}} a^+_{\mathbf{p}_+} a^+_{-\mathbf{p}_-} + \Delta^*_{\mathbf{q}} a_{-\mathbf{p}_-} a_{\mathbf{p}_+} \right\} \qquad (5.201)$$

where, as usual, we use the notation $\mathbf{p}_+ = \mathbf{p} \pm \frac{1}{2}\mathbf{q}$.

Let us now calculate the correction to thermodynamic potential (free energy) due to (5.201). According to [Abrikosov A.A., Gorkov L.P., Dzyaloshinskii I.E. (1963)], correction to thermodynamic potential due to any interaction is expressed via the average value of (Matsubara) S-matrix:

$$\Delta F = -T \ln < S > \qquad (5.202)$$

---

[24]We assume that these fluctuations are static and "smooth" enough in space.
[25]This Hamiltonian can also be interpreted as describing electron interaction with random "source" of Cooper pairs.

where angular brackets denote Gibbs average, while

$$S = T_\tau \exp\left\{ -\int_0^{\frac{1}{T}} d\tau H_{int}(\tau) \right\} \qquad (5.203)$$

Then it is given by the *loop* expansion over connected diagrams:

$$\Delta F = -T\{< S >_c -1\} \qquad (5.204)$$

Diagrams for $< S >_c -1$ are closed loops which are drawn according to the rules of diagram technique (for the given interaction), with additional factor of $\frac{1}{n}$ attributed to each diagram of n-th order of perturbation theory (for topologically nonequivalent diagrams) [Abrikosov A.A., Gorkov L.P., Dzyaloshinskii I.E. (1963)].

In the absence of an external magnetic field[26] Ginzburg–Landau (GL) expansion for the difference of free energies of superconducting an normal states is usually written as [De Gennes P.G. (1966)]:

$$F_s - F_n = A|\Delta(\mathbf{r})|^2 + \frac{B}{2}|\Delta(\mathbf{r})|^4 + C|\nabla\Delta(\mathbf{r})|^2 \qquad (5.206)$$

Introducing Fourier expansion:

$$\Delta(\mathbf{r}) = \sum_\mathbf{q} \Delta_\mathbf{q} e^{i\mathbf{q}\mathbf{r}} \qquad (5.207)$$

and restricting analysis to a single Fourier component, we can write (5.206) as:

$$F_s - F_n = A|\Delta_\mathbf{q}|^2 + \frac{B}{2}|\Delta_\mathbf{q}|^4 + Cq^2|\Delta_\mathbf{q}|^2 \qquad (5.208)$$

Now the task of microscopic theory is reduced to calculation of GL coefficients A, B, C. From general considerations we can only say that $A \sim T - T_c$ [Lifshits E.M., Pitaevskii L.P. (1980); Sadovskii M.V. (2019a); De Gennes P.G. (1966)].

The knowledge of GL coefficients allows us to find the main characteristics of a superconductor at temperatures close to T_c [Lifshits E.M., Pitaevskii L.P. (1980); Sadovskii M.V. (2019a); De Gennes P.G. (1966)].

---

[26]Taking into account that our "field" Δ is charged, and the charge is $2e$ (5.98), interaction with an external magnetic field can be introduced via the standard replacement:

$$\nabla \to \nabla \mp 2ie\mathbf{A} \qquad (5.205)$$

for Δ and Δ^* respectively, where $\mathbf{A}$ is the vector-potential of external field.

In particular, the coherence length $\xi(T)$, which determines characteristic scale of inhomogeneities of the order parameter Δ, i.e., in fact, the typical size of the Cooper pair, is given by:

$$\xi^2(T) = -\frac{C}{A} \tag{5.209}$$

Penetration depth of an external magnetic field is expressed via GL coefficients and electric charge e as:

$$\lambda_L^2(T) = -\frac{c^2}{32\pi e^2}\frac{B}{AC} \tag{5.210}$$

where we have "restored" the velocity of light c. Dimensionless parameter of Ginzburg and Landau is given by:

$$\kappa = \frac{\lambda_L(T)}{\xi(T)} = \frac{c}{4eC}\sqrt{\frac{B}{2\pi}} \tag{5.211}$$

Close to T_c the upper critical magnetic field H_{c2} is determined by:

$$H_{c2} = \frac{\phi_0}{2\pi\xi^2(T)} = -\frac{\phi_0}{2\pi}\frac{A}{C} \tag{5.212}$$

where $\phi_0 = \pi\hbar c/|e|$ is magnetic flux quantum [Lifshits E.M., Pitaevskii L.P. (1980); Sadovskii M.V. (2019a)]. At last, specific heat discontinuity at superconducting transition is given by:

$$c_s - c_n = \frac{T_c}{B}\left(\frac{A}{T - T_c}\right)^2 \tag{5.213}$$

Up to terms of fourth order in $\Delta_\mathbf{q}$ we have:

$$
\begin{aligned}
<S>_c -1 &= \frac{1}{2!}\int_0^{1/T} d\tau_1 \int_0^{1/T} d\tau_2 <T_\tau(H_{int}(\tau_1)H_{int}(\tau_2))>_c \\
&+ \frac{1}{4!}\int_0^{1/T} d\tau_1 ... \int_0^{1/T} d\tau_4 <T_\tau(H_{int}(\tau_1)...H_{int}(\tau_4))>_c
\end{aligned}
\tag{5.214}
$$

Consider now the second order correction to free energy (5.204):

$$\Delta F_2 = -\frac{T}{2}\int_0^{1/T} d\tau_1 \int_0^{1/T} d\tau_2 <T_\tau(H_{int}(\tau_1)H_{int}(\tau_2))>_c \tag{5.215}$$

To calculate this correction, we can use Wick theorem, allowing to reduce the average of the products of Fermi operators to the products of averages of pairs of operators[27] a and a^+. Then, from (5.215), after appropriate calculations, we obtain:

$$\Delta F_2 = -\frac{T}{2}\int_0^{1/T}d\tau_1\int_0^{1/T}d\tau_2\Delta_{\mathbf{q}}^*\Delta_{\mathbf{q}}\left\{\sum_{\mathbf{p}}G(\mathbf{p}_+,\tau_1-\tau_2)G(-\mathbf{p}_-,\tau_1-\tau_2)\right.$$

$$\left.+\sum_{\mathbf{p}}G(\mathbf{p}_+,\tau_2-\tau_1)G(-\mathbf{p}_-,\tau_2-\tau_1)\right\}$$

$$= -T\int_0^{1/T}d\tau_1\int_0^{1/T}d\tau_2|\Delta_{\mathbf{q}}|^2\sum_{\mathbf{p}}G(\mathbf{p}_+,\tau_1-\tau_2)G(-\mathbf{p}_-,\tau_1-\tau_2)$$

$$(5.216)$$

where

$$G(\mathbf{p},\tau_1-\tau_2) = -<T_\tau a_{\mathbf{p}}(\tau_1)a_{\mathbf{p}}^+(\tau_2)> \qquad (5.217)$$

is the Green's function of a free electron in $(\mathbf{p},\tau)$ representation. Expanding this Green's function into Fourier series over τ and calculating "time" integrals, we get:

$$\Delta F_2 = -|\Delta_{\mathbf{q}}|^2 T\sum_{\mathbf{p}}\sum_n G(\mathbf{p}_+,\varepsilon_n)G(-\mathbf{p}_-,-\varepsilon_n) \qquad (5.218)$$

where

$$G(\mathbf{p}\varepsilon_n) = \frac{1}{i\varepsilon_n-\xi(\mathbf{p})}, \qquad \varepsilon_n = (2n+1)\pi T \qquad (5.219)$$

is the standard form of Matsubara Green's function of free electron.

Similarly, for the fourth order correction, we have:

$$\Delta F_4 = -T\frac{1}{4!}12|\Delta_{\mathbf{q}}|^4\int_0^{1/T}d\tau_1...\int_0^{1/T}d\tau_4 G(\mathbf{p},\tau_1-\tau_3)G(-\mathbf{p},\tau_1-\tau_4)$$

$$\times G(\mathbf{p},\tau_2-\tau_4)G(-\mathbf{p},\tau_2-\tau_3) \qquad (5.220)$$

where in Green's function we put $q = 0$, neglecting the contribution of spatial inhomogeneities to fourth order term of GL-expansion. Then, after some calculations, similar to those just done for the second order contribution, we obtain:

$$\Delta F_4 = \frac{T}{2}|\Delta_{\mathbf{q}}|^4\sum_{\mathbf{p}}\sum_n G(\mathbf{p},\varepsilon_n)G(-\mathbf{p},-\varepsilon_n)G(\mathbf{p},\varepsilon_n)G(-\mathbf{p},-\varepsilon_n)$$

$$(5.221)$$

---

[27]We do need to consider here any anomalous averages, as we analyze the normal metal at $T > T_c$, where some random "source" of fluctuating Cooper pairs (5.201) is "operational".

Finally, the correction to free energy, up to fourth order in $\Delta_{\mathbf{q}}$, is given by:

$$\Delta F \approx -|\Delta_{\mathbf{q}}|^2 T \sum_{\mathbf{p}} \sum_n G(\mathbf{p}_+, \varepsilon_n) G(-\mathbf{p}_-, -\varepsilon_n)$$

$$+ \frac{T}{2}|\Delta_{\mathbf{q}}|^4 \sum_{\mathbf{p}} \sum_n G(\mathbf{p}, \varepsilon_n)^2 G(-\mathbf{p}, -\varepsilon_n)^2 \qquad (5.222)$$

Now GL-expansion for the difference of free energies of a superconductor and normal metal will be is obtained from (5.222), if we rewrite it in such a form, that the coefficient before $|\Delta_{\mathbf{q}}|^2$ at $q = 0$ will be zero at $T = T_c$ and negative for $T < T_c$. This type of behavior is easily guaranteed, if we subtract from the r.h.s. of (5.222) the value of ΔF_2 (5.218), taken with the coefficient before $|\Delta_{\mathbf{q}}|^2$ calculated at $T = T_c$ and $q = 0$. Then, the GL-expansion is written as:

$$F_s - F_n = -|\Delta_{\mathbf{q}}|^2 T \sum_{\mathbf{p}} \sum_n G(\mathbf{p}_+, \varepsilon_n) G(-\mathbf{p}_-, -\varepsilon_n)$$

$$+ |\Delta_{\mathbf{q}}|^2 T_c \sum_{\mathbf{p}} \sum_n G(\mathbf{p}, \varepsilon_n) G(-\mathbf{p}, -\varepsilon_n)|_{T=T_c}$$

$$+ \frac{T_c}{2}|\Delta_{\mathbf{q}}|^4 \sum_{\mathbf{p}} \sum_n G(\mathbf{p}, \varepsilon_n)^2 G(-\mathbf{p}, -\varepsilon_n)^2|_{T=T_c} \qquad (5.223)$$

Here we have taken into account that the coefficient B before $|\Delta_{\mathbf{q}}|^4$ is finite at $T = T_c$, so that calculating it (besides neglecting the q-dependence) we can safely put $T = T_c$. With the account of BCS equation for T_c, taken in the form given in (5.115) with $\Delta = 0$, we see that the second term in (5.223) reduces just to $\frac{1}{|\lambda|}|\Delta_{\mathbf{q}}|^2$, so that GL-expansion can be rewritten in more compact form:

$$F_s - F_n = \frac{1}{|\lambda|}|\Delta_{\mathbf{q}}|^2 - |\Delta_{\mathbf{q}}|^2 T \sum_{\mathbf{p}} \sum_n G(\mathbf{p}_+, \varepsilon_n) G(-\mathbf{p}_-, -\varepsilon_n)$$

$$+ \frac{T_c}{2}|\Delta_{\mathbf{q}}|^4 \sum_{\mathbf{p}} \sum_n G(\mathbf{p}, \varepsilon_n)^2 G(-\mathbf{p}, -\varepsilon_n)^2|_{T=T_c} \qquad (5.224)$$

Graphically Eqs. (5.223), (5.224) can be represented by diagrams, shown in Fig. 5.11. Subtraction of the second diagram here precisely guarantees correct behavior of the coefficient A, so that it goes through zero and changes sign at $T = T_c$.

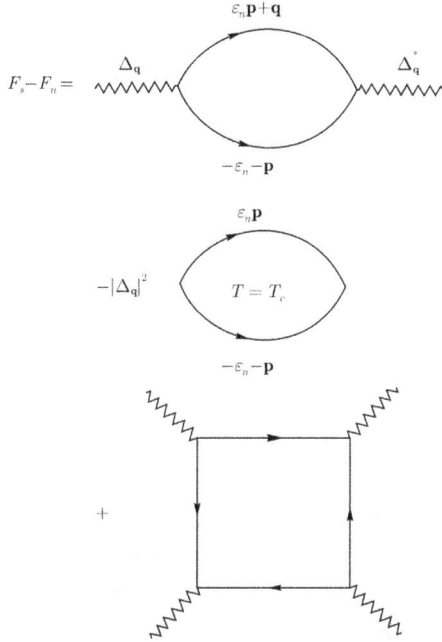

Fig. 5.11 Diagrammatic representation of Ginzburg–Landau expansion.

Let us now start with explicit calculations of GL-coefficients. From (5.223) we write the coefficient A as:

$$A = -T\sum_{\mathbf{p}}\sum_{n} G(\mathbf{p}\varepsilon_n)G(-\mathbf{p},-\varepsilon_n) + T_c\sum_{\mathbf{p}}\sum_{n} G(\mathbf{p}\varepsilon_n)G(-\mathbf{p},-\varepsilon_n)|_{T=T_c}$$

$$= -T\nu_F \int_{-\infty}^{\infty} d\xi \sum_{n} \frac{1}{\varepsilon_n^2 + \xi^2} + T_c\nu_F \int_{-\infty}^{\infty} d\xi \sum_{n} \frac{1}{\varepsilon_n^2 + \xi^2}\bigg|_{T=T_c}$$

$$= -\nu_F \int_{-\infty}^{\infty} \frac{d\xi}{\xi} \left[th\frac{\xi}{2T} - th\frac{\xi}{2T_c} \right] \tag{5.225}$$

For $T \approx T_c$ we have $\frac{1}{\xi}th\frac{\xi}{2T} \approx \frac{1}{\xi}th\frac{\xi}{2T_c} + \frac{T_c-T}{2T_c^2}\frac{1}{ch^2\frac{\xi}{2T_c}}$, so that:

$$A = -\nu_F \frac{T_c - T}{4T_c^2} \int_{-\infty}^{\infty} d\xi \frac{1}{ch^2\frac{\xi}{2T_c}} = \nu_F \frac{T - T_c}{T_c} \tag{5.226}$$

To calculate the coefficient C we have to expand the product of Green's functions in the first term of (5.223) in powers of q:

$$G(\mathbf{p}_+, \varepsilon_n)G(-\mathbf{p}_-, -\varepsilon_n) \approx \frac{1}{\varepsilon_n^2 + \xi^2(\mathbf{p})} - \frac{(\mathbf{qp})^2}{4m^2(\varepsilon_n^2 + \xi^2(\mathbf{p}))^2} - \frac{i\varepsilon_n(\mathbf{qp})}{m(\varepsilon_n^2 + \xi^2(\mathbf{p}))^2}$$

$$-\frac{\xi(\mathbf{p})q^2}{4m^2(\varepsilon_n^2 + \xi^2(\mathbf{p}))^2} - \frac{\varepsilon_n^2 - \xi^2(\mathbf{p})}{2m^2(\varepsilon_n^2 + \xi^2(\mathbf{p}))^3}(\mathbf{qp})^2 \qquad (5.227)$$

so that

$$-T_c \sum_{\mathbf{p}} \sum_n G(\mathbf{p}_+ \varepsilon_n)G(-\mathbf{p}_-, -\varepsilon_n)$$

$$\approx -T_c \sum_{\mathbf{p}} \sum_n \left\{ \frac{1}{\varepsilon_n^2 + \xi^2(\mathbf{p})} - (\mathbf{qp})^2 \frac{3\varepsilon_n^2 - \xi^2(\mathbf{p})}{4m^2(\varepsilon_n^2 + \xi^2(\mathbf{p}))^3} \right\} \qquad (5.228)$$

and we get the expression for C as:

$$C = T_c \nu_F \frac{1}{d} \sum_n \int_{-\infty}^{\infty} d\xi \left(\frac{p}{2m}\right)^2 \frac{3\varepsilon_n - \xi^2}{(\varepsilon_n^2 + \xi^2(\mathbf{p}))^3} \qquad (5.229)$$

where d is the spatial dimensionality. Due to fast convergence of the integral in (5.229) the main contribution here comes from the immediate vicinity of the Fermi level and we can put $p \approx p_F = m v_F$. Then finally:

$$C = T_c \nu_F \frac{v_F^2}{4d} \sum_n \frac{\pi}{|\varepsilon_n|^3} = \nu_F \frac{7\zeta(3)}{16\pi^2 d} \frac{v_F^2}{T_c^2} \equiv \nu_F \xi_0^2 \qquad (5.230)$$

where we have defined the coherence length ξ_0 [De Gennes P.G. (1966)] as:

$$\xi_0^2 = \frac{7\zeta(3)}{16\pi^2 d} \frac{v_F^2}{T_c^2} \qquad (5.231)$$

and $\zeta(3) \approx 1.202...$ ($\zeta(x) - \zeta$ is Riemann zeta-function). In these expressions we again introduced:

$$\nu_F = \begin{cases} \frac{m p_F}{2\pi^2} & d = 3 \\ \frac{m}{2\pi} & d = 2 \end{cases} \qquad (5.232)$$

— density of states of electrons at the Fermi level for a single spin projection.

At last, the value of the coefficient B is immediately obtained from the diagram with four Δ-"tails", shown in Fig. 5.11:

$$B = T_c \sum_{\mathbf{p}} \sum_n G^2(\varepsilon_n \mathbf{p})G^2(-\varepsilon_n, -\mathbf{p}) = \nu_F T_c \sum_n \int_{-\infty}^{\infty} d\xi \frac{1}{(\varepsilon_n^2 + \xi^2(\mathbf{p}))^2}$$

$$= \nu_F T_c \sum_n \frac{\pi}{2|\varepsilon_n|^3} = \nu_F \frac{7\zeta(3)}{8\pi^2 T_c^2} \qquad (5.233)$$

Expressions (5.226), (5.230) and (5.233) give the standard expressions for GL coefficients, obtained by Gorkov (1959) for "clean" superconductors. The use of these in Eqs. (5.209)–(5.213) gives the well known expressions of BCS theory for temperatures close to T_c. Thus we obtain the complete microscopic justification of Ginzburg–Landau theory within BCS model. At the same time, it should be noted that GL approach is much more convenient (and simpler) than the complete microscopic theory.

Now we can proceed with further generalizations. Consider e.g. the so-called "dirty" superconductors with nonmagnetic impurities. GL expansion for this case is obtained by direct generalization of our previous analysis. We only have to take into account scattering by impurities, as we have already done several times. It is not difficult to convince yourself that the proper generalization of GL expansion is described by diagrams shown in Fig. 5.12, where we have introduced "triangular" vertices (which we analyzed in the previous Chapter), taking into account impurity scattering, and all electron lines are assumed to be "dressed" by impurities:[28]

$$ G(\varepsilon_n \mathbf{p}) = \frac{1}{i\varepsilon_n - \xi(\mathbf{p}) + i\gamma \frac{\varepsilon_n}{|\varepsilon_n|}}, \qquad \gamma = \frac{1}{2\tau} = \pi \rho v^2 \nu_F \qquad (5.234) $$

and, as usual, we consider the case of point-like impurities.

An important difference from the analysis given in previous Chapter is that now we have to consider the Cooper channel and loops in Fig. 5.12(a) are defined (in notations similar to that of the previous Chapter, and in Matsubara technique) as:

$$ \Psi(\mathbf{q}\omega_m \varepsilon_n) = -\frac{1}{2\pi i} \sum_{\mathbf{p}\mathbf{p}'} < G(\mathbf{p}_+ \mathbf{p}'_+, -\varepsilon_n + \omega_m) G(-\mathbf{p}'_-, -\mathbf{p}_-, -\varepsilon_n) > $$

$$ (5.235) $$

which is shown in Fig. 5.13 (for the case of interest to us with $\omega_m = 2\varepsilon_n$). Then from an expansion shown in Fig. 5.12 it is not difficult to obtain

---

[28]In the standard approximation without intersecting impurity lines, valid in the limit of weak disorder, when $p_F l \gg 1$, $E_F \tau \gg 1$.

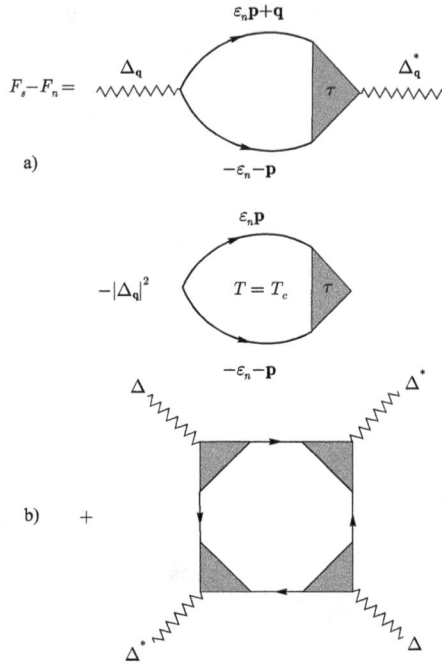

Fig. 5.12 Diagrammatic representation of Ginzburg–Landau expansion for a superconductor with impurities.

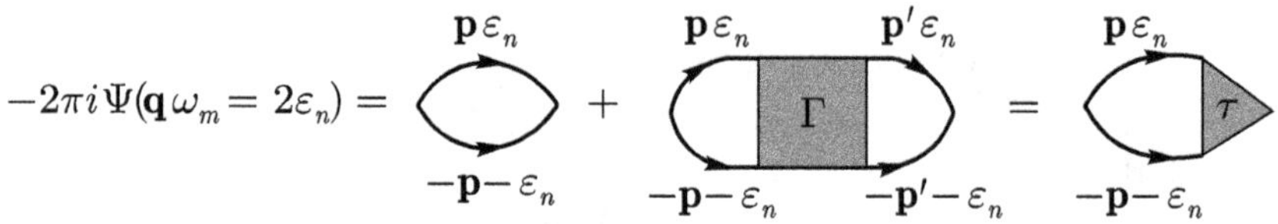

Fig. 5.13 Diagrammatic representation of $\Psi(\mathbf{q}\omega_m = 2\varepsilon_n)$ for impure system.

the following general expressions for GL coefficients A and C:[29]

$$A = \frac{1}{|\lambda|} + 2\pi i T \sum_n \Psi(\mathbf{q} = 0, \omega_m = 2\varepsilon_n) \qquad (5.236)$$

---

[29]Dropping details of calculations we just note that coefficient B in a superconductor with impurities, determined by Fig. 5.12(b), is again given by Eq. (5.233), i.e. is the same as for the "clean" case [Sadovskii M.V. (2000)]. Thus, below we only analyze the coefficients A and C.

$$C = i\pi T \sum_n \frac{\partial^2}{\partial q^2} \Psi(\mathbf{q}, \omega_m = 2\varepsilon_n)|_{q=0, T=T_c} \qquad (5.237)$$

Our analysis is simplified in the case of time-reversal invariance (i.e. in the absence of an external magnetic filed or magnetic impurities), when we can just reverse one of electronic lines (as was already done in the previous Chapter) in the loops in Figs. 5.12, 5.13 and convince ourselves that

$$\Psi(\mathbf{q}\omega_m\varepsilon_n) = \Phi(\mathbf{q}\omega_m\varepsilon_n) \qquad (5.238)$$

where $\Phi(\mathbf{q}\omega_m\varepsilon_n)$ is the obvious Matsubara formalism generalization of the two-particle Green's function (loop), which was analyzed in details in the previous Chapter (cf. (4.88), (4.305)). Thus, in fact we do not need any calculations at all! GL coefficients are determined from $\Phi(\mathbf{q}\omega_m = 2\varepsilon_n)$, the form of which is easily "guessed", returning to the appropriate results given in the previous Chapter:[30]

$$\Phi(\mathbf{q}\omega_m) = -\frac{\nu_F}{i|\omega_m| + iD_0 q^2} \qquad (5.239)$$

where $D_0 = \frac{1}{d}v_F^2\tau = \frac{1}{d}\frac{E_F}{m\gamma}$ is Drude diffusion coefficient, until we discuss the case of weak enough disorder ($p_F l \gg 1, E_F\tau \gg 1$) and can use the "ladder" approximation for the vertex part Γ of impurity scattering.[31]

Using (5.239) and (5.238) in (5.236), we find:

$$A = \frac{1}{|\lambda|} - 2\nu_F \sum_{n\geq 0}^{n^*} \frac{1}{2n+1} = \frac{1}{|\lambda|} - \nu_F \ln \frac{2\gamma}{\pi} \frac{\omega_D}{T}$$

$$= \nu_F \ln \frac{T}{T_c} \approx \nu_F \frac{T - T_c}{T_c} \qquad (5.240)$$

where we have introduced the cut-off (of logarithmically divergent) sum over n at $n^* = \frac{\omega_D}{2\pi T}$. Here we obtained a standard result of BCS theory for $T_c = \frac{2\gamma}{\pi}\omega_D \exp\left(-\frac{1}{|\lambda|\nu_F}\right)$, so that transition temperature does not depend on impurity concentration, in complete accordance with the analysis given in the previous section (Anderson "theorem"). Formally,

---

[30]Note, that we are obviously dealing with *single* spin projection here!

[31]In reality, Eq. (5.239) "works" also in the framework of self-consistent theory of localization [Sadovskii M.V. (2000)], with the only replacement of Drude diffusion coefficient D_0 by the generalized diffusion coefficient $D(|\omega_m|)$.

impurity contribution "drops out" with D_0, as in (5.236) we have to take $q = 0$.

In contrast, for the coefficient C, from (5.237) we obtain:

$$C = -i\pi T \nu_F \sum_n \frac{\partial^2}{\partial q^2} \frac{1}{2i|\varepsilon_n| + iD_0 q^2}\Big|_{q=0}$$

$$= \pi T \nu_F D_0 \sum_n \frac{1}{2\varepsilon_n^2} = \frac{\nu_F D_0}{\pi T} \sum_{n \geq 0} \frac{1}{(2n+1)^2} = \frac{\pi}{8T} \nu_F D_0 \quad (5.241)$$

Writing (5.241) as:

$$C \equiv \nu_F \xi^2 = \frac{\pi}{8T_c} D_0 = \frac{\pi}{8T_c} \frac{1}{3} v_F^2 \tau = \frac{\pi}{24} \frac{v_F}{T_c} v_F \tau = \frac{\pi}{24} \frac{v_F}{T_c} l = 0.13 \frac{v_F}{T_c} l \quad (5.242)$$

where $l = v_F \tau$ is the mean free path, and taking into account the definition (5.231) for the "clean" case, i.e. $\xi_0 = 0.18 \frac{v_F}{T_c}$, we immediately get the main result of the theory of "dirty" superconductors for the coherence length (L.P. Gorkov, 1959):

$$\xi^2 \approx \xi_0 l \qquad \xi \approx \sqrt{\xi_0 l} \qquad (5.243)$$

Thus, the effective size of Cooper pairs in "dirty limit", when $l \ll \xi_0$, is suppressed in comparison with the "clean" case. Then, according to (5.212), this means that disordering (introduction of impurities), while leaving T_c untouched, may lead to a significant growth of the upper critical field H_{c2}, which may be useful for the practical applications![32]

Using (5.242), (5.243) in (5.212) we may obtain the remarkable Gorkov's relation, connecting the temperature derivative (slope of the temperature dependence) of the upper critical field H_{c2} (close to T_c) with conductivity of the system $\sigma = \frac{ne^2}{m} \tau = 2e^2 \nu_F D_0$ and the density of states at the Fermi level ν_F:

$$-\frac{\sigma}{\nu_F} \left(\frac{dH_{c2}}{dT} \right)_{T_c} = \frac{8e^2}{\pi^2 \hbar} \phi_0 \qquad (5.244)$$

where $\phi_0 = \frac{\pi c \hbar}{e}$ is magnetic flux quantum. In the r.h.s. of Eq. (5.244) we have only fundamental constants, while in the l.h.s. can be determined experimentally. Disorder (concentration of impurities) growth, in general, does not change density of states ν_F significantly, while (residual)

---

[32]Of course, we oversimplify the real situation. As we already noted above, T_c may be strongly dependent on disordering [Sadovskii M.V. (2000)].

resistivity grows linearly with impurity concentration (i.e. conductivity is suppressed).[33] Then from (5.244) it follows that the slope of the upper critical field H_{c2} (close to T_c) grows linearly with disorder (impurity concentration). This is confirmed by many experiments on traditional superconductors.

Up to now in our analysis of "dirty" superconductors we were dealing with weak enough disorder. As in traditional superconductors the typical values of coherence length $\xi_0 \sim \frac{v_F}{T_c}$ are orders of magnitude larger than interatomic spacing a, there is no problem with reaching the "dirty" limit $l \ll \xi_0$ at relatively large mean free paths $l \gg a$ (corresponding to weak disorder in the sense of satisfying the inequality $p_F l \gg 1$). An interesting question is what happens to the usual "dirty" limit results with further decrease of the mean free path (growth of disorder) up to $l \sim a$, when, as we have seen in the previous Chapter, Anderson metal–insulator transition takes place [Sadovskii M.V. (2000)]. In fact we understand, that in this limit an expression for the two-particle Green's function of the type of Eq. (5.239) is conserved, but Drude diffusion coefficient has to be replaced by the generalized one $D_0 \to D(|\omega_m|)$, which is determined (for Matsubara frequencies) by an equation of self-consistent theory of localization of the type of (4.286), which (for $d = 3$) takes the form [Sadovskii M.V. (2000)]:

$$\frac{D_E(\omega_m)}{D_0} = 1 - \frac{\lambda}{\lambda_c} + \frac{\pi}{2}\frac{\lambda}{\lambda_c}\left[\frac{D_0}{D_E(\omega_m)}\frac{\omega_m}{2\gamma}\right]^{1/2} \qquad (5.245)$$

where all notations are the same as in the previous Chapter.[34] Similarly to (4.293) and with sufficient (for our purposes) accuracy solution of (5.245) can be written as:

$$D_E(\omega_m) \approx Max\left\{ D_E\frac{\omega_m}{\omega_m + 3D_E\omega_0^2(E)/v_F^2}; \quad D_0\left(\frac{\omega_m}{2\gamma}\right)^{1/3}\right\} \qquad (5.246)$$

where $D_E = \frac{D_0}{p_F \xi_{loc}(E)}$ is the renormalized diffusion coefficient (which drops to zero at the mobility edge), ω_0 is characteristic frequency define in (4.250).

Then we see that GL coefficients A and B are again given by (5.240) and (5.233), while the coefficient C is significantly changed. Calculating it for the vicinity of Anderson transition we have to take into account an important frequency dependence of the generalized diffusion coefficient (4.294), defined (in Matsubara formalism) by the second expression under the brackets in (5.246).

In metallic region, not very close to the mobility edge, we have $D_E(\omega_m) = D_E$,

---

[33] For low enough temperatures, of interest to us in traditional superconductors, we can limit ourselves with the discussion of residual resistivity only.
[34] In particular, for shortness, we use here E instead of E_F.

and coefficient C is determined as:

$$C = -i\pi T\nu_F \sum_{\varepsilon_n} \frac{\partial^2}{\partial q^2} \frac{1}{2i|\varepsilon_n| + iD_E q^2}\bigg|_{q=0}$$

$$= \pi T\nu_F D_E \sum_{\varepsilon_n} \frac{1}{2\varepsilon_n^2} = \frac{\nu_F D_E}{\pi T} \sum_{n\geq 0} \frac{1}{(2n+1)^2} = \frac{\pi}{8T}\nu_F D_E \qquad (5.247)$$

In insulating region, also not very close to the mobility edge, according to (5.246):

$$D_E(\omega_m) = D_E \frac{\omega_m}{\omega_m + 3D_E \omega_0^2/v_F^2} \qquad (5.248)$$

and we obtain (R_{loc} is localization radius, defined by (4.250)):

$$C = \frac{\pi}{2} T\nu_F \sum_{\varepsilon_n} \frac{1}{\varepsilon_n^2} D_E(2|\varepsilon_n|)$$

$$= \frac{\nu_F D_E}{2\pi T} \sum_n \frac{1}{(2n+1)^2 + (2n+1)3D_E\omega_0^2/2\pi T v_F^2}$$

$$= \frac{\nu_F v_F^2}{3\omega_0^2} \left[\psi\left(\frac{1}{2} + \frac{3D_E\omega_0^2}{4\pi T v_F^2} \right) - \psi\left(\frac{1}{2} \right) \right]$$

$$= \nu_F R_{loc}^2 \left[\psi\left(\frac{1}{2} + \frac{D_E}{4\pi T R_{loc}^2} \right) - \psi\left(\frac{1}{2} \right) \right]$$

$$\approx \nu_F R_{loc}^2 \ln \frac{1.78 D_E}{\pi T R_{loc}^2} \qquad (5.249)$$

where the approximate equality is valid until $D_E R_{loc}^{-2} \gg 4\pi T$.

In the immediate vicinity of the mobility edge, both in metallic and insulating regions, we can write (cf. (5.246)):

$$D_E(\omega_m) \approx D_0(\omega_m \tau)^{1/3} \approx (D_0 l)^{2/3} \omega_m^{1/3} \qquad (5.250)$$

so that

$$C = \frac{\pi}{2} T\nu_F \sum_{\varepsilon_n} \frac{1}{\varepsilon_n^2} D_E(2|\varepsilon_n|) \sim (D_0 l)^{2/3} T\nu_F \sum_{\varepsilon_n} \frac{1}{|\varepsilon_n|^{5/3}}$$

$$\sim \left(\frac{D_0 l}{T} \right)^{2/3} \nu_F \sum_{n\geq 0} \frac{1}{(2n+1)^{5/3}}$$

$$\sim \zeta\left(\frac{5}{3} \right) \nu_F \left(\frac{D_0 l}{T} \right)^{2/3} \qquad (5.251)$$

Expression (5.251) is dominating, in comparison to (5.247), in the region where:

$$D_E/T_c \approx D_0 l/\xi_{loc} T_c \leq D_0^{2/3}(l/T_c)^{2/3} \qquad (5.252)$$

Finally we obtain the following behavior of coefficient C on the way from "dirty" metal to Anderson insulator (L.N. Bulaevskii, M.V. Sadovskii, 1984):

$$C \equiv \nu_F \xi^2 \approx \nu_F \begin{cases} \frac{\pi}{8T_c} D_E & \text{for } \xi_{loc}(E) < (\xi_0 l^2)^{1/3} & \text{for } E > E_c \\ \left(\frac{D_0 l}{T_c} \right)^{2/3} \approx (\xi_0 l^2)^{2/3} & \text{for } \xi_{loc}(E) > (\xi_0 l^2)^{1/3} & E \sim E_c \\ R_{loc}^2(E) ln \frac{1.78 D_E}{\pi T_c R_{loc}^2(E)} & \text{for } R_{loc}(E) < (\xi_0 l^2)^{1/3} & E < E_c \end{cases}$$

$$(5.253)$$

where $\xi_0 = 0.18 v_F/T_c$ as usual, l is Drude mean free path.

As Fermi level moves towards the mobility edge E_c in metallic phase, correlation length of localization theory (4.279) ξ_{loc} grows, so that coefficient C at first diminishes along with generalized diffusion coefficient D_E, i.e. with suppression of conductivity of the system in normal state. However, in the vicinity of the Anderson transition, while $\sigma \to 0$, further suppression of coefficient C stops and it remains finite even for $E < E_c$, i.e. in the insulating phase. With further lowering of E within localization region (or with the growth of E_c with disorder) coefficient C is determined by localization length R_{loc}, which is diminishing as E moves deeper into localization region.

The finiteness of GL coefficient C in the vicinity of Anderson transition means that in this region the superconducting (Meissner) response of the system persists. Accordingly, in principle, at temperatures $T < T_c$ the system may transform from Anderson insulator to superconducting state (L.N. Bulaevskii, M.V. Sadovskii, 1984). Of course, analysis based upon the GL expansion and the simplest BCS-like model of superconducting pairing is not sufficient for a complete proof of such an exotic behavior of strongly disordered system. Note that all considerations were based on the concept of T_c being independent of disorder (Anderson "theorem"). We have noted above that this statement is valid in strongly disordered system (up to the Anderson transition) if we neglect disorder influence on pairing interaction itself. In real systems, the growth of disorder leads to the appropriate growth of effective Coulomb repulsion of electrons forming the Cooper pair [Sadovskii M.V. (2000)]. Thus, for more or less typical values of parameters, characterizing a superconductor, transition temperature T_c is completely suppressed long before the Anderson transition. However, under very restrictive conditions (e.g. if the initial value of T_c, when no disorder is present, is high enough) we may hope to find the finite values of T_c even in the immediate vicinity of Anderson transition (or even in insulating phase) [Sadovskii M.V. (2000)]. Unfortunately, in these lectures there is no time and place for further discussion of these interesting possibilities. Further details, as well as discussion of experimental situation can be found in [Sadovskii M.V. (2000)].

It is convenient to rewrite (5.253) using the relation between conductivity and generalized diffusion coefficient (e.g. (4.207)) and expressions (4.265) and (4.268). Then, using the definition of characteristic length ξ, from (5.253) we can easily obtain the following expression for temperature dependent coherence length $\xi(T)$ [De Gennes P.G. (1966)] of disordered superconductor:

$$\xi^2(T) = \frac{T_c}{T_c - T} \begin{cases} \xi_0 l \frac{\sigma}{\sigma + \sigma_c} & \sigma > \sigma^\star \quad (E > E_c) \\ (\xi_0 l^2)^{2/3} & \sigma < \sigma^\star \quad (E \sim E_c) \end{cases} \tag{5.254}$$

where in accordance with (4.265) $\sigma_c = e^2 p_F / (\pi^3 \hbar^2)$, while characteristic conductivity scale $\sigma^\star$ is defined as:

$$\sigma^\star \approx \sigma_c (p_F \xi_0)^{-1/3} \approx \sigma_c \left(\frac{T_c}{E}\right)^{1/3} \tag{5.255}$$

Thus, in the region of very small conductivity $\sigma < \sigma^\star$ the scale of coherence length $\xi(T)$ is determined not by $\xi \sim \sqrt{\xi_0 l}$, as in the usual theory of "dirty" superconductors, but by the new characteristic length $\xi \sim (\xi_0 l^2)^{1/3} \sim (\xi_0/p_F^2)^{1/3}$, which gives an estimate of the Cooper pair size in a superconductor in the vicinity of Anderson transition (L.N. Bulaevskii, M.V. Sadovskii, 1984).

The density of superconducting electrons n_s in GL theory can be defined as [De Gennes P.G. (1966)]:

$$n_s(T) = 8mC\Delta^2(T) = 8mC(-A)/B \tag{5.256}$$

Close to the Anderson transition this can be estimated as:

$$n_s \sim mN(E)\xi^2\Delta^2 \sim mp_F(\xi_0/p_F^2)^{2/3}\Delta^2 \sim n(T_c^{1/2}/E_F^2)^{2/3}(T_c - T) \tag{5.257}$$

where $n \sim p_F^3$ is the total electron density. If we take here $T \sim 0.5\, T_c$, we get the simple estimate:

$$n_s \sim n\left(\frac{T_c}{E_F}\right)^{4/3} \tag{5.258}$$

which, by the order of magnitude, is valid up to $T = 0$. Thus we can see that only a small fraction of electrons in strongly disordered superconductor remains superconducting. However, it confirms possibility of superconductivity in the vicinity of Anderson metal–insulator transition.

The value of conductivity $\sigma^\star$, defined by (5.255), determines the typical scale of conductivity, below which localization effects are significant for superconducting properties. While σ_c is of the order of Mott's "minimal metallic conductivity" [Mott N.F. (1974)], the value of $\sigma^\star$ is even smaller. However, for a superconductor with small enough Cooper pairs (which is typical for strong coupling and high-temperature superconductors) $\sigma^\star$ is more or less of the order of σ_c. Experimentally this can be determined as conductivity scale at which, with further growth of disorder, appear significant deviations from predictions of traditional theory of "dirty" superconductors.

Direct information on the value of coherence length $\xi^2(T)$ can be obtained from the measurements of the upper critical field (5.212). In particular, it is easy to convince oneself, that the use of (5.253) and (5.254) leads to the following generalization of Gorkov's relation (5.244) (L.N. Bulaevskii, M.V. Sadovskii, 1984):

$$-\frac{\sigma}{\nu_F}\left(\frac{dH_{c2}}{dT}\right)_{T_c} \approx \begin{cases} \frac{8e^2}{\pi^2\hbar}\phi_0 & \sigma > \sigma^\star \\ \phi_0\dfrac{\sigma}{\nu_F(\xi_0 l^2)^{2/3}T_c} \approx \phi_0\dfrac{\sigma}{[\nu_F T_c]^{1/3}} & \sigma < \sigma^\star \end{cases} \tag{5.259}$$

We see that for $\sigma < \sigma^\star$, i.e. close to the Anderson metal–insulator transition, the standard relation (5.244) becomes invalid and (assuming weak dependence of ν_F and T_c on disorder) the usual growth of the derivative $(dH_{c2}/dT)_{T_c}$ (the slope of $H_{c2}(T)$-curve) with the growth of disorder "saturates", and this slope becomes more or less independent of conductivity of the system in the normal state. Qualitatively, this behavior of the slope of the upper critical field is observed in a number of strongly disordered superconductors [Sadovskii M.V. (2000)].

5.6 Superconductors in electromagnetic field

In this section we shall consider the microscopic theory of electromagnetic response of superconductors, with the aim to understand important differences from the case of normal metals.

Let us start with a general formulation of the problem [Schrieffer J.R. (1964)]. Consider a superconductor in a weak external electromagnetic field, described by vector and scalar potentials $\mathbf{A}(\mathbf{r}t)$ and $\varphi(\mathbf{r}t)$, which can be combined in a single 4-vector $(x = (\mathbf{r}, t))$:

$$A_\mu(x) = \begin{cases} A_i(x) & \text{for} \quad \mu = i = 1, 2, 3 \\ c\varphi(x) & \text{for} \quad \mu = 0 \end{cases} \tag{5.260}$$

where we again "restored" the velocity of light c. In the first order in A_μ the interaction of electrons with electromagnetic field is given by:

$$H^p = -\frac{1}{c} \int d^3 r j_\mu^p(x) A_\mu(x) = -\frac{1}{c} \int d^3 r [\mathbf{j}^p(x) \mathbf{A}(x) - \rho_e(x) c\varphi(x)] \tag{5.261}$$

This is the interaction Hamiltonian with "paramagnetic" 4-vector of current density, which has the form:

$$j_\mu^p(x) = \begin{cases} \mathbf{j}^p(x) = \frac{e}{2m} i\{\psi^+(x)\nabla\psi(x) - [\nabla\psi^+(x)]\psi(x)\}, \mu = i = 1, 2, 3 \\ \rho_e(x) = -e\psi^+(x)\psi(x) = -\rho(x), \quad \mu = 0 \end{cases} \tag{5.262}$$

The total current density $j_\mu(x)$ in the presence of a vector-potential $\mathbf{A}$ is, as we have already noted above, the sum of paramagnetic and "diamagnetic" contributions:

$$j_\mu(x) = j_\mu^p(x) + j_\mu^d(x) \tag{5.263}$$

where the density of diamagnetic current is given by:

$$j_\mu^d(x) = \begin{cases} \frac{e}{mc}\rho_e(x)\mathbf{A}(x) & \text{for} \quad \mu = i = 1, 2, 3 \\ 0 & \text{for} \quad \mu = 0 \end{cases} \tag{5.264}$$

Then the total interaction of electrons with an external electromagnetic field can be written as:

$$H_{int} = H^p + H^d \tag{5.265}$$

where diamagnetic part of interaction is determined as:

$$H^d = -\frac{e}{2mc^2} \int d^3 r \rho_e(x) \mathbf{A}^2(x) \tag{5.266}$$

Now we can use interaction representation with H_{int} (assuming that $A_\mu \to 0$ for $t \to -\infty$), which defines the following change of the system ground state in time [Abrikosov A.A., Gorkov L.P., Dzyaloshinskii I.E. (1963)]:

$$|\Phi(t)>= T\exp\left\{-i\int_{-\infty}^t dt' H_{int}(t')\right\}|0>\equiv U(t,-\infty)|0> \quad (5.267)$$

Then the value of current density in the state $|\Phi(t)>$ is given by:

$$J_\mu(x) =< \Phi(t)|j_\mu(\mathbf{r}t)|\Phi(t)>=< 0|U^+(t,-\infty)j_\mu(\mathbf{r}t)U(t,-\infty)|0> \quad (5.268)$$

We are interested in first order in A_μ contributions to $J_\mu(x)$, so that after direct calculations we obtain:

$$J_\mu(x) = \frac{e}{mc} < 0|\rho_e(x)|0 > A_\mu(x)[1-\delta_{\mu 0}] - i < 0|[j_\mu^p(\mathbf{r}t),$$
$$\int_{-\infty}^\infty dt' H_{int}(t')]|0> \quad (5.269)$$

All terms of the zeroth order in $J_\mu(x)$ vanish, except $< j_0(x) > -$ the average charge density, which is of no interest to us and can be dropped. Using (5.261)–(5.266) we find that the linear response of the system J_μ to the external potential A_μ is nonlocal and expressed via integral kernel $K_{\mu\nu}$:

$$J_\mu(x) = -\frac{c}{4\pi}\int K_{\mu\nu}(\mathbf{r}t;\mathbf{r}'t')A_\nu(\mathbf{r}'t')d^3r'dt' \quad (5.270)$$

where time integration is done over the whole time axis, while the kernel is given by:

$$K_{\mu\nu}(x,x') = -\frac{4\pi i}{c^2} < 0|[j_\mu^p(x),j_\nu^p(x')]|0 > \theta(t-t')$$
$$-\frac{4\pi e}{mc^2} < 0|\rho_e(x)|0 > \delta(x-x')\delta_{\mu\nu}[1-\delta_{\nu 0}] \quad (5.271)$$

If our system is translationally invariant, the kernel $K_{\mu\nu}$ depends only on the difference of coordinates and it is convenient to use Fourier representation:

$$K_{\mu\nu}(\mathbf{q},t-t') = \int K_{\mu\nu}(x,x')e^{-i\mathbf{q}(\mathbf{r}-\mathbf{r}')}d^3rd^3r'$$
$$= -\frac{4\pi i}{c^2} < 0|[j_\mu^p(\mathbf{q}t),j_\nu^p(-\mathbf{q}t')]|0 > \theta(t-t')$$
$$+\frac{4\pi ne^2}{mc^2}\delta(t-t')\delta_{\mu\nu}(1-\delta_{\nu 0}) \quad (5.272)$$

where n is the total electron density. As the second term in (5.272) (diamagnetic response) is known exactly, we have to calculate only paramagnetic contribution (first term):

$$R_{\mu\nu}(\mathbf{q}t) = -i < 0|[j_{\mu}^{p}(\mathbf{q}t), j_{\nu}^{p}(-\mathbf{q}0)]|0 > \theta(t) \qquad (5.273)$$

This can be found with the help of T-ordered Green's function of the following form:[35]

$$P_{\mu\nu}(\mathbf{q}t) = -i < 0|Tj_{\mu}^{p}(\mathbf{q}t)j_{\nu}^{p}(-\mathbf{q}0)|0 > \qquad (5.274)$$

Comparing spectral representations for time Fourier transforms of (5.273) and (5.274) [Abrikosov A.A., Gorkov L.P., Dzyaloshinskii I.E. (1963)], it is not difficult to convince yourself, that the real parts of $R_{\mu\nu}$ and $P_{\mu\nu}$ coincide, while imaginary parts differ by sign for $\omega < 0$:

$$Re P_{\mu\nu}(\mathbf{q}\omega) = Re R_{\mu\nu}(\mathbf{q}\omega) \qquad (5.275)$$

$$Im P_{\mu\nu}(\mathbf{q}\omega) = sign\omega Im R_{\mu\nu}(\mathbf{q}\omega) \qquad (5.276)$$

which gives us the required expression for $R_{\mu\nu}$ via $P_{\mu\nu}$. As $K_{\mu\nu}$ contains only system parameters in the absence of an external vector-potential A_{μ}, operators j_{μ} and j_{μ}^{p} just coincide and in the following we drop the index p.

Finally, the response of the system to an external vector-potential $A_{\mu}(q) = [\mathbf{A}(q), c\varphi(q)]$ (where $q = [\mathbf{q}, \omega]$) takes the form:

$$J_{\mu}(q) = -\frac{c}{4\pi} K_{\mu\nu}(q) A_{\nu}(q) = -\frac{c}{4\pi} \left[\sum_{i=1}^{3} K_{\mu i}(q) A_i(q) - K_{\mu 0}(q) c\varphi(q) \right]$$
$$(5.277)$$

where:

$$K_{\mu\nu}(q) = \frac{4\pi}{c^2} R_{\mu\nu}(q) + \frac{1}{\lambda_L^2} \delta_{\mu\nu}[1 - \delta_{\nu 0}] \qquad (5.278)$$

and we defined $\lambda_L^2 = \frac{mc^2}{4\pi ne^2}$ — the square of London penetration depth at $T = 0$. Two terms in (5.278) reflect contributions of paramagnetic and diamagnetic currents.

Introduce now the following shortened notation:

$$Q_{\alpha\beta}(q) = -\frac{c}{4\pi} K_{\alpha\beta}(q) \qquad (5.279)$$

---

[35]Similar Green's function was used by us (in Matsubara technique) in (4.108).

Consider first an ideal Fermi-gas (normal metal) [Levitov L.S., Shitov A.V. (2003)]. Let us determine the response (current density **j**) to spatially inhomogeneous static vector-potential[36] **A**. In momentum representation, the linear relation between **j** and **A** can be written as:

$$\mathbf{j_q} = \hat{Q}(\mathbf{q})\mathbf{A_q} \qquad (5.280)$$

In the previous Chapter we have already noted that the static vector-potential does not lead to the appearance of electric current in a normal metal (or ideal Fermi-gas) in the long wavelength limit. This means that $\hat{Q}(\mathbf{q} = 0) = 0$, so that at $\mathbf{q} = 0$ diamagnetic and paramagnetic contributions to (5.278) completely compensate each other. At the same time, in the limit of $\mathbf{q} \to 0$, and for $\omega = 0$, the response contains a small contribution due to Landau diamagnetism. Consider this in more details. Calculations will be done for the case of finite temperatures, using Matsubara formalism. From the analysis made in the previous Chapter concerning (4.108)–(4.120), after the obvious change of notations, it becomes clear that for an ideal Fermi-gas the paramagnetic contribution to the response kernel[37] can be written as:

$$Q^p_{\alpha\beta}(\mathbf{q}=0)=-\lim_{\tau \to 0} 2\frac{e^2}{m^2c}T\sum_n e^{i\varepsilon_n\tau}\int v\frac{p_\alpha p_\beta}{(i\varepsilon_n - \xi(\mathbf{p}))(i\varepsilon_n - \xi(\mathbf{p}))}\frac{d^3p}{(2\pi)^3} \qquad (5.281)$$

Now use the obvious relation, valid for the free electron Green's function:

$$\mathbf{p}G_0^2(\varepsilon_n\mathbf{p}) = m\nabla_{\mathbf{p}}G_0(\varepsilon_n\mathbf{p}) \qquad (5.282)$$

and write (5.281) as:

$$Q^p_{\alpha\beta}(0) = -\frac{e^2}{mc}2T\sum_n \int p_\alpha \nabla_{p_\beta} G_0(\varepsilon_n\mathbf{p})\frac{d^3p}{(2\pi^3)} \qquad (5.283)$$

Integration over d^3p can be done in parts, so that ∇_{p_β} acts on p_α, and leading to:

$$Q^p_{\alpha\beta}(0) = \frac{e^2}{mc}\delta_{\alpha\beta}2T\sum_n \int G_0(\varepsilon_n\mathbf{p})\frac{d^3p}{(2\pi)^3} = \frac{ne^2}{mc}\delta_{\alpha\beta} \qquad (5.284)$$

which totally compensates diamagnetic part of (5.278).

---

[36] We assume the gauge choice: $div\mathbf{A} = 0, \varphi = 0$.

[37] For an impure metal everything can be done in a similar way, but we have to take into account the finite damping in denominators of Green's functions due to impurity scattering. Vertex corrections due to impurity scattering in "current" vertices just vanish due to angular integrations (in the model with point-like impurities).

Consider now the case of $\mathbf{q} \neq 0$. Expansion of $\hat{Q}(\mathbf{q})$ in powers of $\mathbf{q}$ starts from the second order: $Q(\mathbf{q}) = aq^2 + O(q^4)$, which is easily checked by an expansion of diagram shown in Fig. 5.14. Then we can write:

$$\mathbf{j} = \hat{Q}\mathbf{A} = -a\nabla^2\mathbf{A} \qquad (5.285)$$

On the other hand, as we know from electrodynamics, the current may be related to magnetization of the system as:

$$\mathbf{j} = c \, rot\mathbf{M} = c \, rot\chi_0\mathbf{B} = \chi_0 c \, rot \, rot\mathbf{A} \qquad (5.286)$$

so that using $rot \, rot\mathbf{A} = grad \, div\mathbf{A} - \nabla^2\mathbf{A}$ and gauge condition $div\mathbf{A} = 0$ we may write:

$$\mathbf{j} = -\chi_0 c\nabla^2\mathbf{A} \qquad (5.287)$$

where χ_0 is the magnetic susceptibility of electron gas. Then, comparing (5.285) with (5.287), we find: $\chi_0 = \frac{a}{c}$, where coefficient a can be calculated from the diagram shown in Fig. 5.14. In fact, from this diagram we have:

$$Q_{xx} = -\lim_{\tau \to 0} 2\frac{e^2}{c}T\sum_n e^{i\varepsilon_n\tau} \int \frac{v_x(\mathbf{p}+\mathbf{q})v_x(\mathbf{p})}{(i\varepsilon_n - \xi(\mathbf{p}+\mathbf{q}))(i\varepsilon_n - \xi(\mathbf{p}))} \frac{d^3p}{(2\pi)^3}$$
$$(5.288)$$

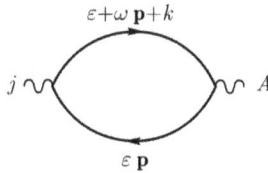

Fig. 5.14 Diagram determining the response of an ideal Fermi-gas to an external vector-potential.

where $v_x(\mathbf{p}+\mathbf{q}) = \partial\xi(\mathbf{p}+\mathbf{q})/\partial p_x$ and $v_x(\mathbf{p}) = \partial\xi(\mathbf{p})/\partial p_x$ are appropriate velocity projections. Let us assume that the wave vector $\mathbf{q}$ is directed along z-axis. In contrast to spin susceptibility, orbital susceptibility is determined not only by electrons from the vicinity of the Fermi surface, but by all electrons inside it, as Landau quantization influence all electrons with $E < E_F$, and the finite value of susceptibility χ_0 appears due to the difference between the usual integral over continuous spectrum and the sum over quantized Landau levels. Let us limit ourselves to the case of free electrons, when $\xi(\mathbf{p}) = \frac{p^2}{2m} - E_F$. Expand Q_{xx} in powers of $q = q_z$ and take into account $v_x(\mathbf{p} + \mathbf{q}) = v_x(\mathbf{p}) = p_x/m$, so that:

$$Q_{xx} = -\lim_{\tau \to 0} \frac{e^2 q^2}{c} T \sum_n e^{i\varepsilon_n \tau} \int \frac{v_x^2}{i\varepsilon_n - \xi(\mathbf{p})} \frac{\partial^2}{\partial p_z^2} \frac{1}{i\varepsilon_n - \xi(\mathbf{p})} \frac{d^3 p}{(2\pi)^3} \tag{5.289}$$

Now:

$$\frac{\partial^2}{\partial p_z^2} \frac{1}{i\varepsilon_n - \xi(\mathbf{p})} = \frac{1}{m} \frac{1}{(i\varepsilon_n - \xi(\mathbf{p}))^2} + 2\frac{p_z^2}{m^2} \frac{1}{(i\varepsilon_n - \xi(\mathbf{p}))^3}$$

and we immediately obtain:

$$\chi_0 = -\lim_{\tau \to 0} \frac{e^2}{c^2} T \sum_n e^{i\varepsilon_n \tau} \int \left\{ \frac{1}{m} \frac{v_x^2}{(i\varepsilon_n - \xi(\mathbf{p}))^3} + 2\frac{v_x^2 v_z^2}{(i\varepsilon_n - \xi(\mathbf{p}))^4} \right\} \frac{d^3 p}{(2\pi)^3} \tag{5.290}$$

Angular averaging gives $<v_x^2> = v^2/3$ and $<v_x^2 v_z^2> = v^4/15$, and after the double partial integration in the first term of (5.290):

$$\int_0^\infty \frac{p^4}{(i\varepsilon_n + E_F - p^2/2m)^3} dp = \frac{3m^2}{2} \int_0^\infty \frac{dp}{i\varepsilon_n + E_F - p^2/2m} \tag{5.291}$$

and after triple partial integration in the second term,

$$\int_0^\infty dp \frac{p^6}{(i\varepsilon_n + E_F - p^2/2m)^4} = -\frac{5m^3}{2} \int_0^\infty \frac{dp}{i\varepsilon_n + E_F - p^2/2m} \tag{5.292}$$

Then, summation over n in (5.290) gives just the Fermi distribution [Abrikosov A.A., Gorkov L.P., Dzyaloshinskii I.E. (1963)] and from the first term in (5.290) we get $-\frac{e^2 p_F}{4\pi^2 mc^2}$, while from the second we have $+\frac{e^2 p_F}{6\pi^2 mc^2}$ so that the sum of these contributions gives the final result:

$$\chi_0 = -\frac{e^2}{12\pi^2 mc^2} \int_0^\infty dp n(\xi(\mathbf{p})) = -\frac{e^2 p_F}{12\pi^2 mc^2} \tag{5.293}$$

which is the usual diamagnetic susceptibility of free electron gas [Sadovskii M.V. (2019a)]. If we remember a similar result for Pauli (spin) susceptibility:

$$\chi_p = 2\mu_B^2 \nu_F = \frac{e^2}{4m^2 c^2} \frac{m p_F}{\pi^2} \tag{5.294}$$

we obtain the well known result for total susceptibility [Sadovskii M.V. (2019a)]:

$$\chi_0 = -\frac{1}{3}\chi_p \tag{5.295}$$

Let us return now to superconductors and calculate again the response to static vector-potential. We shall see that in contrast to the normal metal (free Fermi-gas), there will be no complete compensation of paramagnetic and diamagnetic contributions in this case. Thus, even if we take $\mathbf{q} = 0$ in (5.280), we get the finite response, which is written as:

$$\mathbf{j}(\mathbf{r}) = -\frac{n_s e^2}{mc}\mathbf{A}(\mathbf{r}) \tag{5.296}$$

and which is called London equation — basic equation of electrodynamics of superconductors. Here n_s represents (by definition) the density of superconducting (superfluid) electrons. Eq. (5.296), "obviously" breaks gauge invariance. The physical reason for the absence of total compensation of diamagnetic and paramagnetic contributions in response kernel $Q(0)$ is precisely due to the fact (already noted above), that gauge invariance (charge conservation) in superconductors is *spontaneously broken*. This is the symmetry (of the ground state!) which is lowered when the system goes superconducting. In a normal metal the zero value of $Q(0)$ is guaranteed by the validity of the Ward identity, which is directly related to charge conservation and gauge invariance [Sadovskii M.V. (2019b)]. It is not so (simple) for superconductors, and special care is needed to "restore" gauge invariance of electromagnetic response absent in a simple BCS-approach.

Now, let us directly calculate $Q(0)$ for a superconductor. Diamagnetic contribution is obviously the same as before and can be written as:

$$Q_{\alpha\beta}^{(2)}(0) = -\frac{ne^2}{mc}\delta_{\alpha\beta} \tag{5.297}$$

But in calculations of paramagnetic contribution, in contrast to the case of the normal metal, we have to take into account contributions to $Q(0)$ both from normal and anomalous Green's functions. Paramagnetic contribution $Q_{\alpha\beta}^{(1)}(\omega, \mathbf{q})$ is again obtained (for finite temperatures, in Matsubara formalism) by averaging the product of two current operators and is given by the following expression (σ, σ' are spin indices):

$$Q^{(1)}_{\alpha\beta}(\omega_m, \mathbf{q}) = \frac{1}{2c} \sum_{\sigma\sigma'} \int_{-\beta}^{\beta} e^{i\omega_m \tau} \int d^3 r e^{i\mathbf{q}\mathbf{r}}$$

$$\times < T_\tau \psi^+_\sigma(\mathbf{r}\tau) \hat{j}_\alpha \psi_\sigma(\mathbf{r}\tau) \psi^+_{\sigma'}(00) \hat{j}_\beta \psi_{\sigma'}(00) > \qquad (5.298)$$

where $\hat{j} = -i\frac{e}{m}\nabla_\mathbf{r}$. Now we have to consider all the pairing of ψ-operators, taking into account both normal and anomalous averages. Then we obtain:

$$< T_\tau \psi^+_\sigma(\mathbf{r}\tau) \hat{j}_\alpha \psi_\sigma(\mathbf{r}\tau) \psi^+_{\sigma'}(00) \hat{j}_\beta \psi_{\sigma'}(00) >$$
$$= -\hat{j}_\alpha G_{\sigma\sigma'}(\mathbf{r}\tau) \hat{j}_\beta G_{\sigma'\sigma}(-\mathbf{r}, -\tau) - \hat{j}_\alpha F^+_{\sigma\sigma'}(\mathbf{r}\tau) \hat{j}_\beta F_{\sigma'\sigma}(-\mathbf{r}, -\tau) \quad (5.299)$$

so that $Q_{\alpha\beta}(\omega_m \mathbf{q})$ takes the form:

$$Q^{(1)}_{\alpha\beta}(\omega_m, \mathbf{q}) = \frac{1}{2c} \sum_{\sigma\sigma'} \int_{-\beta}^{\beta} e^{i\omega_m \tau} \int d^3 r e^{i\mathbf{q}\mathbf{r}} \{ 2\hat{j}_\alpha G(\mathbf{r}\tau) \hat{j}_\beta G(-\mathbf{r}, -\tau)$$

$$+ 2\hat{j}_\alpha F(\mathbf{r}\tau) \hat{j}_\beta F^*(-\mathbf{r}, -\tau) \} \qquad (5.300)$$

Note that the signs of loops containing G and F are here the same.[38] Rewriting (5.300) in momentum representation we have:

$$Q_{\alpha\beta}(\omega_m, \mathbf{q}) = -2\frac{e^2}{c} T \sum_n \int \frac{d^3 p}{(2\pi)^3} v_\alpha \left(\mathbf{p} - \frac{\mathbf{q}}{2}\right) v_\beta \left(\mathbf{p} + \frac{\mathbf{q}}{2}\right)$$

$$\times \left\{ G\left(\varepsilon_n \mathbf{p} - \frac{\mathbf{q}}{2}\right) G\left(\varepsilon_n + \omega_m \mathbf{p} + \frac{\mathbf{q}}{2}\right)\right.$$

$$\left. + F\left(\varepsilon_n \mathbf{p} - \frac{\mathbf{q}}{2}\right) F^*\left(\varepsilon_n + \omega_m \mathbf{p} + \frac{\mathbf{q}}{2}\right) \right\} \qquad (5.301)$$

Then, for $\omega_m = 0$, $\mathbf{q} = 0$ we have:

$$Q^{(1)}_{\alpha\alpha}(0) = -\frac{2e^2}{3m^2 c} T \sum_n \int \frac{d^3 p}{(2\pi)^3} p^2 \{ G^2(\varepsilon_n \mathbf{p}) + |F(\varepsilon_n \mathbf{p})|^2 \} \quad (5.302)$$

Combining (5.297) and (5.302) together, using the definition of n_s, following from (5.296), and also the explicit form of Gorkov's functions (5.111), (5.112), we obtain:

$$n - n_s = -\frac{2\nu_F p^2_F}{3m} T \sum_n \int_{-\infty}^{\infty} d\xi \frac{\Delta^2 + \xi^2 - \varepsilon^2_n}{(\varepsilon^2_n + \xi^2 + \Delta^2)^2} \qquad (5.303)$$

---

[38]In the next Chapter we shall see that the situation is opposite for loops with scalar (not vector or "current") vertices.

This expression is not fully satisfactory. Total electron density is determined by all electronic states, including those far from the Fermi surface. Thus, the usual integration over the linearized spectrum ξ, used in Eq. (5.303), is not, strictly speaking, justified. At the same time it is clear that density of superconducting electrons n_s is determined by the close vicinity of the Fermi surface, as only there an important transformation of electronic spectrum takes place in BCS theory. Deep inside the Fermi sphere (i.e. for $\xi \sim E_F$) nothing happens at all. Thus, the correct procedure is to subtract from (5.303) the same expression for the normal metal, i.e. with $\Delta = 0$. Then the contribution of deep states just drops out. This will also guarantee the obvious requirement of n_s being zero for $\Delta \to 0$. Finally we obtain:

$$n_s = -\frac{2\nu_F p_F^2}{3m} T \sum_n \int_{-\infty}^{\infty} d\xi \left[\frac{\xi^2 - \varepsilon_n^2}{(\xi^2 + \varepsilon_n^2)^2} - \frac{\Delta^2 + \xi^2 - \varepsilon_n^2}{(\varepsilon_n^2 + \xi^2 + \Delta^2)^2} \right] \quad (5.304)$$

Now we can perform ξ-integrations using the standard integrals:

$$\int_{-\infty}^{\infty} \frac{dx}{(x^2 + a^2)^2} = \frac{\pi}{2a^3}, \qquad \int_{-\infty}^{\infty} \frac{dx\, x^2}{(x^2 + a^2)^2} = \frac{\pi}{2a} \quad (5.305)$$

Then we obtain:

$$n_s = \frac{2\nu_F p_F^2}{3m} T \sum_n \frac{\pi \Delta^2}{(\varepsilon_n^2 + \Delta^2)^{3/2}} \quad (5.306)$$

Let us analyze now the different limiting cases.

- The case of $T \to 0$. Here we can replace summation by integration and obtain:

$$n_s = \frac{2\nu_F p_F^2}{3m} \frac{\pi \Delta^2}{2\pi} \int_{-\infty}^{\infty} \frac{d\varepsilon}{(\varepsilon^2 + \Delta^2)^{3/2}} = \frac{2\nu_F p_F^2}{3m} = \frac{p_F^2}{3\pi^2} = n \quad (5.307)$$

  We see that at $T = 0$ superfluid density is equal to total density of electrons, as it should be in translationally invariant system.

- The case of $T \to T_c$. Now we can neglect Δ^2 in the denominator of (5.306). The remaining sum over frequencies is already known to us, it is expressed via ζ-functions, and we obtain:

$$\frac{n_s}{n} = \frac{7\zeta(3)\Delta^2}{4\pi^2 T_c^2} = 2 \left(1 - \frac{T}{T_c} \right) \quad (5.308)$$

  so that for $T \to T_c$ superfluid density goes to zero.

Impurity scattering suppresses superfluid density n_s, but it remains finite. To understand it in more details, we calculate $Q(0)$ for an impure system. It is sufficient to calculate the loop diagrams, built upon the impurity averaged Gorkov's functions, which are obtained from those of the "clean" superconductor, given by (5.111) and (5.112), via the replacement (5.186). Vertex corrections (diagrams with impurity lines connecting different Green's functions in the loop) can be dropped, as they vanish due to angular integration (vector nature of vertices, describing interaction with external electromagnetic field). Thus, an expression for n_s can be immediately written as (cf. (5.303)):

$$\frac{n - n_s}{n} = -T \int_{-\infty}^{\infty} \frac{\xi^2 + \tilde{\Delta}^2 - \tilde{\varepsilon}_n^2}{(\xi^2 + \tilde{\Delta}^2 + \tilde{\varepsilon}_n^2)^2} \tag{5.309}$$

where $\tilde{\varepsilon}_n$ and $\tilde{\Delta}$ were defined in (5.185), (5.186). Let us again subtract from this expression its value with $\Delta = 0$ and in the absence of impurities, to exclude contribution of deep levels under the Fermi sphere. Then, after the integration over ξ, we obtain the following generalization of (5.306):

$$\frac{n_s}{n} = \pi T \sum_n \frac{\tilde{\Delta}^2}{(\tilde{\varepsilon}_n^2 + \tilde{\Delta}^2)^{3/2}} \tag{5.310}$$

or, with the account of (5.185), (5.186):

$$\frac{n_s}{n} = \pi T \sum_n \frac{\Delta^2}{(\varepsilon_n^2 + \Delta^2)^{3/2} \left(1 + \frac{1}{2\tau\sqrt{\varepsilon_n^2 + \Delta^2}}\right)} \tag{5.311}$$

For small impurity concentration, when $\Delta_0 \tau \gg 1$, this expression reduces to (5.306), while in the "dirty" limit, when $\Delta_0 \tau \ll 1$, it gives:

$$\frac{n_s}{n} = 2\pi\tau T \sum_n \frac{\Delta^2}{(\varepsilon_n^2 + \Delta^2)^2} = \pi\tau\Delta \, th\frac{\Delta}{2T} \tag{5.312}$$

Thus in a "dirty" superconductor, even for $T \to 0$, we have:

$$\frac{n_s(T \to 0)}{n} = \pi\tau\Delta_0 \ll 1 \tag{5.313}$$

i.e. only a small fraction of electrons is superconducting.

Let us continue the general discussion after the return to our initial notations (5.270)–(5.278) and putting $c = 1$ for shortness. In isotropic system, electromagnetic response kernel reduces to $K_{\mu\nu} = K\delta_{\mu\nu}$. The previous discussion can be summarized as follows. To obtain superconducting response it is necessary (and sufficient) to satisfy the following requirements:[39]

$$\lim_{q\to 0}\lim_{\omega\to 0} K(\mathbf{q}\omega) = \lim_{\omega\to 0}\lim_{q\to 0} K(\mathbf{q}\omega) = K(0,0) \neq 0 \qquad (5.316)$$

where, according to (5.306), we have:

$$K(0,0) = \frac{1}{\lambda_L^2}\frac{n_s}{n} = \frac{4\pi n_s e^2}{m} = \frac{1}{\lambda_L^2} 2\pi T \sum_n \frac{\Delta^2}{(\varepsilon_n^2 + \Delta^2)^{3/2}} \qquad (5.317)$$

with $\lambda_L^2 = \frac{mc^2}{4\pi n e^2}$ — the usual definition of the square of London penetration depth at zero temperature $T = 0$.

From (5.270) we get:

$$J_\mu(\mathbf{q}\omega) = -\frac{1}{4\pi}K_{\mu\nu}(\mathbf{q}\omega)A_{\mathbf{q}\omega}^\nu \qquad (5.318)$$

so that, taking into account the definition of electric field:

$$\mathbf{E} = -\frac{\partial\mathbf{A}}{\partial t} \qquad \mathbf{E}_{\mathbf{q}\omega} = i\omega\mathbf{A}_{\mathbf{q}\omega} \qquad (5.319)$$

we can introduce conductivity as:

$$\sigma_{\mu\nu}(\mathbf{q}\omega) = -\frac{1}{4\pi i\omega}K_{\mu\nu}(\mathbf{q}\omega) \qquad \sigma(\mathbf{q}\omega) = -\frac{1}{4\pi i\omega}K(\mathbf{q}\omega) \qquad (5.320)$$

where the second equality is valid for isotropic case. In most cases we are interested in the limit $\mathbf{q} \to 0$ (response to homogeneous field), which is assumed in what follows.

---

[39]There are two contributions to $K_{\mu\nu}(\mathbf{q}\omega)$: paramagnetic one $K_{\mu\nu}^p$ and diamagnetic $K_{\mu\nu}^d$. Diamagnetic contribution is proportional to the total density of electrons and is the same as in the normal state. In normal state the total current induced by static vector-potential is negligible (and determines only the the small contribution of Landau diamagnetism). Thus, with high accuracy we have: $K_{\mu\nu}^n(\mathbf{q}0) = K_{\mu\nu}^{np}(\mathbf{q}0) + K_{\mu\nu}^d \approx 0$, $K_{\mu\nu}^{np}(\mathbf{q}0) \approx -K_{\mu\nu}^d$ (exact equality holds for $\mathbf{q} \to 0$). Then the current density in a superconductor is given by:

$$J_\mu(\mathbf{q}\omega) = -\frac{1}{4\pi}\{K_{\mu\nu}^{sp}(\mathbf{q}\omega) - K_{\mu\nu}^{np}(\mathbf{q}0)\}A_{\mathbf{q}\omega}^\nu \qquad (5.314)$$

Thus, as noted above, we have only to calculate the paramagnetic response. The difference of current densities in superconducting and normal states can be written as:

$$J_\mu^s(\mathbf{q}\omega) - J_\mu^n(\mathbf{q}\omega) = \{K_{\mu\nu}^s(\mathbf{q}\omega) - K_{\mu\nu}^n(\mathbf{q}\omega)\}A_{\mathbf{q}\omega}^\nu$$
$$= \{K_{\mu\nu}^{sp}(\mathbf{q}\omega) - K_{\mu\nu}^{np}(\mathbf{q}\omega)\}A_{\mathbf{q}\omega}^\nu \qquad (5.315)$$

It is convenient to write:

$$\sigma(\omega) = \sigma_s(\omega) + \sigma_{exc}(\omega) \tag{5.321}$$

where

$$\sigma_s(\omega) = -\frac{K(0)}{4\pi i \omega} \tag{5.322}$$

is conductivity of superconducting condensate, while

$$\sigma_{exc}(\omega) = -\frac{1}{4\pi}\frac{K(\omega) - K(0)}{i\omega} \tag{5.323}$$

is conductivity due to single-particle excitations. We see that (5.322), with the use of (5.317), is reduced to:

$$\sigma_s(\omega) = \frac{n_s e^2}{m}\frac{i}{\omega + i\delta}, \qquad \delta \to +0 \tag{5.324}$$

so that:

$$Re\,\sigma_s(\omega) = \frac{n_s e^2}{m}\pi\delta(\omega) \tag{5.325}$$

which corresponds to dissipationless contribution of condensate into conductivity (i.e. to superconductivity itself).

Let us consider now in detail $\sigma_{exc}(\omega)$, i.e. conductivity due to single-particle excitations in superconductor, which, in particular, determines the absorption of electromagnetic energy by a superconductor at finite frequencies (optical properties of a superconductor). In fact, we have to return to (5.300), (5.301) and repeat all calculation for the case of a finite frequency of external field $i\omega_m \to \omega + i\delta$ and $\mathbf{q} \to 0$. We have:

$$Q_{\alpha\alpha}(\omega_m, \mathbf{q}) = -2\frac{e^2}{3m^2}T\sum_n \int \frac{d^3p}{(2\pi)^3}\left(\mathbf{p} + \frac{\mathbf{q}}{2}\right)^2$$
$$\times \left\{G\left(\varepsilon_n \mathbf{p} - \frac{\mathbf{q}}{2}\right)G\left(\varepsilon_n + \omega_m \mathbf{p} + \frac{\mathbf{q}}{2}\right)\right.$$
$$\left. + F\left(\varepsilon_n \mathbf{p} - \frac{\mathbf{q}}{2}\right)F^*\left(\varepsilon_n + \omega_m \mathbf{p} + \frac{\mathbf{q}}{2}\right)\right\} \tag{5.326}$$

Now substitute here explicit expressions for Gorkov's functions, perform summation over frequencies and introduce integration over the linearized spectrum $\xi_p \equiv \xi(\mathbf{p})$ in the vicinity of the Fermi surface, where $p \approx p_F$ and can be taken out of integral. These calculations are rather cumbersome and we just skip them.[40] Finally, after the usual continuation $i\omega_m \to \omega + i\delta$, we get:

$$Q_{\alpha\alpha}(\omega_m, \mathbf{q}) = -\frac{2e^2}{3m^2}\nu_F p_F^2 \int_{-\infty}^{\infty} d\xi_p \left\{ \frac{\varepsilon_p \varepsilon_{p+q} + \xi_p \xi_{p+q} + \Delta^2}{\varepsilon_p \varepsilon_{p+q}} \right\} \{n(\varepsilon_p) - n(\varepsilon_{p+q})\}$$

$$\times \left\{ \frac{1}{\varepsilon_p - \varepsilon_{p+q} + \omega + i\delta} + \frac{1}{\varepsilon_p - \varepsilon_{p+q} - \omega - i\delta} \right\}$$

$$+\frac{2e^2}{3m^2}\nu_F p_F^2 \int_{-\infty}^{\infty} d\xi_p \left\{ \frac{\varepsilon_p \varepsilon_{p+q} - \xi_p \xi_{p+q} - \Delta^2}{\varepsilon_p \varepsilon_{p+q}} \right\} \{1 - n(\varepsilon_p) - n(\varepsilon_{p+q})\}$$

$$\times \left\{ \frac{1}{\varepsilon_p + \varepsilon_{p+q} + \omega + i\delta} + \frac{1}{\varepsilon_p + \varepsilon_{p+q} - \omega - i\delta} \right\} \tag{5.327}$$

where we have introduced the usual notation $\varepsilon_p = \sqrt{\xi_p^2 + \Delta^2}$ for electron spectrum in BCS model (5.68), $n(\varepsilon_p)$ is Fermi distribution with this spectrum. In the limit of $T \to 0$ only the second term contributes, as $n(\varepsilon_p) \to 0$ for $T \to 0$ due to the gap in BCS spectrum. At the same time we are interested in the limit of $\mathbf{q} \to 0$, when actually goes to zero numerator (so-called coherence factor) of the expression, standing in the first brackets of this term. Thus we obtain zero, and all conductivity, in fact, is reduced to (5.322), (5.324). In particular, the real part of conductivity is concentrated in δ-function contribution at zero frequency (5.325), with $n_s = n$, i.e. all electrons contribute to this dissipationless motion of condensate.

To understand what has happened remember, that we are analyzing the response at finite frequency of a superconductor without impurity (or any other) scattering, leading to current dissipation. The velocity of i-th electron in an external field is determined as: $m\mathbf{v}_i = \mathbf{p}_i - \frac{e}{c}\mathbf{A}(\mathbf{r}_i)$. Then the total current is equal to:

$$\mathbf{J} = e\sum_i < \mathbf{v}_i > = \frac{e}{m}\sum_i < \mathbf{p}_i > -\frac{e^2}{mc}\sum_i \mathbf{A}(\mathbf{r}_i) \tag{5.328}$$

The first term here is proportional to the applied electric field $\mathbf{E}(\mathbf{r}t) = \mathbf{E}e^{i\mathbf{q}\mathbf{r} - i\omega t}$, so that:

$$\mathbf{J}(\mathbf{r}t) = \sigma(\omega)\mathbf{E}(\mathbf{r}t) - \frac{e^2}{mc}n\mathbf{A}(\mathbf{r}t) \tag{5.329}$$

where in writing the second term we assumed the homogeneity of the system. With the account of $\mathbf{E} = -\frac{1}{c}\frac{\partial \mathbf{A}(\mathbf{r}t)}{\partial t}$, we have $\mathbf{A} = \frac{ic}{\omega}\mathbf{E}$. Then London equation

---

[40]In the next Chapter we shall perform similar calculations (for another problem) in all details.

(5.296) with $n_s = n$ (at $T = 0$) directly follows from (5.329), if $\sigma(\omega) = 0$. It is precisely what was obtained in (5.327).

Thus, consistent analysis of conductivity in a superconductor should be done with the account of e.g. impurity scattering (S.B. Nam, 1967). These calculations are also very cumbersome, and we shall use instead much simplified arguments, which give correct answer in the "dirty" limit [Mahan G.D. (1981)]. Let us use the definition of conductivity (cf. (5.320), (5.279)) in the following form:

$$\sigma(\omega) = \lim_{q \to 0} \frac{Q(q\omega)}{i\omega}, \qquad Re\sigma(\omega) = \frac{1}{\omega} ImQ(0\omega) \qquad (5.330)$$

and consider the imaginary part of (5.327) for $\omega > 0$. Besides, let us take into account that in "dirty" superconductor momenta $\mathbf{p}$ and $\mathbf{p} + \mathbf{q}$ are not well defined quantum number for an electron. Then we may assume that in this limit both $\xi_{\mathbf{p}}$ and $\xi_{\mathbf{p}+\mathbf{q}}$ in Eq. (5.327) can be considered as *independent* variables, so that we can write:

$$
\begin{aligned}
Re\sigma(\omega) &= \frac{C_0}{\omega} \int_{-\infty}^{\infty} d\xi_p \int_{-\infty}^{\infty} d\xi_p' \delta(\omega - \varepsilon_p - \varepsilon_{p'}) \frac{\varepsilon_p \varepsilon_{p'} - \xi_p \xi_{p'} - \Delta^2}{\varepsilon_p \varepsilon_{p'}} \\
&= \frac{C_0}{\omega} \int_{\Delta}^{\infty} d\varepsilon \int_{\Delta}^{\infty} d\varepsilon' N(\varepsilon) N(\varepsilon') \delta(\omega - \varepsilon - \varepsilon') \left(1 - \frac{\Delta^2}{\varepsilon \varepsilon'} \right)
\end{aligned}
$$
$$(5.331)$$

where in the last equality we have changed integration variable from ξ to BCS spectrum (5.68) ε and, accordingly, introduced BCS density of states, defined by Eq. (5.79). In Eq. (5.331) we also introduced a constant C_0, which will be determined from matching with conductivity of a normal metal (when $\Delta \to 0$). From (5.331), performing δ-function integration, we obtain:

$$
\begin{aligned}
Re\sigma(\omega) &= \frac{C_0}{\omega} \theta(\omega - 2\Delta) \int_{\Delta}^{\omega-\Delta} d\varepsilon N(\varepsilon) N(\omega - \varepsilon) \left\{ 1 - \frac{\Delta^2}{\varepsilon(\omega - \varepsilon)} \right\} \\
&= \frac{C_0}{\omega} \theta(\omega - 2\Delta) \int_{\Delta}^{\omega-\Delta} d\varepsilon \frac{\varepsilon(\omega - \varepsilon) - \Delta^2}{(\varepsilon^2 - \Delta^2)^{1/2}[(\omega - \varepsilon)^2 - \Delta^2]^{1/2}}
\end{aligned}
$$
$$(5.332)$$

Introducing new integration variable x via $2\varepsilon = \omega + x(\omega - 2\Delta)$, we have:

$$Re\sigma(\omega) = \frac{1}{2} \frac{C_0}{\omega} \theta(\omega - 2\Delta)(\omega - 2\Delta) \int_{-1}^{1} dx \frac{1 - \alpha x^2}{[(1 - x^2)(1 - \alpha^2 x^2)]^{1/2}}$$
$$(5.333)$$

where $\alpha = \frac{\omega-2\Delta}{\omega+2\Delta}$. The integral in (5.333) is expressed via elliptic functions as:

$$Re\sigma(\omega) = \frac{C_0}{\omega}[(\omega + 2\Delta)E(\alpha) - 4\Delta K(\alpha)]\theta(\omega - 2\Delta) \qquad (5.334)$$

In the normal state ($\Delta = 0$) electromagnetic absorption is determined by C_0 (which is seen from (5.331), calculated for $\Delta = 0$). Accordingly, the ratio of optical conductivities in a superconductor and normal metal is defined as:

$$\frac{Re\sigma_s(\omega)}{Re\sigma_n(\omega)} \equiv \frac{\sigma_{1s}(\omega)}{\sigma_{1n}(\omega)} = \frac{1}{\omega}[(\omega+2\Delta)E(\alpha)-4\Delta K(\alpha)]\theta(\omega-2\Delta) \quad (5.335)$$

This is the so-called Mattis–Bardeen formula (D. Mattis, J. Bardeen, 1958), which gives beautiful agreement with experiment, as you can see in Fig. 5.15.

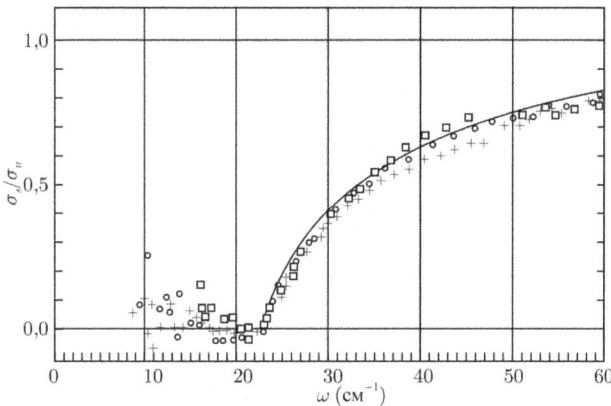

Fig. 5.15 Real part of optical conductivity of lead in superconducting state at $T = 2K$. Shown are experimental data for different samples and theoretical curve (full line) (L. Palmer, M. Tinkham, 1968).

In a wide interval of frequencies of an external electromagnetic field the qualitative behavior of the real part of conductivity (optical absorption) in superconductors is given in Fig. 5.16, where we show the results of more detailed calculations of optical conductivity in BCS model, with the account of impurity scattering with fixed value of $\gamma = \frac{1}{2\tau} = \Delta$. In the normal phase (for $T \geq T_c$) conductivity dependence on frequency is given by the usual Drude expression (4.210) with mean free time, determined (e.g. for low enough temperatures) by impurity scattering. This gives characteristic behavior shown in Fig. 5.16 by full curve.

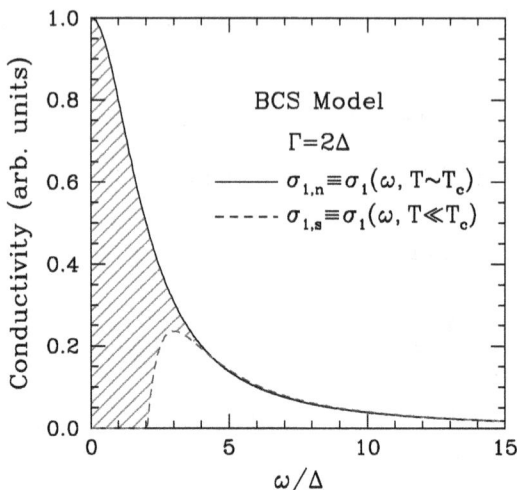

Fig. 5.16   Real part of optical conductivity in normal and superconducting states with the account of finite impurity scattering rate $\Gamma = \frac{1}{\tau} = 2\Delta$.

After the superconducting transition, a $\delta(\omega)$-contribution appears in conductivity (5.325), which is due to superconducting response of Cooper pairs condensate, while the finite absorption appearing at $\omega > 2\Delta$ corresponds to excitation of single electrons through BCS gap. At $T = 0$ these electrons are created by external field "breaking" of Cooper pairs (in accordance with the physical meaning of Δ as a binding energy of an electron in Cooper pair). Conductivity of an arbitrary system has to satisfy the following exact *sum rule* [Nozieres P., Pines D. (1966)]:

$$\int_0^\infty d\omega \, Re\sigma(\omega) = \frac{\pi n e^2}{2m} = \frac{\omega_p^2}{8} \qquad (5.336)$$

where $\omega_p^2 = \frac{4\pi n e^2}{m}$ is the square of plasma frequency. For superconductors this sum rule means that the dashed area below Drude conductivity curve of a normal metal in Fig. 5.16, after the superconducting transition is transformed into the amplitude of δ-function contribution in (5.325), while the remaining area under conductivity curve at $\omega > 2\Delta$ guarantees the validity of the sum rule (5.336) together with this contribution of superconducting condensate.

This relatively simple analysis based on BCS model leads to the correct results for the response of a superconductor to the *transverse* electromagnetic field, but it leads to wrong results if applied to calculations of the response to a *longitudinal* field.[41] This is deeply connected with gauge non invariant expressions for response functions obtained above and spontaneous breaking of gauge invariance by BCS ground state. The physical reason for the difficulty with consistent description of longitudinal response is due to the fact, that longitudinal gauge (gradient) transformations are directly connected with collective excitations (of electron density) in superconductors. In the model with only short-range interactions these excitations correspond to the so-called Bogoliubov sound, which is the Goldstone mode, appearing due to spontaneous breaking of gauge invariance. The account of long-range Coulomb interactions leads to transformation of Bogoliubov sound into the usual plasma oscillations. Gauge invariant formulation of electromagnetic response of superconductors can be obtained if we generalize BCS scheme by inclusion of these collective excitations [Schrieffer J.R. (1964)].

---

[41] We always assumed above the transverse gauge $div\mathbf{A} = 0$, $\phi = 0$!

Chapter 6

Electronic Instabilities and Phase Transitions

6.1 Phonon spectrum instability

Let us consider renormalization of phonon spectrum due to electron–phonon interaction in *one-dimensional* metal. We shall use the general approach, described by Eqs. (3.16), (3.18) and (3.91). It will be shown, that this renormalization leads to phonon spectrum instability, which we already mentioned shortly in the Chapter on electron–phonon interaction. Now we shall discuss this instability in more details.

For $d = 1$ polarization operator of the free electron gas at $T = 0$ is defined by the following expression:

$$\Pi(k\omega) = -2i \int \frac{dp}{2\pi} \int \frac{d\varepsilon}{2\pi} G_0(\varepsilon p) G_0(\varepsilon + \omega p + k) \qquad (6.1)$$

where the factor of 2 is due to spin. Calculating the integral over ε as was already done previously, we obtain:

$$\Pi(k\omega) = \frac{1}{\pi} \int dp \frac{n(\xi_p) - n(\xi_{p+k})}{\omega - \xi_{p+k} + \xi_p + i\delta(sign\xi_{p+k} - sign\xi_p)} \qquad (6.2)$$

where we again introduced the notation $\xi_p \equiv \xi(p) = \frac{p^2}{2m} - \mu$. Nonzero contributions to integral in (6.2) come from two regions:

(1) $\xi_p > 0,$ $\xi_{p+k} < 0$
(2) $\xi_p < 0,$ $\xi_{p+k} > 0$

in all other cases we have $n(\xi_p) - n(\xi_{p+k}) = 0$. To be specific, consider the case of $k > 0$. Then these regions correspond to:

(1) $-p_F - k < p < -p_F$
(2) $p_F - k < p < p_F$

These inequalities determine also the correct sign of $\pm i\delta$-contribution in the denominator of (6.2). Then (6.2) can be rewritten as:

$$\Pi(k\omega) = -\frac{1}{\pi}\int_{-p_F-k}^{-p_F}\frac{dp}{\omega - \frac{k^2}{2m} - \frac{pk}{m} - i\delta} + \frac{1}{\pi}\int_{p_F-k}^{p_F}\frac{dp}{\omega - \frac{k^2}{2m} - \frac{pk}{m} + i\delta}$$

$$= \frac{m}{\pi k}\ln\left[\frac{\left(\frac{k^2}{2m} - \frac{kp_F}{m} - \omega + i\delta\right)\left(\frac{k^2}{2m} - \frac{kp_F}{m} + \omega + i\delta\right)}{\left(\frac{k^2}{2m} + \frac{kp_F}{m} + \omega - i\delta\right)\left(\frac{k^2}{2m} + \frac{kp_F}{m} - \omega - i\delta\right)}\right] \quad (6.3)$$

Consider the behavior of $\Pi(k\omega)$ for $\omega = 0$ and close to $k = 2p_F$. Define $k = 2p_F + q$ (i.e. $q = k - 2p_F$, $|q| \ll 2p_F$). Then:

$$\Pi(k = 2p_F + q,\ \omega = 0) = -\frac{m}{\pi p_F}\ln\frac{4p_F}{|q|} = -\frac{m}{\pi p_F}\ln\frac{4p_F}{|k - 2p_F|} \quad (6.4)$$

In fact, we have mentioned this result previously in (3.89), in our preliminary discussion of the "giant" Kohn anomaly.

Consider now renormalization of phonon spectrum. Let us write down the phonon Green's function as (cf. (3.18) and (3.91)):

$$D^{-1}(k\omega) = D_0^{-1}(\omega k) - g^2\Pi(k\omega) \quad (6.5)$$

Close to $k = 2p_F$, using (6.4),[1] we have:

$$\frac{\omega^2 - \omega_{2p_F}^2}{\omega_{2p_F}^2} + \frac{mg^2}{\pi p_F}\ln\frac{4p_F}{|k - 2p_F|} = 0 \quad (6.6)$$

or

$$\omega^2 = \omega_{2p_F}^2\left\{1 - \frac{mg^2}{\pi p_F}\ln\frac{4p_F}{|k - 2p_F|}\right\} \quad (6.7)$$

Now we can see that for k, close enough to $2p_F$, the second term overcomes the first one, so that phonon frequency becomes imaginary ($\omega^2 < 0$). Suppression of phonon frequency at $k \sim 2p_F$ is usually called the appearance of the "soft" mode (lattice "softening"), as shown in Fig. 6.1, and respective instability of the phonon spectrum leads to spontaneous deformation of the lattice. This means (as we shall see below) that formation of (atomic) density modulation with the period $\frac{2\pi}{2p_F} = \frac{\pi}{p_F}$ becomes thermodynamically advantageous, leading to the

---

[1]Strictly speaking we should have used here $\Pi(\omega, k \approx 2p_F)$ at $\omega \neq 0$, but neglect of this frequency dependence does not change qualitative conclusions.

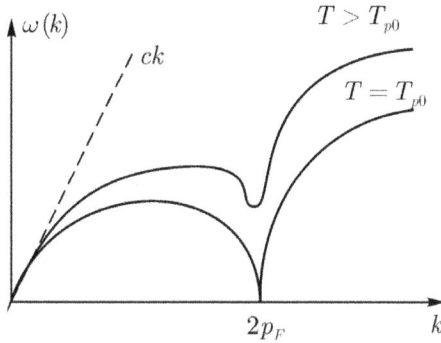

Fig. 6.1   Phonon "soft" phonon mode due to "giant" Kohn anomaly.

appearance of *static* density wave $\sim Ree^{i2p_Fx} \sim \cos(2p_Fx + \phi)$. This is called *Peierls* structural transition. Atomic displacements in this density wave directly lead to modulation of electronic charge density: $\rho(x) = \rho_0 + \rho_1 \cos(2p_Fx + \phi)$, or to formation of the so-called *charge density wave* (CDW). This is illustrated in Fig. 6.2.

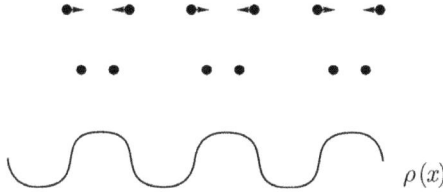

Fig. 6.2   Atomic displacements in one-dimensional chain leading to new period and appearance of charge density wave (CDW) due to Peierls transition. The case of period doubling.

In fact, the instability of phonon spectrum appears at some finite temperature $T = T_{p0}$, when the square of the frequency $\omega^2(k = 2p_F)$ becomes zero for the first time. To understand it in more details we consider the case of $T \neq 0$. As usual we shall use Matsubara technique. For phonon Green's function we again write down the Dyson equation:

$$D^{-1}(k\omega_m) = D_0^{-1}(\omega_m k) - g^2\Pi(\omega_m k) \qquad (6.8)$$

where

$$D_0(k\omega_m) = \frac{\omega_k^2}{(i\omega_m)^2 - \omega_k^2}, \qquad \omega_m = 2\pi mT \qquad (6.9)$$

Polarization operator is given by:

$$\Pi(k\omega_m) = 2T \sum_n \int \frac{dp}{2\pi} G_0(\varepsilon_n p) G_0(\varepsilon_n - \omega_m, p - k) \qquad (6.10)$$

We are interested in the region of $k \sim 2p_F$, so that we introduce:

$$k = 2p_F + q, \qquad |q| \ll 2p_F \qquad (6.11)$$

As everything of interest to us is determined by electrons from relatively narrow vicinity of the Fermi level, in future estimates we use the linearized spectrum of electrons, shown in Fig. 6.3. Then we have:

$$\xi_{p-k} = -\xi_p + v_F q \qquad \text{for} \qquad p \sim +p_F$$
$$\xi_{p+k} = -\xi_p + v_F q \qquad \text{for} \qquad p \sim -p_F \qquad (6.12)$$

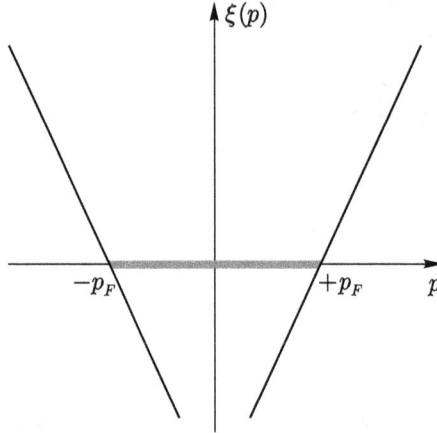

Fig. 6.3 Linearized (close to the Fermi level) spectrum of electrons in one-dimensional metal. Fermi "sphere" is represented by a straight line, electronic states with momenta from dashed region $(-p_F, \ p_F)$ are filled. Fermi "surface" consists of two points $\pm p_F$.

In fact, this is valid for any form of electronic spectrum in one-dimensional metal close to two "ends" of the Fermi line (Fermi "points"). Accordingly, we write the polarization operator as:

$$\Pi(q\omega_m) = TN(E_F) \sum_n \int_{-\infty}^{\infty} \frac{d\xi_p}{2\pi i} 2\pi i \frac{1}{i\varepsilon_n - \xi_p} \frac{1}{i(\varepsilon_n - \omega_m) + \xi_p - v_F q}$$
$$= -2\pi TN(E_F) \sum_n \theta[\varepsilon_n(\varepsilon_n - \omega_m)] \frac{sign\varepsilon_n}{2\varepsilon_n - \omega_m + iv_F q} \qquad (6.13)$$

where spin degeneracy is included in $N(E_F)$. Here, the factor of $\theta[\varepsilon_n(\varepsilon_n - \omega_m)]$ guarantees $sign\varepsilon_n = sign(\varepsilon_n - \omega_m)$, which, in turn, places the poles in ξ_p into different half-planes of appropriate complex variable, and guarantees nonzero value of integral over ξ_p, which is trivially calculated using Cauchy theorem.

Multiplying both numerator and denominator by $sign\varepsilon_n = sign(\varepsilon_n - \omega_m)$, we obtain:

$$\Pi(q\omega_m) = -2\pi T N(E_F) \sum_n \frac{\theta[\varepsilon_n(\varepsilon_n - \omega_m)]}{|\varepsilon_n| + |\varepsilon_n - \omega_m| + iv_F q \, sign\varepsilon_n}$$

$$= -2\pi T N(E_F) \sum_n \frac{\theta[\varepsilon_n(\varepsilon_n - \omega_m)]}{|2\varepsilon_n - \omega_m| + iv_F q \, sign\varepsilon_n} \tag{6.14}$$

where we have taken into account $|\varepsilon_n| = \varepsilon_n sign\varepsilon_n; |\varepsilon_n - \omega_m| = (\varepsilon_n - \omega_m) sign\varepsilon_n; |2\varepsilon_n - \omega_m| > 0$. Then:

$$\Pi(q\omega_m) = -2\pi T N(E_F) \sum_{n \geq 0} \frac{1}{2\varepsilon_n + \omega_m + iv_F q}$$

$$-2\pi T N(E_F) \sum_{n \geq 0} \frac{1}{2\varepsilon_n + \omega_m - iv_F q} \tag{6.15}$$

The sums entering here are formally divergent and we have to introduce a cut-off, remembering that electron–phonon interaction (coupling "constant" g) is effectively suppressed for frequencies $\sim E_F$, i.e. of the order of conduction band width.[2] Thus we have to perform summation up to $\varepsilon_n \sim E_F$! Then, adding and subtracting to the sums in (6.15) $2\pi T N(E_F) \sum_{n \geq 0}^{N^*} \frac{1}{2\varepsilon_n}$, with $N^* = \left[\frac{E_F}{2\pi T}\right]$, we can write:

$$\Pi(\omega_m q) = -4\pi T N(E_F) \sum_{n \geq 0}^{N^*} \frac{1}{2\varepsilon_n}$$

$$-2\pi T N(E_F) \sum_{n=0}^{\infty} \left[\frac{1}{2\varepsilon_n + \omega_m + iv_F q} - \frac{1}{2\varepsilon_n} \right]$$

$$-2\pi T N(E_F) \sum_{n=0}^{\infty} \left[\frac{1}{2\varepsilon_n + \omega_m - iv_F q} - \frac{1}{2\varepsilon_n} \right] \tag{6.16}$$

---

[2]Do not mix it with *pairing* interaction in superconductors, which, as we have seen above, is cut-off at frequencies $\sim \omega_D$.

where in convergent sums we can already take the infinite upper limit. Using now:[3]

$$4\pi T N(E_F) \sum_{n\geq 0}^{N^*} \frac{1}{2\varepsilon_n} = N(E_F) \ln \frac{2\gamma E_F}{\pi T}, \qquad (6.17)$$

we get:

$$\Pi(q\omega_m) = -N(E_F) \ln \frac{2\gamma E_F}{\pi T}$$
$$-\frac{1}{2}N(E_F) \sum_{n=0}^{\infty} \left[\frac{1}{n + \frac{1}{2} + \frac{\omega_m}{4\pi T} + \frac{iv_F q}{4\pi T}} - \frac{1}{n + \frac{1}{2}} \right]$$
$$-\frac{1}{2}N(E_F) \sum_{n=0}^{\infty} \left[\frac{1}{n + \frac{1}{2} + \frac{\omega_m}{4\pi T} - \frac{iv_F q}{4\pi T}} - \frac{1}{n + \frac{1}{2}} \right] \qquad (6.18)$$

Using now the definition of $\psi(x)$-function:

$$\psi(x) = -C - \sum_{n=0}^{\infty} \left[\frac{1}{n + x} - \frac{1}{n + 1} \right] \qquad (6.19)$$

where $C = \ln\gamma = 0.577...$, we can write down the final expression for polarization operator as:

$$\Pi(q\omega_m) = -N(E_F) \ln \frac{2\gamma E_F}{\pi T} + \frac{1}{2}N(E_F) \left[\psi\left(\frac{1}{2} + \frac{\omega_m}{4\pi T} + \frac{iv_F q}{4\pi T} \right) \right.$$
$$+ \psi\left(\frac{1}{2} + \frac{\omega_m}{4\pi T} - \frac{iv_F q}{4\pi T} \right) - 2\psi\left(\frac{1}{2} \right) \right]$$
$$= -N(E_F) \ln \frac{E_F}{2\pi T} + \frac{1}{2}N(E_F) \left[\psi\left(\frac{1}{2} + \frac{\omega_m}{4\pi T} + \frac{iv_F q}{4\pi T} \right) \right.$$
$$+ \psi\left(\frac{1}{2} + \frac{\omega_m}{4\pi T} - \frac{iv_F q}{4\pi T} \right) \right] \qquad (6.20)$$

where we also used $\psi\left(\frac{1}{2} \right) = -\ln 4\gamma$.

Let us define dimensionless coupling constant for electron–phonon interaction as:[4]

$$\lambda = g^2 N(E_F) \qquad (6.21)$$

---

[3] We use here: $2\sum_{n=0}^{N} \frac{1}{2n+1} = \ln(4\gamma N)$.

[4] It should not be mixed with dimensional constant λ of pairing interaction used in the previous Chapter!

Then the equation for phonon Green's function (6.8) is rewritten as:

$$
\begin{aligned}
\omega_{2p_F}^2 D^{-1}(q\omega_m) &= (i\omega_m)^2 - \omega_{2p_F}^2 \left\{ 1 - \lambda \ln \frac{E_F}{2\pi T} \right. \\
&+ \frac{\lambda}{2} \left[\psi\left(\frac{1}{2} + \frac{\omega_m}{4\pi T} + \frac{iv_F q}{4\pi T} \right) + \psi\left(\frac{1}{2} + \frac{\omega_m}{4\pi T} - \frac{iv_F q}{4\pi T} \right) \right] \right\} \\
&= (i\omega_m)^2 - \omega_{2p_F}^2 \left\{ 1 - \lambda \ln \frac{2\gamma E_F}{\pi T} \right. \\
&+ \frac{\lambda}{2} \left[\psi\left(\frac{1}{2} + \frac{\omega_m}{4\pi T} + \frac{iv_F q}{4\pi T} \right) + \psi\left(\frac{1}{2} + \frac{\omega_m}{4\pi T} - \frac{iv_F q}{4\pi T} \right) - 2\psi\left(\frac{1}{2} \right) \right] \right\}
\end{aligned}
$$

$$(6.22)$$

Then we define the phase transition temperature T_{p0}, as temperature at which the frequency of phonons with $k = 2p_F$ (i.e. $q = 0$) becomes zero. Putting in (6.22) $q = 0$ and $\omega_m = 0$, we obtain the equation:

$$
D^{-1}(0,0) = 1 - \lambda \ln \frac{2\gamma E_F}{\pi T_{p0}} = 0 \tag{6.23}
$$

which gives BCS-like expression for the transition temperature (at which the crystal lattice becomes unstable):

$$
T_{p0} = \frac{2\gamma}{\pi} E_F e^{-\frac{1}{\lambda}} \tag{6.24}
$$

Writing (6.23) as:

$$
1 - \lambda \ln \frac{E_F}{2\pi T_{p0}} + \lambda \psi\left(\frac{1}{2} \right) = 0 \tag{6.25}
$$

and subtracting this expression from the r.h.s. of (6.22), we obtain:

$$
\begin{aligned}
\omega_{2p_F}^2 D^{-1}(\omega_m q) &= (i\omega_m)^2 - \omega_{2p_F}^2 \lambda \left\{ \ln \frac{T}{T_{p0}} \right. \\
&+ \frac{1}{2} \left[\psi\left(\frac{1}{2} + \frac{\omega_m}{4\pi T} + \frac{iv_F q}{4\pi T} \right) + \psi\left(\frac{1}{2} + \frac{\omega_m}{4\pi T} - \frac{iv_F q}{4\pi T} \right) - 2\psi\left(\frac{1}{2} \right) \right] \right\}
\end{aligned}
$$

$$(6.26)$$

To find dispersion relation for the soft mode we have to perform her the usual analytic continuation $i\omega_m \to \omega$ and use the expansion of ψ-function (which is easily obtained from (6.19)):

$$\psi\left(\frac{1}{2} - \frac{i\omega}{4\pi T} \pm \frac{iv_F q}{4\pi T}\right) \approx -C - \sum_{n=0}^{\infty}\left[\frac{1}{n+1/2} - \frac{1}{n+1}\right]$$

$$+ \sum_{n=0}^{\infty}\frac{1}{(n+1/2)^2}\left[-\frac{i\omega}{4\pi T} \pm i\frac{v_F q}{4\pi T}\right]$$

$$- \sum_{n=0}^{\infty}\frac{1}{(n+1/2)^3}\left[-\frac{i\omega}{4\pi T} \pm i\frac{v_F q}{4\pi T}\right]^2$$

$$= \psi(1/2) + \frac{\pi}{8}\left(-\frac{i\omega}{T} \pm iv_F q\right) + 7\zeta(3)\left[\frac{\omega^2}{16\pi^2 T^2} + \frac{v_F^2 q^2}{16\pi^2 T^2} \mp \frac{\omega v_F q}{8\pi^2 T^2}\right] \quad (6.27)$$

Then we obtain the equation for the soft mode dispersion (spectrum):

$$\omega_{2p_F}^2 D^{-1}(q\omega) = \omega^2 - \omega_{2p_F}^2 \lambda\left\{\ln\frac{T}{T_{p0}} - \frac{i\pi}{8T}\omega + \frac{7\zeta(3)}{16\pi^2}\frac{v_F^2}{T^2}q^2\right.$$

$$\left. + \frac{7\zeta(3)}{16\pi^2 T^2}\omega^2\right\} = 0 \quad (6.28)$$

Considering $T \sim T_{p0}$ and introducing characteristic "coherence" length $\xi_0(T)$:

$$\xi_0^2(T) = \frac{7\zeta(3)}{16\pi^2}\frac{v_F^2}{T^2} \quad (6.29)$$

we rewrite (6.28) in the following form (D. Allender, J. Bray, J. Bardeen, 1974; B. Patton, L. Sham, 1974):

$$\omega^2 - \omega_{2p_F}^2 \lambda\left\{\frac{T - T_{p0}}{T_{p0}} - \frac{i\pi}{8T}\omega + \xi_0^2(T)q^2 + \frac{7\zeta(3)\omega^2}{16\pi^2 T^2}\right\} = 0 \quad (6.30)$$

Neglecting damping (as a first approximation), we obtain the dispersion of the soft mode as:

$$\omega^2 \approx \lambda\omega_{2p_F}^2\left\{\frac{T - T_{p0}}{T_{p0}} + \xi_0^2(T)q^2\right\} \quad (6.31)$$

This demonstrates the qualitative picture described above: for $T \to T_{p0}$ (in the region of $T > T_{p0}$) we observe suppression (softening) of phonon frequency for $k \sim 2p_F$, so that at $T = T_{p0}$ frequency square at $k = 2p_F$ is zero and becomes negative in the region of $T < T_{p0}$ (corresponding to lattice instability).

Returning to (6.26), let us write:

$$D(q\omega_m = 0) = -\frac{1}{\lambda}\frac{1}{\frac{T-T_{p0}}{T_{p0}} + \frac{7\zeta(3)}{16\pi^2}\frac{v_F^2}{T_{p0}^2}q^2} = -\frac{1}{\lambda}\frac{1}{\frac{T-T_{p0}}{T_{p0}} + \xi_0^2(T_{p0})q^2}$$
(6.32)

As noted above in connection with (3.118), (3.119), correlation function of atomic displacements differs from Green's function only by sign and the factor of $(\rho\omega_{2p_F}^2)^{-1}$ (where ρ is the density of the medium), so that:

$$C(q) = \int dx e^{iqx} <u(x)u(0)> = \frac{T_{p0}}{\lambda\rho\omega_{2p_F}^2}\frac{1}{\frac{T-T_{p0}}{T_{p0}} + \xi_0^2(T_{p0})q^2}$$
(6.33)

where the factor of $T = T_{p0}$ appeared due to "summation" over Matsubara frequencies (cf. (3.119)), where we have left only the term with $\omega_m = 0$, which corresponds to high temperatures (most strongly fluctuating contribution, corresponding to classic limit). Shortly speaking, we write:

$$C(q) = \frac{A}{(T - T_{p0}) + Bq^2} \quad \text{for} \quad A = \frac{T_{p0}^2}{\lambda\rho\omega_{2p_F}^2}, \quad B = T_{p0}\xi_0^2(T_{p0})$$
(6.34)

Then in coordinate representation we have:

$$C(x) = \int_{-\infty}^{\infty}\frac{dq}{2\pi}C(q)e^{-iqx} = \frac{A}{\pi}\int_0^{\infty}dq\frac{\cos qx}{T - T_{p0} + Bq^2}$$
$$= \frac{A}{2\sqrt{B(T - T_{p0})}}\exp\left\{-\frac{|x|}{\xi(T)}\right\}$$
(6.35)

where

$$\xi^2(T) = \xi_0^2(T_{p0})\frac{T_{p0}}{T - T_{p0}}$$
(6.36)

It is not difficult to understand that we, in fact, obtained just the usual "mean-field" description of second-order phase transition, taking place at $T = T_{p0}$, and (6.34), (6.35), (6.36) correspond to the picture of non-interacting fluctuations of atomic displacements at $T \sim T_{p0}$, described by Ornstein–Zernike correlator [Sadovskii M.V. (2019a)].

The previous discussion illustrates the general microscopic method of investigation of second-order phase transitions with the help of diagram technique. First we have to study fluctuations of the order parameter in harmonic ("Gaussian") approximation. In the case just discussed, the order parameter is, in fact, the Fourier component of atomic displacements, corresponding to wave-vector $k = 2p_F$. If phase transition really happens, there must appear the temperature, at which the system becomes "soft" enough and static ($\omega_m = 0$) correlation function (in momentum representation) diverges. This condition defines the transition temperature in "mean field" approach. Then we have to consider long wave fluctuations of the order parameter and their interactions in the vicinity of critical temperature. To do this we have to surpass (quadratic) Gaussian approximation. In the model under discussion we have to take into account "anharmonicities" due to electron–phonon interaction, e.g. diagrams for free energy with four external "tails" of order parameter fluctuations (to be considered below). Finally this will lead to Ginzburg–Landau type of free energy expansion, which is used to study thermodynamics of the transition. This scheme was already realized above for superconductivity.

In fact, for one-dimensional systems our analysis is oversimplified. We have already seen (cf. discussion around (3.128) and (3.132)), that for $d = 1$ phase transition (long-range order) of the "mean field" type is impossible. In the following we shall return to discussion of this problem and demonstrate, that in the temperature region $T \sim T_{p0}$ well developed fluctuations of short-range order appear in the system, while stabilization of the true long-range order takes place only after we take into account three-dimensional interactions in a system of one-dimensional atomic chains (i.e. in three-dimensional anisotropic crystal).

Consider now one-dimensional metallic chain with random impurities. In this case polarization operator is determined by diagrams, shown in Fig. 6.4, and can be written analytically as:

$$\Pi(\omega_m k) = 2T \sum_n \int \frac{dp}{2\pi} G(\varepsilon_n p) G(\varepsilon_n - \omega_m p - k) \mathcal{T}(\varepsilon_n, \omega_m, k) \quad (6.37)$$

where

$$G(\varepsilon_n \xi_p) = \frac{1}{i\tilde{\varepsilon}_n - \xi_p}, \qquad \tilde{\varepsilon}_n = \varepsilon_n \left[1 + \frac{1}{2\tau|\varepsilon_n|} \right], \qquad \frac{1}{2\tau} = \pi \rho v^2 \nu_F \tag{6.38}$$

and "triangular" vertex in "ladder" approximation of Fig. 6.4(b) is determined by the equation:

$$\mathcal{T}(\varepsilon_n, \varepsilon_{n-m}, q) = 1 - \frac{1}{2\pi\tau} \int_{-\infty}^{\infty} d\xi_p \frac{\mathcal{T}(\varepsilon_n, \varepsilon_{n-m}, q)}{(\xi_p - i\tilde{\varepsilon}_n)(\xi_p - v_F q + i\tilde{\varepsilon}_{n-m})} \tag{6.39}$$

where, as usual, we changed integration over dp to that over $d\xi_p$, introduced the notation $\varepsilon_{n-m} = \varepsilon_n - \omega_m$ and taken into account that

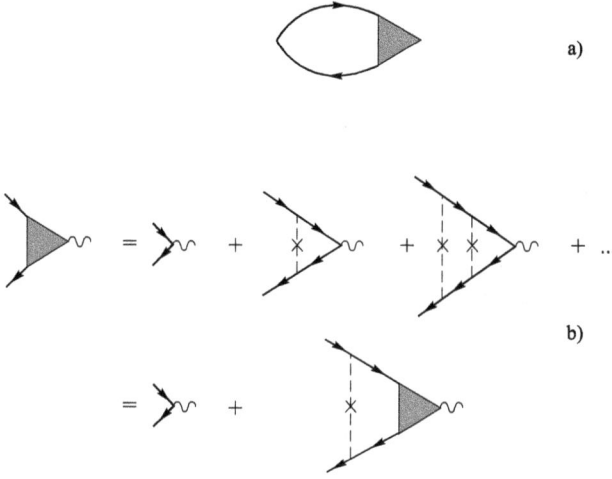

Fig. 6.4   Polarization operator for impure system (a) and appropriate "triangular" vertex (b).

$\rho v^2 \nu_F = \frac{1}{2\pi\tau}$. Then we immediately obtain:

$$\mathcal{T}^{-1}(\varepsilon_n, \varepsilon_{n-m}, q) = 1 + \frac{1}{\tau}\theta(\varepsilon_n\varepsilon_{n-m})\frac{\varepsilon_n}{|\varepsilon_n|}\frac{1}{\tilde{\varepsilon}_n + \tilde{\varepsilon}_{n-m} + iv_F q} \qquad (6.40)$$

To simplify calculations we assume that impurity lines in the "ladder" shown in Fig. 6.4(b) do not scatter electrons from one end of the Fermi line to the other. In general case in (6.40) we have to write symmetrized sum of terms with $\pm q$.

As a result, polarization operator is written as:[5]

$$\Pi(q\omega_m) = TN(E_F)\sum_n \int d\xi_p \frac{\mathcal{T}(\varepsilon_n, \varepsilon_{n-m}, q)}{(i\tilde{\varepsilon}_n - \xi_p)(i\tilde{\varepsilon}_{n-m} + \xi_p - v_F q)}$$

$$= -2\pi TN(E_F)\sum_n \theta(\varepsilon_n\varepsilon_{n-m})\frac{\varepsilon_n}{|\varepsilon_n|}\frac{1}{\tilde{\varepsilon}_n + \tilde{\varepsilon}_{n-m} + iv_F q}$$

$$\times \left\{1 + \frac{1}{\tau}\theta(\varepsilon_n\varepsilon_{n-m})\frac{\varepsilon_n}{|\varepsilon_n|}\frac{1}{\tilde{\varepsilon}_n + \tilde{\varepsilon}_{n-m} + iv_F q}\right\}^{-1} \qquad (6.41)$$

which reduces to:

$$\Pi(q\omega_m) = -2\pi TN(E_F)\sum_n \theta[\varepsilon_n(\varepsilon_n - \omega_m)]\frac{sign\varepsilon_n}{2\varepsilon_n - \omega_m + iv_F q + \frac{2}{\tau}sign\varepsilon_n} \qquad (6.42)$$

---

[5]In contrast to similar calculations in previous Chapters, here it is more convenient to perform ξ_p-integration first.

Further transformations can be done similarly to those done before, during the transition from (6.13) to (6.22), so that we obtain:

$$\Pi(q\omega_m) = -2\pi T N(E_F) \sum_n \frac{1}{|2\varepsilon_n - \omega_m| + \frac{2}{\tau} - iv_F q sign\varepsilon_n}$$

$$= -2\pi T N(E_F) \sum_{n>0} \frac{1}{2\varepsilon_n + \omega_m + \frac{2}{\tau} - iv_F q}$$

$$-2\pi T N(E_F) \sum_{n>0} \frac{1}{2\varepsilon_n + \omega_m + \frac{2}{\tau} + iv_F q}$$

$$= N(E_F) \ln \frac{2\gamma E_F}{\pi T} + \frac{1}{2} N(E_F) \left\{ \psi \left(\frac{1}{2} + \frac{\omega_m}{4\pi T} + \frac{1}{2\pi T\tau} + \frac{iv_F q}{4\pi T} \right) \right.$$

$$\left. + \psi \left(\frac{1}{2} + \frac{\omega_m}{4\pi T} + \frac{1}{2\pi T\tau} - \frac{iv_F q}{4\pi T} \right) - 2\psi \left(\frac{1}{2} \right) \right\} \tag{6.43}$$

Then we get the following equation for phonon Green's function:

$$\omega_{2p_F}^2 D^{-1}(q\omega_m) = (i\omega_m)^2 - \omega_{2p_F}^2 \left\{ 1 - \lambda \ln \frac{2\gamma E_F}{\pi T} \right.$$

$$+ \frac{\lambda}{2} \left[\psi \left(\frac{1}{2} + \frac{\omega_m}{4\pi T} + \frac{1}{2\pi T\tau} + \frac{iv_F q}{4\pi T} \right) \right.$$

$$\left. + \psi \left(\frac{1}{2} + \frac{\omega_m}{4\pi T} + \frac{1}{2\pi T\tau} - \frac{iv_F q}{4\pi T} \right) - 2\psi \left(\frac{1}{2} \right) \right] \right\}$$

$$= (i\omega_m)^2 - \omega_{2p_F}^2 \lambda \left\{ \ln \frac{T}{T_{p0}} \right.$$

$$+ \frac{1}{2} \left[\psi \left(\frac{1}{2} + \frac{\omega_m}{4\pi T} + \frac{1}{2\pi T\tau} + \frac{iv_F q}{4\pi T} \right) \right.$$

$$\left. + \psi \left(\frac{1}{2} + \frac{\omega_m}{4\pi T} + \frac{1}{2\pi T\tau} - \frac{iv_F q}{4\pi T} \right) - 2\psi \left(\frac{1}{2} \right) \right] \right\} \tag{6.44}$$

Then, similarly to (6.23), i.e. from the condition $D^{-1}(0,0) = 0$, we obtain the equation, determining temperature T_p of Peierls transition in a system with impurities (L.N. Bulaevskii, M.V. Sadovskii, 1974):

$$\ln \frac{T_p}{T_{p0}} + \psi \left(\frac{1}{2} + \frac{1}{2\pi T_p \tau} \right) - \psi \left(\frac{1}{2} \right) = 0 \tag{6.45}$$

which is formally identical to Eq. (5.197), discussed above and determining the critical temperature of a superconductor with magnetic impurities. Thus, normal (nonmagnetic) impurities (disordering) strongly suppress the temperature of Peierls structural transition.

6.2 Peierls dielectric

As we already noted, Peierls instability of phonon spectrum leads to the appearance of spontaneous deformation of the lattice (chain) with the wave vector $Q = 2p_F$. Let us now consider the description of "condensed" phase, which exists at temperatures $T < T_{p0}$. Hamiltonian of our system can be written in the following form:

$$H = \sum_p \xi_p a_p^+ a_p + \sum_l \omega_k b_k^+ b_k + \sum_{pk} g_k a_{p+k}^+ a_p (b_k + b_{-k}^+) \qquad (6.46)$$

where we have defined coupling constant as:

$$g_k = g\sqrt{\frac{\omega_k}{2}} \qquad (6.47)$$

as we are using notations of [Abrikosov A.A., Gorkov L.P., Dzyaloshin-skii I.E. (1963)].

Formation of the Peierls superstructure is described by introduction of the following anomalous average [Bogoliubov N.N. (1991b)], which breaks translational symmetry of initial lattice:

$$\Delta = g_{2p_F} < b_{2p_F} + b_{-2p_F}^+ > \neq 0 \qquad (6.48)$$

where angular brackets denote thermodynamic average. Appearance of such anomalous average can be interpreted as Bose-condensation of phonons into a state with (quasi) momentum $Q = 2p_F$. In coordinate representation (6.48) describes potential field of Peierls deformation:

$$V(x) = \Delta e^{i2p_F x} + \Delta^* e^{-i2p_F x} = 2|\Delta| \cos(2p_F x + \phi) \qquad (6.49)$$

where $|\Delta|$ is the modulus, while ϕ — the phase of appropriate order parameter: $\Delta = |\Delta| e^{i\phi}$.

To find the spectrum of an electron moving in the field defined by (6.49) we have, in fact, to solve the usual problem of electron motion in one-dimensional periodic field. This is well known from any course on solid state theory. Let us show how this can be done within Green's functions formalism. For generality, consider electron motion in periodic field characterized by an arbitrary wave vector Q:

$$V(x) = \Delta e^{iQx} + \Delta^* e^{-iQx} = 2|\Delta| \cos(Qx + \phi) \qquad (6.50)$$

Then, limiting ourselves to first order in V, we can describe everything by the system of equations for Green's functions of Gorkov type, shown

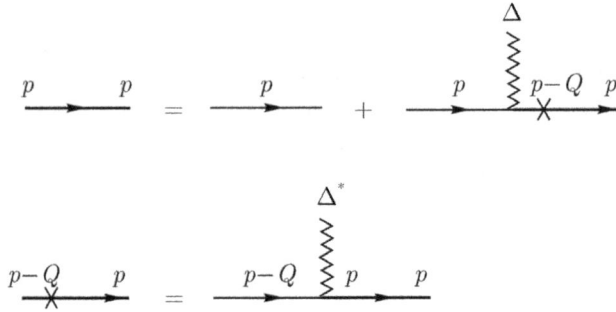

Fig. 6.5 Diagrammatic representation of "Gorkov equations" for an electron moving in periodic field.

diagrammatically in Fig. 6.5. For $T = 0$ we can write this system analytically as:

$$G(\varepsilon p) = G_0(\varepsilon p) + G_0(\varepsilon p)\Delta F(\varepsilon p) \qquad (6.51)$$

$$F(\varepsilon p) = G_0(\varepsilon p - Q)\Delta^* G(\varepsilon p) \qquad (6.52)$$

where normal and anomalous Green's functions (in momentum-coordinate representation) are defined as:

$$G(tp) = -i < Ta_p(t)a_p^+(0) > \qquad (6.53)$$

$$F(tp) = -i < Ta_p(t)a_{p-Q}^+(0) > \qquad (6.54)$$

Anomalous Green's function F (6.54) describes here an elementary *Umklapp* scattering process $p - Q \to p$, which appears in periodic field (6.50). The system of equations (6.51), (6.52) is easily rewritten as:

$$(\varepsilon - \xi_p)G(\varepsilon p) - \Delta F(\varepsilon p) = 1 \qquad (6.55)$$

$$(\varepsilon - \xi_{p-Q})F(\varepsilon p) - \Delta^* G(\varepsilon p) = 0 \qquad (6.56)$$

which gives the following solutions:

$$G(\varepsilon p) = \frac{\varepsilon - \xi_{p-Q}}{(\varepsilon - \xi_p)(\varepsilon - \xi_{p-Q}) - |\Delta|^2} \qquad (6.57)$$

$$F(\varepsilon p) = \frac{\Delta^*}{(\varepsilon - \xi_p)(\varepsilon - \xi_{p-Q}) - |\Delta|^2} \qquad (6.58)$$

Zero of denominators (pole) is determined here by an equation:

$$(\varepsilon - \xi_p)(\varepsilon - \xi_{p-Q}) - |\Delta|^2 = 0 \qquad (6.59)$$

which gives the standard result for the spectrum of "new" quasiparticles:

$$\varepsilon_{1,2}(p) = \frac{1}{2}(\xi_p + \xi_{p-Q}) \pm \sqrt{\frac{1}{4}(\xi_p - \xi_{p-Q})^2 + |\Delta|^2}, \qquad \xi_p = \frac{p^2}{2m} - \mu \qquad (6.60)$$

i.e. the usual "band" spectrum in "two-wave" approximation [Ziman J.M. (1972)], which is shown in Fig. 6.6. Naturally, all our analysis is symmetric with respect to $Q \rightarrow -Q$.

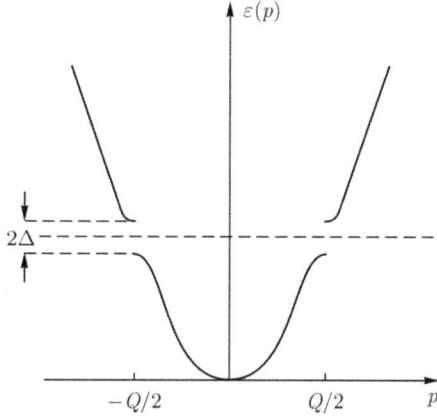

Fig. 6.6 Electron spectrum in periodic potential field, characterized by wave-vector Q ("two-wave" approximation).

For $Q = 2p_F$ Fermi level μ is precisely in the middle of the band gap of the width 2Δ, so that our system is *dielectric*. For one-dimensional case free electron spectrum always satisfies "nesting" condition:

$$\xi_{p-Q} = \xi_{p-2p_F} = -\xi_p \qquad (6.61)$$

which is easily seen from Fig. 6.7. Then, the spectrum (6.60) reduces to BCS-like:

$$\varepsilon_{1,2}(p) = \pm\sqrt{\xi_p^2 + |\Delta|^2} \qquad (6.62)$$

and ξ_p can be taken in a linearized form (close to Fermi level).

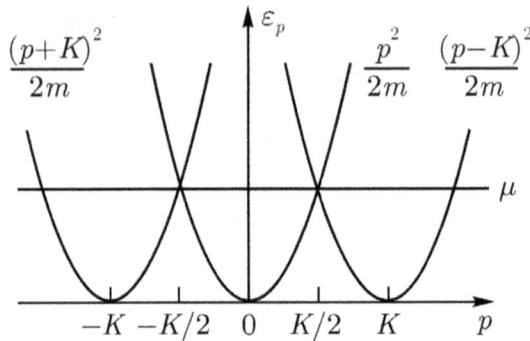

Fig. 6.7   Graphic illustration of validity of "nesting" property for one-dimensional spectrum of free electrons: $K = Q = 2p_F$.

Often we have to deal with electronic spectrum in tight-binding approximation [Ziman J.M. (1972)]. For example, in a model with transfer integral J being nonzero only for nearest neighbors (in a chain), we have the spectrum:

$$\varepsilon_p = -2J\cos pa, \qquad \xi_p = \varepsilon_p - \mu = -2J\cos pa \qquad (6.63)$$

where a is lattice constant (distance between nearest atoms in the chain), and the second equality is valid for half-filled band (one conduction electron per atom), when $p_F = \frac{\pi}{2a}$, and Fermi level is exactly in the middle of the band ($E_F = 2J$, if the origin of the energy scale is at the "bottom" of the band, but $E_F = \mu = 0$, if it is placed at the Fermi level). In this case $Q = 2p_F = \frac{\pi}{a}$, which corresponds to Peierls transition with period doubling (lattice dimerization). Here we again have "nesting" condition valid:

$$\xi_{p-Q} = -\xi_p, \qquad \varepsilon_{p-\frac{\pi}{a}} = -\varepsilon_p \qquad (6.64)$$

The form of the spectrum after Peierls transition for this case is shown in Fig. 6.8. This is a typical example of *commensurate* Peierls transition, when the period of Peierls superstructure and initial period of the chain form a rational relation (e.g. the new period is equal to rational number of initial periods). The case of free electron spectrum in the chain (Fig. 6.7) is a good model for *incommensurate* Peierls transition, when the period of the new superstructure is unrelated to the period of initial chain (incommensurate with it). In particular, this is due to Fermi momentum in this case being determined only by electron concentration, unrelated to the period of initial lattice. In the following we shall deal only with this last case.

Let us rewrite the system of equations (6.51), (6.52) in Matsubara technique:

$$G(\varepsilon_n p) = G_0(\varepsilon_n p) + G_0(\varepsilon_n p)\Delta F(\varepsilon_n p) \qquad (6.65)$$

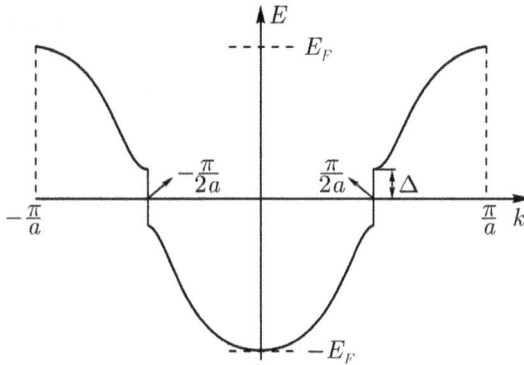

Fig. 6.8 Electron spectrum in Peierls dielectric in tight-binding approximation, the case of period doubling.

$$F(\varepsilon_n p) = G_0(\varepsilon_n p - Q)\Delta^* G(\varepsilon_n p) \tag{6.66}$$

or (if "nesting" condition is valid):

$$G(\varepsilon_n p) = \frac{1}{i\varepsilon_n - \xi_p} + \frac{1}{i\varepsilon_n - \xi_p}\Delta F(\varepsilon_n p) \tag{6.67}$$

$$F(\varepsilon_n p) = \frac{1}{i\varepsilon_n + \xi_p}\Delta^* G(\varepsilon_n p) \tag{6.68}$$

which may be rewritten also as:

$$(i\varepsilon_n - \xi_p)G(\varepsilon_n p) - \Delta F(\varepsilon_n p) = 1 \tag{6.69}$$

$$(i\varepsilon_n + \xi_p)F(\varepsilon_n) - \Delta^* G(\varepsilon_n p) = 0 \tag{6.70}$$

which (almost!) coincides with (5.108) and (5.109). Solution of this system is:

$$G(\varepsilon_n p) = \frac{i\varepsilon_n + \xi_p}{(i\varepsilon_n)^2 - \xi_p^2 - |\Delta|^2} \tag{6.71}$$

$$F(\varepsilon_n p) = \frac{\Delta^*}{(i\varepsilon_n)^2 - \xi_p^2 - |\Delta|^2} \tag{6.72}$$

which again is almost the same as (5.111) and (5.112). There is only some difference in signs of (5.112) and (6.72).

As in superconductivity theory, for Peierls dielectric we can introduce matrix Green's function:

$$\hat{G}^{-1}(\varepsilon_n \mathbf{p}) = \begin{pmatrix} i\varepsilon_n + \xi_p & -\Delta^* \\ -\Delta & i\varepsilon_n - \xi_p \end{pmatrix} \tag{6.73}$$

or $(F^- \equiv F)$

$$\hat{G}(\varepsilon_n \mathbf{p}) = \begin{pmatrix} G & F^- \\ F^+ & \tilde{G} \end{pmatrix} = \begin{pmatrix} G_{++} & G_{+-} \\ G_{-+} & G_{--} \end{pmatrix} \tag{6.74}$$

where notations $\pm$ correspond to "ends" of Fermi line (Fermi "points") $(\pm p_F)$ and to obvious electron transitions in our system. Often we speak just about $\pm$ ("right" or "left") electrons. In addition to (6.69) and (6.70) here we also have the obvious equations:

$$(i\varepsilon_n + \xi_p)\tilde{G}(\varepsilon_n p) - \Delta^* F(\varepsilon_n p) = 1 \tag{6.75}$$

$$(i\varepsilon_n - \xi_p)F^+(\varepsilon_n) - \Delta\tilde{G}(\varepsilon_n p) = 0 \tag{6.76}$$

From (6.71), after analytic continuation $i\varepsilon_n \to \varepsilon + i\delta$, we can easily (similarly to (5.78)) calculate the density of electronic states close to Fermi level, which takes the same form (5.79) as in BCS theory:

$$\frac{N(\varepsilon)}{N(E_F)} = \begin{cases} \frac{|\varepsilon|}{\sqrt{\varepsilon^2 - |\Delta^2|}} & \text{for} \quad |\varepsilon| > |\Delta| \\ 0 & \text{for} \quad |\varepsilon| < |\Delta| \end{cases} \tag{6.77}$$

Characteristic form of the density of states is shown in Fig. 6.9. Despite almost complete coincidence of all expressions, obtained here, with those of BCS theory, from previous discussion it is quite clear that energy gap $|\Delta|$ is now of *dielectric* nature.

Up to now we have done only a first part of our task — we still have to write down equations for self-consistent determination of Δ, which was introduced above "by hand". As Δ is determined by the anomalous average (6.48), we shall write down (Matsubara) equations of motion for operators b_Q and b_Q^+ perform Gibbs averaging. With the help of Hamiltonian (6.46), in a standard way, we get:

$$\left(-\frac{\partial}{\partial \tau} - \omega_Q\right) < b_{\pm Q}(\tau) > = -g_Q \sum_p < a_{p \mp Q}^+ a_p >$$

$$= -g_Q \sum_p F^{\mp}(p\tau = -0) \tag{6.78}$$

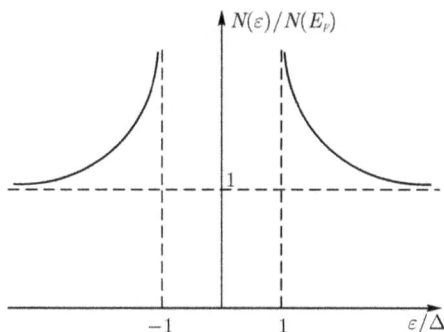

Fig. 6.9   BCS-like density of electronic states in Peierls insulator.

$$\left(-\frac{\partial}{\partial \tau} + \omega_Q\right) < b^+_{\pm Q}(\tau) > = g_Q \sum_p < a^+_{p\pm Q} a_p >$$

$$= -g_Q \sum_p F^{\pm}(p\tau = -0) \quad (6.79)$$

From these expressions we see that "phonon" anomalous average (6.48) are proportional to "electronic" anomalous averages $< a^+_{p\pm Q} a_p >$.[6] Thus, sometimes it is said that we are dealing here with *electron-hole pairing*.

After Fourier transformation over Matsubara "time", the system of equations (6.78), (6.79) reduces to:

$$(i\omega_m - \omega_Q) < b_{\pm Q} >_{\omega_m} = -g_Q \sum_p \sum_n F^{\mp}(p\varepsilon_n) \quad (6.81)$$

$$(i\omega_m + \omega_Q) < b^+_{\pm Q} >_{\omega_m} = g_Q \sum_p \sum_n F^{\pm}(p\varepsilon_n) \quad (6.82)$$

From these equations we obtain:

$$< b_Q + b^+_{-Q} >_{\omega_m} = -\frac{g_Q}{\omega_m^2 + \omega_Q^2} 2\omega_Q T \sum_p \sum_n F^-(\varepsilon_n p) \quad (6.83)$$

---

[6]It is clear now that charge density wave (CDW) appears in a system, as its order parameter is $< a^+_{p\pm Q} a_p >$. For the Fourier component of charge density we have ($Q = 2p_F$):

$$< \rho_q >= \rho_0 \delta(q) + \rho_1 \delta(q \pm Q), \qquad \rho_1 \sim \sum_p < a^+_{p\pm Q} a_p > \quad (6.80)$$

which gives $< \rho(x) >= \rho_0 + \rho_1 \cos(2p_F x + \phi)$.

For $\omega_m = 0$ (Bose-condensate!) we have:

$$< b_Q + b_{-Q}^+ >_{\omega_m=0} = -\frac{\lambda \omega_Q}{g_Q} \int d\xi_p T \sum_n F^-(\varepsilon_n p) \qquad (6.84)$$

where, as usual, we have changed variables from p to ξ_p integration and taken into account definition (6.21). Then, from (6.48) and (6.84) we get:

$$\Delta = g_Q < b_Q + b_{-Q}^+ >_{\omega_m=0} = -\frac{\lambda}{\omega_Q} \int_{-E_F}^{E_F} d\xi_p T \sum_n F^-(\varepsilon_n p) \qquad (6.85)$$

where we have introduced the cut-off in divergent integral at energies of the order of $\pm E_F$, similar to frequency summation cut-off discussed above. Substituting here (6.72) and performing standard calculations, similar to those made above in case of superconducting transition, we obtain BCS-type gap equation for Δ:

$$1 = \lambda \int_0^{E_F} d\xi \frac{1}{\sqrt{\xi^2 + \Delta^2(T)}} th \frac{\sqrt{\xi^2 + \Delta^2(T)}}{2T} \qquad (6.86)$$

From this equation it follows immediately, that the temperature of Peierls transition and the value of the gap at $T = 0$ are determined by standard relations:

$$T_{p0} = \frac{2\gamma}{\pi} E_F e^{-\frac{1}{\lambda}}, \qquad \Delta_0 = \frac{\pi}{\gamma} T_{p0} \qquad (6.87)$$

In particular, the value of transition temperature T_{p0} naturally coincides with (6.24), obtained from our analysis of instability of the "normal" phase.

6.3 Peierls dielectric with impurities

It is instructive to analyze the condensed phase of Peierls insulator with the account of scattering by non magnetic impurities (disorder).[7] Acting in a standard way and in obvious notations, we write down the system of Gorkov equations with impurity scattering (for the case

---

[7]It also gives us an opportunity to perform in detail all calculations, which are practically the same as dropped above in our discussion of superconductors with magnetic impurities.

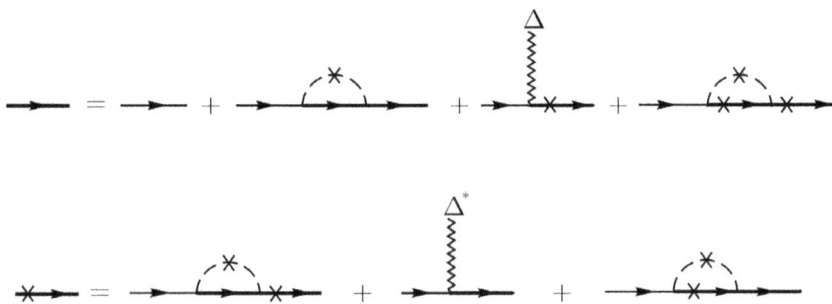

Fig. 6.10 Gorkov equations for Peierls dielectric with impurity scattering.

of weak disorder, $p_F l \gg 1$), shown diagrammatically in Fig. 6.10 (cf. Fig. 5.8 and the following discussion for superconductors):

$$G(\varepsilon_n p) = G_0(\varepsilon_n p) + G_0(\varepsilon_n p)\Sigma(\varepsilon_n pp)G(\varepsilon_n p) + G_0(\varepsilon_n p)\Delta F(\varepsilon_n p)$$
$$+ G_0(\varepsilon_n p)\Sigma(\varepsilon_n pp - Q)F(\varepsilon_n p) \qquad (6.88)$$

$$F(\varepsilon_n p) = G_0(\varepsilon_n p - Q)\Sigma(\varepsilon_n p - Qp - Q)F(\varepsilon_n p)$$
$$+ G_0(\varepsilon_n p - Q)\Delta^* G(\varepsilon_n p)$$
$$+ G_0(\varepsilon_n p - Q)\Sigma(\varepsilon_n p - Qp)G(\varepsilon_n p) \qquad (6.89)$$

Solution of this system, with the account of "nesting" condition $\xi_{p-Q} = -\xi_p$ (for $Q = 2p_F$), has the following form (cf. (5.177), (5.178), (5.179), (5.180)):

$$G(\varepsilon_n p) = [i\tilde{\varepsilon}_n + \xi_p]Det^{-1} \qquad (6.90)$$

$$F(\varepsilon_n p) = \tilde{\Delta}^* Det^{-1} \qquad (6.91)$$

where

$$i\tilde{\varepsilon}_n = i\varepsilon_n - \Sigma(\varepsilon_n pp) \equiv i\varepsilon_n - \Sigma_n(\varepsilon_n) \qquad (6.92)$$

$$\tilde{\Delta}_n = \Delta + \Sigma(\varepsilon_n p - Qp) \equiv \Delta + \Sigma_a(\varepsilon_n) \qquad (6.93)$$

$$Det = (i\tilde{\varepsilon}_n)^2 - \xi_p^2 - |\tilde{\Delta}_n|^2 \qquad (6.94)$$

and, similarly to the case of superconductors, here we also have:

$$\Sigma_n(\varepsilon_n) = \rho v^2 \nu_F \int_{-\infty}^{\infty} d\xi_p G(\varepsilon_n \xi_p) = -\frac{\Gamma}{2} \frac{i\tilde{\varepsilon}_n}{\sqrt{\tilde{\varepsilon}_n^2 + |\tilde{\Delta}_n|^2}} \qquad (6.95)$$

$$\Sigma_a(\varepsilon_n) = \rho v^2 \nu_F \int_{-\infty}^{\infty} d\xi_p F(\varepsilon_n \xi_p) = -\frac{\Gamma}{2} \frac{\tilde{\Delta}_n}{\sqrt{\tilde{\varepsilon}_n^2 + |\tilde{\Delta}_n|^2}} \qquad (6.96)$$

where $\Gamma = \frac{1}{\tau} = 2\pi \rho v^2 \nu_F$. Then from (6.92)–(6.96) we have:

$$i\tilde{\varepsilon}_n = i\varepsilon_n + \frac{\Gamma}{2} \frac{i\tilde{\varepsilon}_n}{\sqrt{\tilde{\varepsilon}_n^2 + |\tilde{\Delta}_n|^2}} \qquad (6.97)$$

$$\tilde{\Delta}_n = \Delta_n - \frac{\Gamma}{2} \frac{\tilde{\Delta}_n}{\sqrt{\tilde{\varepsilon}_n^2 + |\tilde{\Delta}_n|^2}} \qquad (6.98)$$

Note the opposite signs in (6.97) and (6.98), making this result different from (5.182), (5.183), (5.185) and similar to (5.192), (5.193) (the case of magnetic impurities in superconductors). This follows directly from difference between (6.71), (6.72) and (5.111), (5.112), already noted above.

Introducing

$$u_n = \frac{\tilde{\varepsilon}_n}{\tilde{\Delta}_n} \qquad (6.99)$$

from (6.97) and (6.98) we find:

$$\frac{\varepsilon_n}{\Delta} = u_n \left\{ 1 - \frac{\Gamma}{\Delta} \frac{1}{\sqrt{u_n^2 + 1}} \right\} \qquad (6.100)$$

This equation determines u_n as a function of ε_n/Δ and Γ/Δ. Knowing u_n, we can find:

$$\tilde{\varepsilon}_n = \varepsilon_n + \frac{1}{2}\Gamma \frac{u_n}{\sqrt{u_n^2 + 1}} \qquad (6.101)$$

$$\tilde{\Delta}_n = \Delta + \frac{1}{2}\Gamma \frac{1}{\sqrt{u_n^2 + 1}} \qquad (6.102)$$

Similarly to (6.85), order parameter (gap function) Δ is determined now by the equation:

$$\Delta = -\lambda T \sum_n \int d\xi_p F^+(\varepsilon_n \xi_p) \qquad (6.103)$$

or

$$1 = -\lambda T \sum_n \int d\xi_p \frac{\frac{\tilde{\Delta}_n}{\Delta}}{(i\varepsilon_n)^2 - \xi_p^2 - |\tilde{\Delta}_n|^2} \tag{6.104}$$

We deliberately drop limits of integration here! Let us now subtract in both r.h.s. and l.h.s. of this equation an expression, standing in the r.h.s. for $\Delta \to 0$ and in the absence of impurities. Then we obtain:

$$1 + \lambda T \sum_n \int_{-\infty}^{\infty} d\xi_p \frac{1}{(i\varepsilon_n)^2 - \xi_p^2}$$

$$= -\lambda T \sum_n \int_{-\infty}^{\infty} d\xi_p \left\{ \frac{\frac{\tilde{\Delta}_n}{\Delta}}{(i\varepsilon_n)^2 - \xi_p^2 - |\tilde{\Delta}_n|^2} - \frac{1}{(i\varepsilon_n)^2 - \xi_p^2} \right\} \tag{6.105}$$

Now we can take infinite limits of integration, due to fast convergence of all integrals!

Performing elementary integrations, we get:

$$1 - \lambda T \sum_n \frac{\pi}{|\varepsilon_n|} = 1 - \lambda \ln \frac{2\gamma E_F}{\pi T} = \lambda \ln \frac{T_{p0}}{T}$$

$$= -\lambda T \sum_n \left\{ \frac{\frac{\tilde{\Delta}_n}{\Delta}}{\sqrt{\tilde{\varepsilon}_n^2 + |\tilde{\Delta}_n|^2}} - \frac{1}{|\varepsilon_n|} \right\} \tag{6.106}$$

where in the sum over n in l.h.s. we introduced cut-off at $|\varepsilon_n| \sim E_F$, as was done before in (6.16), while in r.h.s. we can sum over all n up to infinity. As a result, using (6.99), we immediately obtain:

$$\ln \frac{T_{p0}}{T} = \pi T \sum_n \left\{ \frac{1}{|\varepsilon_n|} - \frac{\Delta^{-1}}{\sqrt{1 + u_n^2}} \right\} \tag{6.107}$$

where u_n is determined from Eq. (6.100). For $\Delta \to 0$, from (6.100) we have:

$$u_n \Delta \to \varepsilon_n + \Gamma \frac{\varepsilon_n}{|\varepsilon_n|} \tag{6.108}$$

so that

$$u_n = \Delta^{-1}(\varepsilon_n + \Gamma sign \varepsilon_n) \tag{6.109}$$

Then

$$\sqrt{1 + u_n^2} \to |u_n| = \frac{1}{\Delta}[|\varepsilon_n| + \Gamma] \tag{6.110}$$

and from (6.107) we obtain the following equation for transition temperature:

$$\ln \frac{T_{p0}}{T_p} = 2\pi T_p \sum_{n \geq 0} \left\{ \frac{1}{\varepsilon_n} - \frac{1}{\varepsilon_n + \Gamma} \right\} \qquad (6.111)$$

which, with the use of (6.19), reduce again to already known to us Eqs. (5.197), (6.45):

$$\ln \frac{T_{p0}}{T_p} = \psi \left(\frac{1}{2} + \frac{\Gamma}{2\pi T_p} \right) - \psi \left(\frac{1}{2} \right) \qquad (6.112)$$

For small Γ, similarly to (5.200), we have:

$$T_p \approx T_{p0} - \frac{\pi}{4}\Gamma = T_{p0} - \frac{\pi}{4\tau} \qquad (6.113)$$

For the critical disorder Γ_c, at which T_p vanishes, we obtain:

$$\Gamma_c = \frac{1}{\tau_c} = \frac{\pi}{2\gamma} T_{p0} = \frac{\Delta_{00}}{2} \qquad (6.114)$$

where Δ_{00} denotes the gap at $T = 0$ and in the absence of impurities (6.87).

Consider now the case of $T = 0$. Introduce notations:

$$\Delta_0 = \Delta(T = 0; \Gamma), \qquad \Delta_{00} = \Delta(T = 0; \Gamma = 0) \qquad (6.115)$$

For $T \to 0$ in equations, discussed above, we have to make the obvious change: $T \sum_n \ldots \to \int \frac{d\varepsilon}{2\pi} \ldots$. In particular, Eq. (6.104) reduces to:

$$1 = -\lambda \int \frac{d\varepsilon}{2\pi} \int d\xi_p \frac{\frac{\tilde{\Delta}_\varepsilon}{\Delta}}{\tilde{\varepsilon}_\varepsilon^2 - \xi_p^2 - |\tilde{\Delta}_\varepsilon|^2} \qquad (6.116)$$

where $\tilde{\varepsilon}_\varepsilon$ and $\tilde{\Delta}_\varepsilon$ are determined by analytic continuation of (6.97)–(6.107):

$$\tilde{\varepsilon}_\varepsilon = \varepsilon + \frac{\Gamma}{2} \frac{u}{\sqrt{1 - u^2}} \qquad (6.117)$$

$$\tilde{\Delta}_\varepsilon = \Delta_0 - \frac{\Gamma}{2} \frac{1}{\sqrt{1 - u^2}} \qquad (6.118)$$

$$\frac{\varepsilon}{\Delta_0} = u \left\{ 1 - \frac{\Gamma}{\Delta} \frac{1}{\sqrt{1 - u^2}} \right\} \qquad (6.119)$$

Subtracting from both r.h.s. and l.h.s. of (6.116) an expression, standing at the r.h.s., but taken with $\tilde{\Delta}_\varepsilon \to \Delta_0$ and $\tilde{\varepsilon} \to \varepsilon$, we obtain (taking infinite integration limits in fast converging integrals):

$$1 + \lambda \int_0^\infty \frac{d\xi_p}{\sqrt{\xi_p^2 + \Delta_0^2}} = 1 + \lambda \ln \frac{2E_F}{\Delta_0} = -\lambda \ln \frac{\Delta_{00}}{\Delta_0}$$

$$= -\lambda \int_{-\infty}^\infty \frac{d\varepsilon}{2\pi} \int_{-\infty}^\infty \left\{ \frac{\frac{\tilde{\Delta}_\varepsilon}{\Delta_0}}{\tilde{\varepsilon}^2 - \xi_p^2 - \tilde{\Delta}_\varepsilon^2} - \frac{1}{\varepsilon^2 - \xi_p^2 - \Delta_0^2} \right\}$$

$$= \frac{\lambda}{2} \int_{-\infty}^\infty d\varepsilon \left\{ \frac{\frac{\tilde{\Delta}_\varepsilon}{\Delta_0}}{\sqrt{\tilde{\Delta}_\varepsilon^2 - \tilde{\varepsilon}^2}} - \frac{1}{\sqrt{\Delta_0^2 - \varepsilon^2}} \right\} \tag{6.120}$$

so that for the gap function at $T = 0$ we have the following equation:

$$\ln \frac{\Delta_0}{\Delta_{00}} = \int_0^\infty d\varepsilon \left\{ \frac{\Delta_0^{-1}}{\sqrt{1 - u^2}} - \frac{1}{\sqrt{\Delta_0^2 - \varepsilon^2}} \right\}$$

$$= \int_0^\infty dx \left\{ \frac{1}{\sqrt{1 - u^2(x)}} - \frac{1}{\sqrt{1 - x^2}} \right\} \tag{6.121}$$

In the first integral in (6.121) we change integration over x to that over u, which is defined in (6.119). Then, taking into account $x = \infty \leftrightarrow u_\infty = \infty$, $x = 0 \leftrightarrow u_0 = 0$ for $\frac{\Gamma}{\Delta_0} \leq 1$, and $x = 0 \leftrightarrow u_0 = \sqrt{\frac{\Gamma^2}{\Delta_0^2} - 1}$ from $\frac{\Gamma}{\Delta_0} > 1$, after calculating elementary integrals, we get:

$$\ln \frac{\Delta_0}{\Delta_{00}}$$

$$= \begin{cases} -\frac{\pi}{4} \frac{\Gamma}{\Delta_0}, & \frac{\Gamma}{\Delta_0} \leq 1 \\ -\frac{1}{2} arctg \left(\frac{\Gamma^2}{\Delta_0^2} \right)^{-1/2} + \frac{\Delta_0}{\Gamma} \left(\frac{\Gamma^2}{\Delta_0^2} \right)^{1/2} - \ln \left[\frac{\Gamma}{\Delta_0} + \left(\frac{\Gamma^2}{\Delta_0^2} \right)^{1/2} \right], & \frac{\Gamma}{\Delta_0} \geq 1 \end{cases} \tag{6.122}$$

Then, it follows that $\Delta = 0$ for $\Gamma > \Gamma_c = \Delta_{00}/2 = \pi T_{p0}/2\gamma$ (cf. (6.114)).

Electronic density of states is given by:

$$\frac{N(\varepsilon)}{N(E_F)} = -\frac{1}{\pi} \int_{-\infty}^\infty d\xi_p Im \frac{\tilde{\varepsilon}_\varepsilon + \xi_p}{\tilde{\varepsilon}_\varepsilon^2 - \xi_p^2 - \tilde{\Delta}_\varepsilon^2}$$

$$= Im \frac{\tilde{\varepsilon}_\varepsilon}{\sqrt{\tilde{\Delta}_\varepsilon^2 - \tilde{\varepsilon}_\varepsilon^2}} = Im \frac{u}{\sqrt{1 - u^2}} \tag{6.123}$$

where $u = u\left(\frac{\varepsilon}{\Delta_0}, \frac{\Gamma}{\Delta_0}\right)$ is determined from (6.119). It is clear that density of states (6.123) is nonzero for $|u| > 1$. Energy gap in the spectrum is defined as the region of ε, where (6.123) is equal to zero. Then, from (6.123) it can be seen, that half-width of the gap is defined as $Max\varepsilon = \varepsilon_g$, for which (6.119) still has a real solution for $u\left(\frac{\varepsilon}{\Delta_0}\right)$ with $|u| < 1$. Thus defined value of ε_g depends on the ratio $\frac{\Gamma}{\Delta_0}$. For $\Gamma = 0$ we obviously have $\varepsilon_g = \Delta_{00}$. Maximizing the r.h.s. of (6.119), we find $u_g\left(\frac{\Gamma}{\Delta_0}\right) = u\left(\frac{\Gamma}{\Delta_0}, \frac{\varepsilon_g}{\Delta_0}\right)$:

$$Max\, u\left[1 - \frac{\Gamma}{\Delta_0}\frac{1}{\sqrt{1 - u^2}}\right] \equiv Max\mathcal{F}(u) = Max\frac{\varepsilon}{\Delta_0} = \frac{\varepsilon_g}{\Delta_0} \quad (6.124)$$

From $\mathcal{F}'(u_g) = 0$ we obtain:

$$1 - \frac{\Gamma}{\Delta_0}\frac{1}{\sqrt{1 - u_g^2}} - \frac{u_g^2}{(1 - u_g)^{3/2}}\frac{\Gamma}{\Delta_0} = 0 \quad (6.125)$$

so that

$$(1 - u_g^2)^{3/2} = \frac{\Gamma}{\Delta_0}, \qquad u_g^2 < 1 \quad (6.126)$$

and:

$$u_g = \sqrt{1 - \left(\frac{\Gamma}{\Delta_0}\right)^{2/3}} \quad (6.127)$$

Substituting (6.127) into (6.119), we find finally:

$$\varepsilon_g = \Delta_0\left\{1 - \left(\frac{\Gamma}{\Delta_0}\right)^{2/3}\right\}^{3/2} \quad (6.128)$$

Thus we have $\varepsilon_g = 0$ for $\frac{\Gamma}{\Delta_0} \geq 1$, when density of states becomes "gapless", despite the fact that order parameter $\Delta_0 \neq 0$. Appropriate region of parameters (at $T = 0$) is determined by an inequality:

$$\Delta_0 \leq \Gamma \leq \frac{\Delta_{00}}{2} \quad (6.129)$$

For $\Gamma = \Delta_0$ from (6.122) it follows that $\Delta_0 = \exp\left(-\frac{\pi}{4}\right)\Delta_{00}$, so that appropriate $\Gamma = 2\exp\left(-\frac{\pi}{4}\right)\Gamma_c \approx 0.91\Gamma_c$, which defines rather narrow gapless region on the phase diagram.[8]

---

[8]All the results obtained here are directly related also to the problem of magnetic impurities in superconductors, discussed in the previous Chapter. Historically, first these were obtained during the analysis of precisely this problem of superconductivity theory (A.A. Abrikosov, L.P. Gorkov, 1960).

6.4 Ginzburg–Landau expansion for Peierls transition

For temperatures $T > T_{p0}$ we can expand free-energy of the system, undergoing Peierls transition, in powers of the order-parameter:

$$\Delta_Q = g_Q < b_Q + b_Q^+ >, \qquad Q \sim 2p_F \qquad (6.130)$$

which is quite similar to GL-expansion in superconductors.

From the very beginning, we shall work in static approximation, taking $\omega_m = 0$. Our aim is to obtain microscopic derivation of coefficients in an expansion for the difference of free-energies of "condensed" and "normal" phases, which is written as:

$$F(\Delta_Q;\ T) - F(0;\ T) = a(Q)|\Delta_Q|^2 + b|\Delta_Q|^4 + \cdots \qquad (6.131)$$

The form of expansion (6.131) directly follows from general Landau theory of second order phase transitions [Sadovskii M.V. (2019a)].

Contributions due to (6.130) appear from the phonon part of Hamiltonian (6.46):

$$H_{ph} = \sum_Q \omega_Q b_Q^+ b_Q \qquad (6.132)$$

and its electron–phonon interaction part:

$$H_{int} = \sum_{pQ} \Delta_Q a_{p+Q}^+ a_p \qquad (6.133)$$

Restricting ourselves to a contribution of a single mode with fixed Q, we have:

$$< H_{ph} >= \omega_Q < b_Q^+ b_Q + b_{-Q}^+ b_{-Q} >= \omega_Q \frac{|\Delta_Q|^2}{2g_Q^2} = \frac{N(E_F)}{\lambda}|\Delta_Q|^2$$

$$(6.134)$$

Then, using the standard loop expansion of free-energy [Abrikosov A.A., Gorkov L.P., Dzyaloshinskii I.E. (1963)] in powers of H_{int}, we immediately obtain an expansion shown in Fig. 6.11. It is clear that diagram shown in Fig. 6.11(a) defines the term $a(Q)|\Delta_Q|^2$, while that of Fig. 6.11(b) gives $b|\Delta_Q|^4$. During the calculations we have to remember [Abrikosov A.A., Gorkov L.P., Dzyaloshinskii I.E. (1963)], that contribution of diagram in Fig. 6.11(a) has to be multiplied by an extra factor of $1/2$, and that of Fig. 6.11(b) by $1/4$ accordingly. Besides, during calculation of the loops, we have to take into account contributions of both Fermi "points", giving an extra factor of 2.

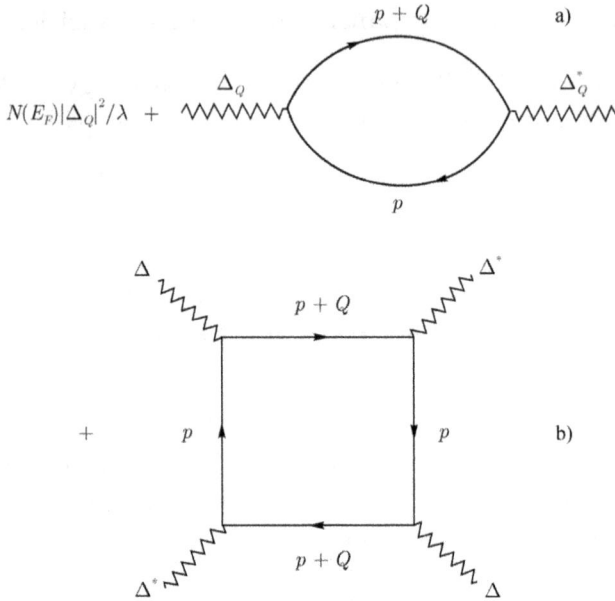

Fig. 6.11 Diagrammatic representation of Ginzburg–Landau expansion for Peierls transition.

In fact, all calculations, leading to $a(Q)|\Delta|^2$-contribution, were already performed in (6.10)–(6.32). Thus we can immediately write (taking into account $Q = 2p_F + q$, $\omega_m = 0$ and $\xi_{p-2p_F} = -\xi_p$):

$$\frac{1}{2}[\text{Fig. 6.11(a)}] = -T\sum_n \int \frac{dp}{2\pi} G_0(\varepsilon_n\xi_p)G_0(\varepsilon_n, -\xi_p + v_Fq)$$

$$= N(E_F)\left\{-\ln\frac{2\gamma E_F}{\pi T} + \frac{1}{2}\left[\psi\left(\frac{1}{2} + \frac{iv_Fq}{4\pi T}\right)\right.\right.$$

$$\left.\left.+\psi\left(\frac{1}{2} - \frac{iv_Fq}{4\pi T}\right) - 2\psi\left(\frac{1}{2}\right)\right]\right\} \tag{6.135}$$

Then we have:

$$a(q) = N(E_F)\left\{\frac{1}{\lambda} - \ln\frac{2\gamma E_F}{\pi T} + \frac{1}{2}\left[\psi\left(\frac{1}{2} + \frac{iv_Fq}{4\pi T}\right) + \psi\left(\frac{1}{2} - \frac{iv_Fq}{4\pi T}\right)\right.\right.$$

$$\left.\left.-2\psi\left(\frac{1}{2}\right)\right]\right\} \tag{6.136}$$

During calculations of b-coefficient, we neglect its dependence on q, so that we have:

$$b = \frac{1}{4}[\text{Fig. } 6.11(\text{b})] = \frac{1}{2}T\sum_n \int \frac{dp}{2\pi} G_0^2(\varepsilon_n \xi_p) G_0^2(\varepsilon_n, -\xi_p) \quad (6.137)$$

or

$$b = \frac{1}{2}TN(E_F)\int \frac{d\xi_p}{2\pi i} 2\pi i \sum_n \frac{1}{(i\varepsilon_n - \xi_p)^2} \frac{1}{(i\varepsilon_n + \xi_p)^2}$$

$$= -TN(E_F)2\pi i \sum_n \frac{\text{sign}\varepsilon_n}{(2i\varepsilon_n)^3} = \frac{N(E_F)}{16\pi^2 T^2} \sum_{n=0}^{\infty} \frac{1}{(n+1/2)^3}$$

$$= \frac{7\zeta(3)}{16\pi^2 T^2} N(E_F) \quad (6.138)$$

Let us return to the analysis of $a(q)$, defined by (6.136). Making expansion in powers of q (as during the derivation of (6.28)), we obtain:

$$a(q)$$
$$= N(E_F)\left\{ \ln\frac{T}{T_{p0}} + \frac{1}{2}\psi\left(\frac{1}{2} + \frac{iv_F q}{4\pi T}\right) + \frac{1}{2}\psi\left(\frac{1}{2} - \frac{iv_F q}{4\pi T}\right) - \psi\left(\frac{1}{2}\right) \right\}$$

$$\approx N(E_F)\left\{ \frac{T - T_{p0}}{T_{p0}} + \frac{7\zeta(3)}{16\pi^2 T^2} v_F^2 q^2 \right\}$$

$$= N(E_F)\left\{ \frac{T - T_{p0}}{T_{p0}} + \xi_0^2(T)q^2 \right\} \quad (6.139)$$

where we used also (6.29), defining the "coherence length".

Finally, Ginzburg–Landau expansion for Peierls transition is written as:

$$F(\Delta_Q; T) - F(0; T) = N(E_F)\frac{T - T_{p0}}{T_{p0}}|\Delta_Q|^2$$

$$+ N(E_F)\xi_0^2(T_{p0})|\Delta_Q|^2(Q - 2p_F)^2 + \frac{7\zeta(3)}{16\pi^2 T^2}N(E_F)|\Delta_Q|^4 \quad (6.140)$$

By the way, from this expression it is clear that correlation length, introduced above in (6.36), is in fact the correlation length of fluctuations of the order-parameter at $T \sim T_{p0}$ in Gaussian approximation.

In the presence of random impurities, all calculation for $a(q)$ are quite similar to those done during the derivation of (6.43) (it is sufficient to consider only the case of $\omega_m = 0$). As a result, we get:

$$a(q) = N(E_F)\left\{ \ln\frac{T}{T_{p0}} + \frac{1}{2}\psi\left(\frac{1}{2} + \frac{1}{2\pi T\tau} + \frac{iv_F q}{4\pi T}\right) \right.$$

$$\left. + \frac{1}{2}\psi\left(\frac{1}{2} + \frac{1}{2\pi T\tau} - \frac{iv_F q}{4\pi T}\right) - \psi\left(\frac{1}{2}\right) \right\} \quad (6.141)$$

For $v_F q \ll 4\pi T$ we have:

$$a(q) \approx N(E_F) \left\{ \frac{T - T_p}{T_p} + \frac{B}{16\pi^2 T^2} v_F^2 q^2 \right\} \tag{6.142}$$

where T_p is determined by the well known equation (6.45)

$$\ln \frac{T_p}{T_{p0}} + \psi \left(\frac{1}{2} + \frac{1}{2\pi T_p \tau} \right) - \psi \left(\frac{1}{2} \right) = 0 \tag{6.143}$$

The constant B in (6.142) is determined as:

$$B = \sum_{n=0}^{\infty} \frac{1}{\left(n + \frac{1}{2} + \frac{1}{2\pi T \tau} \right)^3} = -\frac{1}{2} \psi^{(2)} \left(\frac{1}{2} + \frac{1}{2\pi T \tau} \right) \tag{6.144}$$

Accordingly, in the impure case we have the following asymptotic expressions for "coherence length":

$$\xi_0^2(T) = -\frac{v_F^2}{32\pi^2 T^2} \psi^{(2)} \left(\frac{1}{2} + \frac{1}{2\pi T \tau} \right)$$

$$\approx \begin{cases} \frac{7\zeta(3) v_F^2}{16\pi^2 T^2} & \text{for} \quad \frac{1}{\tau} \ll 4\pi T \\ v_F^2 \tau^2 & \text{for} \quad \frac{1}{\tau} \gg 4\pi T \end{cases} \tag{6.145}$$

Naturally, we everywhere assume that $\tau > \tau_c = \frac{\gamma}{\pi T_{p0}} = 2\Delta_{00}^{-1}$. As to GL-coefficient b in the system with impurities, its calculation is to cumbersome and we drop it.

6.5   Charge and spin density waves in multi-dimensional systems. Excitonic insulator

A natural question may arise — why are we dealing in so much details with Peierls transition? We analyzed the one-dimensional problem, and we actually learned above that long-range order (phase transition) is, strictly speaking, broken by fluctuations and impossible in such a system. In this sense, our approach, based on mean-field description of this phase transition seems to be unjustified. However, in reality, Peierls transition is experimentally observed in a number of *quasi-one-dimensional* systems, where even small interaction of electrons (or order parameters) on neighboring "chains" of atoms (effects of three-dimensionality) leads to stabilization of this transition.[9]

What is more important for us at the moment, is the fact, that theoretical scheme, described in detail above, is almost completely valid

---

[9]Discussion of some early experiments and theoretical studies can be found in a nice review paper by L.N. Bulaevskii, Physics Uspekhi **115**, 263 (1975).

for the description of structural and magnetic phase transitions in two-dimensional and three-dimensional systems, possessing the "nesting" property of an energy spectrum (Fermi surface, or parts of it) [Khomskii D.I. (2010)]. Similar description is also applicable to a model of so-called "excitonic insulator" — one of the basic models in the theory of metal–insulator transitions. The thing is, that in the case, when electronic spectrum $\varepsilon(\mathbf{p})$ satisfies "nesting" condition:

$$\varepsilon(\mathbf{p} + \mathbf{Q}) = -\varepsilon(\mathbf{p}) \qquad (6.146)$$

where $\mathbf{Q}$ is some (nesting) vector in reciprocal space, response functions of appropriate multi-dimensional system (i.e. polarization operator, loop diagram, etc.) are described by practically the same expressions, as in one-dimensional case. Accordingly, such systems become unstable to formation of charge density (CDW) waves (deformation of the lattice) with wave-vector $\mathbf{Q}$, if electron–phonon interaction is a dominating one. If dominating interaction is electron–electron one (repulsion), as a rule, in such systems the similar instability occurs in "spin-channel", leading to the formation of spin-density wave (SDW).

Condition (6.146) is valid on the whole Fermi surface, for example, in the case of tight-binding spectrum with transfer integral J, different from zero only between nearest neighbors in the square lattice:

$$\varepsilon(\mathbf{p}) = -2J(\cos p_x a + \cos p_y a) \qquad (6.147)$$

Eq. (6.146) in this case is satisfied for $\mathbf{Q} = (\pi/a, \pi/a)$ (a is lattice spacing), and the Fermi surface, for the case of half-filled band (one conduction electron per atom), is just a square, shown in Fig. 6.12(a). We see that after the "translation" by vector $\mathbf{Q}$ the opposing sides of this Fermi surface just coincide, which leads to the appearance of logarithmic singularities of "one-dimensional" type in response functions, and Fermi surface becomes completely "closed" by dielectric gap.

Another possibility is that after the translation by vector $\mathbf{Q}$ only finite parts ("patches") of the Fermi surface coincide, as it is shown in Fig. 6.12(b). Then the energy spectrum can also become unstable, but dielectric gap is "opened" only on these parts of the Fermi surface, while the remaining parts remain "metallic". Such situation is realized in some "layered"-compounds of transition metals (e.g. $NbSe_2$, TaS_2 etc.) [White R.M., Geballe T.H. (1979)]. In some of these systems (e.g.

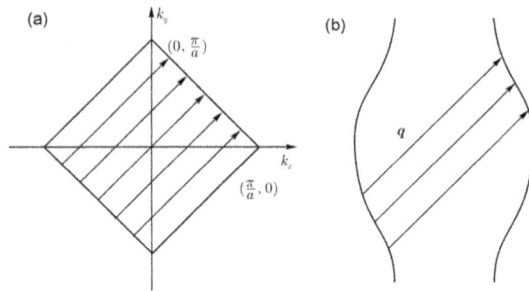

Fig. 6.12 "Nesting" property of Fermi surface in two-dimensional system: (a) half-filled band in tight-binding approximation with $Q = (\pi/a, \pi/a)$, (b) general case.

in $T - TaS_2$) we observe almost complete metal–insulator transition (due to strong "nesting"), so that the energy gap almost completely "closes" the Fermi surface. In other systems (e.g. in $H - NbSe_2$) Fermi surface is only partially "closed", so that the system remains metallic due to "open" parts. However, CDW transition is signalled in anomalies of temperature dependence of resistivity, and in some thermodynamic characteristics (e.g. in specific heat).

Similar in many respects is a remarkable model of "excitonic insulator" (L.V. Keldysh, Yu.V. Kopaev, 1964). This model is based on the model of electronic spectrum, shown in Fig. 6.13. Here we have overlapping bands of electron and holes. Such a band structure (with small band overlap) is typical for so-called semi-metals. From Fig. 6.13 it is seen that in both cases, shown as (a) and (b), Fermi surface consists of electron and hole "pockets" and we have "nesting" condition of the following form:

$$\varepsilon_1(\mathbf{p}) = -\varepsilon_2(\mathbf{p}) \qquad \text{for spectrum (a)} \qquad (6.148)$$

$$\varepsilon_1(\mathbf{p}) = -\varepsilon_2(\mathbf{p} + \mathbf{Q}) \qquad \text{for spectrum (b)} \qquad (6.149)$$

Thus, we have instability of the spectrum at zero wave-vector in case (a), or at finite vector $\mathbf{Q}$ in case (b), which is due to formation of *electron-hole* pairs, induced by natural attraction of electrons and holes (as for the usual Wannier–Mott excitons), and their "Bose-condensation"(sometimes it is called Bose-condensation of "excitons"). Qualitatively, situation here is very similar to Cooper pairing in superconductors, the major difference, of course, is that electron-hole pairs are electrically neutral (non charged), but in the spectrum we again obtain an energy gap at the Fermi level, which is formed due to this phase transition, as it is shown (for case (a)) in Fig. 6.14. As a result, the system becomes "excitonic" insulator. In the case of instability at the finite wave-vector $\mathbf{Q}$, determined by band structure of the type shown in Fig. 6.13(b), a charge density wave (CDW) or a spin density wave (SDW) forms in a system.

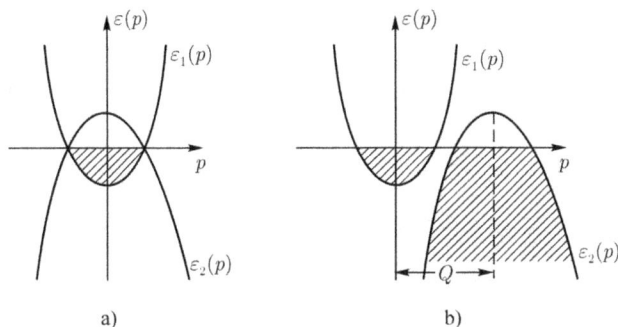

a)   b)

Fig. 6.13 Electronic spectrum of initial semi-metal in the model of an excitonic insulator: (a) directly overlapping bands, (b) indirect overlap with extrema of electron and hole bands at different points of the Brillouin zone, connected by vector $\mathbf{Q}$.

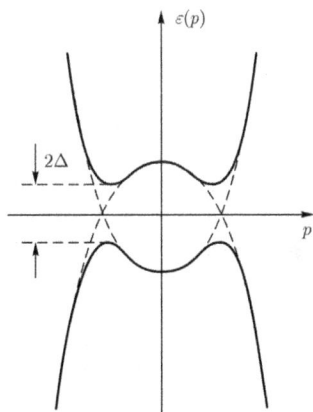

Fig. 6.14 Electronic spectrum of excitonic insulator in the model with direct overlap of electron and hole bands.

Mathematical analysis of this model is very similar to that of BCS and even more to our analysis of Peierls transition. Thus, we limit ourselves to only schematic discussion.[10] For definiteness, let us discuss the case of the spectrum, shown in Fig. 6.13(a), when both spectra of electrons and holes are very simple:

$$\varepsilon_{1,2}(\mathbf{p}) = \pm\left(\frac{p^2}{2m_{1,2}} - \frac{p_F^2}{2m_{1,2}}\right) \tag{6.150}$$

Excitonic instability is determined (similarly to Cooper instability) by the sum of ladder diagrams in particle-hole channel, shown in Fig. 6.15(a). All calculations are practically the same as those done during the derivation of (5.15)–(5.17), and we obtain the vertex, defined by integral equation of Fig. 6.15(b), expression analogous to (5.17):

$$\Gamma(q = 0, \omega) = \frac{\lambda}{1 + \lambda\left(\ln\left|\frac{2\omega_c}{\omega}\right| - i\frac{\pi}{2}\right)} \tag{6.151}$$

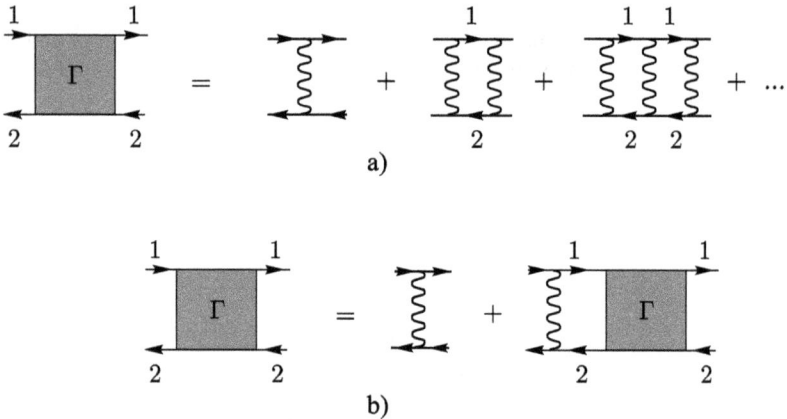

a)

b)

Fig. 6.15 Vertex part describing "excitonic" instability: (a) ladder approximation for particle-hole interaction, (b) integral equation for vertex-part.

Here $\omega_c \sim E_F$ is the cut-off frequency for logarithmic divergence, q and ω are sums of momenta and energies of $e - h$ pair. Dimensionless

[10]Detailed analysis of excitonic instability can be found in a review paper by Yu.V. Kopaev, P.N. Lebedev, Physical Institute Proceedings **86**, 3 (1975) and in [Ginzburg V.L., Kirzhnits D.A. (1982)].

coupling constant $\lambda < 0$ (attraction!). For the case of screened Coulomb $e - h$-interaction, it can be shown that:

$$\lambda = \frac{me^2}{2\pi p_F} \ln \frac{\kappa_D^2}{2p_F^2} \qquad (6.152)$$

Attraction in particle-hole channel leads to the appearance of the pole of (6.151) at imaginary frequency $\omega = i\Omega$, where (cf. (5.19)):

$$\Omega = 2\omega_c e^{-\frac{1}{|\lambda|}} \qquad (6.153)$$

indicating instability of the system towards pairing of electrons and holes (i.e. formation of "excitons") from different bands, close to over-lapping e and h Fermi surfaces.

To find the excitation spectrum in "condensed" phase, we have to act as in BCS theory, or in the case of Peierls transition. Let us write down an interaction Hamiltonian for particles from different bands as:

$$H_{int} = \sum_{\mathbf{ppq'}} V(\mathbf{q})a_{1\sigma}^+(\mathbf{p} + \mathbf{q})a_{2\sigma'}^+(\mathbf{p'} - \mathbf{q})a_{2\sigma'}(\mathbf{p'})a_{1\sigma}(\mathbf{p}) \qquad (6.154)$$

where we have explicitly written band and spin indices. Separating "most highly divergent" contributions to the scattering amplitude (vertex part), corresponding to scattering of electrons from band 1 on holes from band 2, we can transform this Hamiltonian to quadratic form, introducing appropriate anomalous averages:

$$H_{int} \sim a_{1\sigma}^+(\mathbf{p'})a_{2\sigma'}^+(\mathbf{p})a_{2\sigma'}(\mathbf{p'})a_{1\sigma}(\mathbf{p})$$
$$\rightarrow < a_{2\sigma'}^+(\mathbf{p})a_{1\sigma}(\mathbf{p}) > a_{1\sigma}^+(\mathbf{p'})a_{2\sigma'}(\mathbf{p'}) \qquad (6.155)$$

Then it can be diagonalized by Bogoliubov $u - v$ transformation, similarly to the case of BCS theory [Sadovskii M.V. (2019a)]. But instead, we can also write down appropriate Gorkov equations for Green's functions for both bands $G_1(\varepsilon\mathbf{p})$ and $G_2(\varepsilon\mathbf{p})$, introducing again anomalous Green's functions F and F^+, "mixing" particles from different bands:

$$F(\mathbf{r}_1 t_1; \mathbf{r}_2 t_2) = -i < T\psi_2(\mathbf{r}_2 t_2)\psi_1^+(\mathbf{r}_1 t_1) > \qquad (6.156)$$

$$F^+(\mathbf{r}_1 t_1; \mathbf{r}_2 t_2) = -i < T\psi_1(\mathbf{r}_1 t_1)\psi_2^+(\mathbf{r}_2 t_2) > \qquad (6.157)$$

Then we can write down equations of motion, e.g. for G_1, and leave in the "r.h.s." only contributions of the type of (6.155). Similarly we can deal with F. Finally, we obtain Gorkov equations:

$$(\varepsilon - \varepsilon_1(\mathbf{p}))G_1(\varepsilon\mathbf{p}) - \Delta F(\varepsilon\mathbf{p}) = 1 \qquad (6.158)$$

$$(\varepsilon - \varepsilon_2(\mathbf{p}))F(\varepsilon\mathbf{p}) - \Delta^* G_1(\varepsilon\mathbf{p}) = 0 \qquad (6.159)$$

where

$$\Delta(\mathbf{p}) = i \int \frac{d\mathbf{p}'d\varepsilon}{(2\pi)^4} V(\mathbf{p} - \mathbf{p}')F(\mathbf{p}'\varepsilon) \qquad (6.160)$$

Solution of Eqs. (6.158), (6.159) is:

$$G_1(\varepsilon\mathbf{p}) = \frac{\varepsilon - \varepsilon_2(\mathbf{p})}{\varepsilon^2 - (\varepsilon_1(\mathbf{p}) + \varepsilon_2(\mathbf{p}))\varepsilon + \varepsilon_1(\mathbf{p})\varepsilon_2(\mathbf{p}) - |\Delta(\mathbf{p})|^2} \qquad (6.161)$$

$$F(\varepsilon\mathbf{p}) = \frac{\Delta^*(\mathbf{p})}{\varepsilon^2 - (\varepsilon_1(\mathbf{p}) + \varepsilon_2(\mathbf{p}))\varepsilon + \varepsilon_1(\mathbf{p})\varepsilon_2(\mathbf{p}) - |\Delta(\mathbf{p})|^2} \qquad (6.162)$$

and we immediately obtain excitation spectrum:

$$E_{1,2}(\mathbf{p}) = \frac{1}{2}[\varepsilon_1(\mathbf{p}) + \varepsilon_2(\mathbf{p})] \pm \sqrt{\frac{1}{4}[\varepsilon_1(\mathbf{p}) - \varepsilon_2(\mathbf{p})]^2 + |\Delta(\mathbf{p})|^2} \quad (6.163)$$

Solution of the gap equation gives (as in BCS theory) $\Delta = \Omega$, where Ω is defined in (6.153).

Thus we obtain *insulating* spectrum, shown in Fig. 6.14, and mechanism of its formation, introduced above, is considered as one of most important mechanisms of metal–insulator transitions. What are the basic properties of excitonic insulator? Naively, we have direct analogy with BCS superconductor — a Bose condensate of neutral electron-hole pairs forms in the ground state. So it was thought initially that such an insulator may possess anomalous properties due to possible superfluidity of $e - h$ pairs (which may signal itself e.g. in "super thermal conduction"). However, special studies has shown, that apparently no "superfluidity" is realized in this model, and we are dealing with more or less "usual" insulator. Unfortunately, experimentally, the state of excitonic insulator is observed in rather rare cases. This is apparently due to the fact, that exact "nesting" properties of the type of (6.148), (6.149), with complete matching of electron and hole Fermi surfaces, are difficult to obtain (in three-dimensional systems), while deviations from these conditions suppress excitonic instability. Besides, similarly to the case of Peierls transition, normal impurities (disorder) also destroy excitonic phase. However, excitonic instability is considered as a microscopic reason for the formation of different types of charge and spin ordering in a number of real systems.

Up to now we considered attractive interaction. However, even in the case of repulsive interaction, systems with "nesting" properties of Fermi surfaces may acquire instabilities in "spin channel", leading not to CDW, but to SDW transition (A.W. Overhauser, 1965). To understand this, let us return to Eqs. (2.14), (2.15) for magnetic susceptibility in a system with Hubbard interaction:

$$\chi(\mathbf{q}\omega) = \frac{\chi_0(\mathbf{q}\omega)}{1 + U\Pi_0(\mathbf{q}\omega)} = \frac{\chi_0(\mathbf{q}\omega)}{1 - \frac{4U}{g^2\mu_B^2}\chi_0(\mathbf{q}\omega)} \tag{6.164}$$

Let us remind that, in accordance with our definitions $\Pi_0(\mathbf{q}0) < 0$, and due to $g = 2$ we can write $\chi_0 = -\mu_B^2\Pi_0$.

In one-dimensional, and also in higher-dimensional systems with "nesting", magnetic susceptibility $\chi_0 \sim \Pi_0(\mathbf{q})$ possesses logarithmic singularity at $\mathbf{q} = \mathbf{Q}$, where $\mathbf{Q}$ is vector of "nesting". Accordingly, in case of repulsion $(U > 0)$, total susceptibility (6.164) has a pole (divergence), of the same type as in charge channel (dielectric permeability, charge response function) in case of attraction (excitonic instability). In this case, instability signals a tendency to magnetic ordering — formation of SDW with the wave vector $\mathbf{Q}$. Instability appears again even for the case of arbitrarily weak repulsion U. Of course, in case of incomplete "nesting" (or even in its absence) this instability may require strong enough repulsion, so that we satisfy inequality:

$$U|\Pi_0(\mathbf{Q}, 0)| > 1 \tag{6.165}$$

This condition also applies to the theory of itinerant ferromagnetism, when $\mathbf{Q} = 0$. Then we have $\Pi_0(\mathbf{q} \to 0, 0) = -N(E_F)$, so that (6.165) reduces to:

$$UN(E_F) > 1 \tag{6.166}$$

giving the well known Stoner criterion of ferromagnetic instability.

Formal analysis here is again similar to that used in our discussion of Peierls transition. For simplicity, let us again consider one-dimensional case and Hubbard interaction $U \sum_i n_{i\uparrow} n_{i\downarrow}$. In momentum representation Hubbard Hamiltonian can be written as (L is the length of a system):

$$H = \sum_{k\sigma}(\varepsilon_k - \mu)a_{k\sigma}^+ a_{k\sigma} + \frac{U}{L}\sum_{kk'q} a_{k\uparrow}^+ a_{k+q\uparrow} a_{k'\downarrow}^+ a_{k'-q\downarrow} \tag{6.167}$$

Spin density at the point x is defined as:

$$S^z(x) = \frac{1}{2}[c_\uparrow^+(x)c_\uparrow(x) - c_\downarrow^+(x)c_\downarrow(x)]$$

$$= \frac{1}{2L}\sum_{kk'}[c_{k\uparrow}^+ c_{k'\uparrow} - c_{k\downarrow}^+ c_{k'\downarrow}]e^{-i(k-k')x} \qquad (6.168)$$

As susceptibility (6.164) is divergent (in this case at $q = 2p_F$), let us leave in (6.168) only terms with $k - k' = \pm 2p_F$. Then we have:[11]

$$S^z(x) = \frac{1}{2L}\sum_{k}[< c_{k\uparrow}^+ c_{k+2p_F\uparrow} > - < c_{k\downarrow}^+ c_{k+2p_F\downarrow} >]e^{i2p_F x} + c.c.$$

$$(6.169)$$

Defining:

$$< S > = |S|e^{i\phi} = \frac{1}{L}\sum_{k}[< c_{k\uparrow}^+ c_{k+2p_F\uparrow} > - < c_{k\downarrow}^+ c_{k+2p_F\downarrow} >] \quad (6.170)$$

and introducing appropriate anomalous averages into Hamiltonian (6.167) (with only terms with $q = \pm 2p_F$ left), we obtain:

$$H = \sum_{k\sigma}\{\varepsilon_k a_{k\sigma}^+ a_{k\sigma} + (\Delta a_{k+2p_F\sigma}^+ a_{k\sigma} + h.c.)\} \qquad (6.171)$$

where (N is the number of atoms in our chain)

$$\Delta = \frac{U}{N} < S^z > = \frac{U}{N}\frac{1}{2}Re(< S > e^{i2p_F x}) = \frac{U}{N}|S|\cos(2p_F x + \phi)$$

$$(6.172)$$

All these relations are quite similar to Eqs. (6.46)–(6.50), and the structure of solutions is absolutely clear. In particular, electronic spectrum in such system takes the form:

$$E_k = \mu \pm \sqrt{(\varepsilon_k - \mu)^2 + |\Delta|^2} \qquad (6.173)$$

and is the same as shown in Fig. 6.6.

Magnetic structure (SDW), appearing in the system, can be represented by two static waves of electrons with spins $\uparrow$ and $\downarrow$, being in antiphase, so that charge density remains homogeneous, while spin density oscillates with period $2\pi/2p_F$ (order parameter (6.172)). This is called sinusoidal SDW. In principle, another types of solutions are

---

[11]Term with $k' = k - 2p_F$ becomes complex conjugate to the first term in (6.169), after the change of summation index $k \to k + 2p_F$.

possible, e.g. so-called helicoidal SDW, where the modulus of spin remains constant along the chain direction, while its direction rotates in
orthogonal plane, as shown in Fig. 6.16:

$$< S^x >= |S| \cos(2p_F x + \phi), \qquad < S^y >= |S| \sin(2p_F x + \phi) \quad (6.174)$$

Usually, such structure has slightly lower energy.

Fig. 6.16 "Helicoidal" spin density wave.

Of course, thermodynamics of these transitions can also be analyzed
in a standard way. In mean-field we obtain critical temperature, at
which SDW vanishes and the gap in electronic spectrum disappears.
Appropriate expressions are very similar to those obtained in BCS theory, or in the theory of Peierls transition. In particular, for transition
temperature we usually obtain something similar to (6.24), with dimensionless coupling constant $\lambda = UN(E_F)$.

Analogous treatment can be used for higher-dimensional systems
with "nesting". Experimentally SDW of this type were observed in a
number of quasi-one-dimensional organic compounds. But probably
most notorious is SDW transition in chromium. Magnetism of Cr is
explained by this type of a model (or more precisely, in its two-band
analogue, of the type of excitonic insulator), as in Cr we have the Fermi
surface consisting of electron and hole "pockets", possessing "nesting"
property.

Another useful model (with possible practical applications) deals
with two-dimensional electrons on a square lattice in a tight-binding approximation (6.147), with only nearest neighbor transfers. We already
noted that in case of half-filled band, Fermi surface of such system is
represented by the square, shown in Fig. 6.12(a), so that we have complete "nesting" with vector $Q = (\pi/a, \pi/a)$. Accordingly, in the case
of repulsion we shall obtain here SDW with $Q = (\pi/a, \pi/a)$, which
will "close" the whole Fermi surface by the energy gap Δ, so that the
system will become dielectric. At the same time, it is easy to understand that SDW with such wave vector corresponds to the "usual" (two

sublattice) antiferromagnet (with "checkerboard" ordering of oppositely directed spins on the square lattice). One example of such a system is well known, that is the insulating state of La_2CuO_4, where spins of Cu order precisely in the same way in CuO_2 plane. This system is especially interesting as small doping by holes makes it a typical high-temperature superconductor. Thus, the models of the type discussed above, may be of use here.

6.6 Pseudogap

6.6.1 *Fluctuations of Peierls short-range order*

Return again to one-dimensional system with Peierls CDW. As we noted several times before, in purely one-dimensional case long-range order (at finite temperatures) is impossible. So let us discuss qualitatively, what happens in the temperature region $T < T_{p0}$, if we take into account fluctuations, destroying long-range order. Write down one-dimensional GL-expansion (6.140) as:

$$F(\Delta_Q;\ T) - F(0;\ T) = a(T)|\Delta_Q|^2 + c(T)(Q - 2p_F)^2|\Delta_Q|^2 + b(T)|\Delta_Q|^4 \tag{6.175}$$

where coefficients $a(T)$, $c(T)$ are:

$$a(T) = N(E_F)\frac{T - T_{p0}}{T_{p0}}, \quad T_{p0} = \frac{2\gamma}{\pi}E_F e^{-\frac{1}{\lambda}} \tag{6.176}$$

$$c(T) = N(E_F)\xi_0^2(T), \quad \xi_0^2(T) = \frac{7\zeta(3)v_F^2}{16\pi^2 T^2} \tag{6.177}$$

while for coefficient $b(T)$ we can write the following interpolation formula (P.A. Lee, T.M. Rice, P.W. Anderson, 1973):

$$b(T) = \left\{ b_0 + (b_1 - b_0)\frac{T}{T_{p0}} \right\} \frac{N(E_F)}{T_{p0}^2}, \quad b_0 = \frac{\gamma^2}{2\pi^2}, \quad b_1 = \frac{7\zeta(3)}{16\pi^2} \tag{6.178}$$

Then, GL-expansion (6.175), which was formally obtained for the region of $T \sim T_{p0}$, can be applied also for qualitative analysis of low-temperature region. In particular, for $T \to 0$ we get:

$$F(\Delta \sim \Delta_0) \approx -\frac{1}{2}N(E_F)\Delta_0^2 + 2N(E_F)(|\Delta| - \Delta_0)^2 + \cdots \tag{6.179}$$

where $\Delta_0 = \frac{\pi}{\gamma}T_{p0} = 2E_F e^{-\frac{1}{\lambda}}$ coincides with the gap function at $T = 0$, obtained in mean-field approximation. For the temperature dependence of the order parameter, in a standard way [Sadovskii M.V. (2019a)] we obtain:

$$\Delta(T) = \begin{cases} 0 & \text{for} \quad T > T_{p0} \\ \left(-\frac{a}{2b}\right)^{1/2} = \left(\frac{N(E_F)}{2b}\right)^{1/2}\left(\frac{T_{p0}-T}{T_{p0}}\right)^{1/2} & \text{for} \quad T \leq T_{p0} \\ \Delta_0 = \frac{\pi}{\gamma}T_{p0} & \text{for} \quad T \ll T_{p0} \end{cases}$$

$$(6.180)$$

Thus, GL-expansion (6.175)–(6.178) qualitatively reproduces results of mean-field approximation (microscopic theory) in the whole temperature interval. But this approximation, as is well known, does not take into account fluctuations of the order-parameter, which, in fact, just destroy long-range order in one-dimensional systems.

In principle, fluctuations of the order parameter $\{\Delta_Q\}$ may be arbitrary. However, the probability of a given fluctuation Δ_Q is given by [Sadovskii M.V. (2019a)]:

$$\mathcal{P}(\Delta_Q) \sim \exp\left\{-\frac{1}{T}[F(\Delta_Q,T) - F(0,T)]\right\} \qquad (6.181)$$

Then, the statistical sum over all fluctuations is described by the functional integral [Sadovskii M.V. (2019b)] of the following form:

$$Z = \int \{\delta\Delta_Q\} \exp\left\{-\frac{1}{T}[F(\Delta_Q,T) - F(0,T)]\right\} \qquad (6.182)$$

Accordingly, the free energy of the system as a whole is given by $F = -T\ln Z$.

One-dimensional model of GL type (6.175) allows practically exact treatment (D.J. Scalapino, M. Sears, R.A. Ferrell, 1972). We shall not deal with this problem, but only quote most important qualitative results. The absence of long-range order is equivalent to vanishing value of thermodynamic average of the order-parameter:

$$< \Delta_Q >= \frac{1}{Z} \int \{\delta\Delta_Q\}\Delta_Q \exp\left\{-\frac{1}{T}[F(\Delta_Q,T) - F(0,T)]\right\} = 0$$

$$(6.183)$$

However, mean-square fluctuation of the order parameter is obviously nonzero: $< |\Delta_Q|^2 > \neq 0$. Accordingly, two-point correlation function of fluctuations of the order parameter, in our model, is given by:

$$< \Delta(x)\Delta(x') >= 2 < |\Delta|^2 > \exp\left\{-\frac{|x-x'|}{\xi(T)}\right\} \cos 2p_F(x-x')$$

(6.184)

Behavior of parameters, determining (6.184), is shown[12] in Figs. 6.17, 6.18, where $t = \frac{T}{T_{p0}}$, and the value of Δt defines the width of the "Ginzburg" critical region [Sadovskii M.V. (2019a)], which in this model is given by:

$$\Delta t = \frac{\Delta T}{T_{p0}} = \left(\frac{bT_{p0}}{a'^2\xi_0}\right)^{2/3} \sim \left(\frac{N(E_F)}{T_{p0}^2}\frac{T_{p0}}{v_F}\frac{1}{N(E_F)}\right)^{2/3}$$

$$\sim \left(\frac{1}{N(E_F)v_F}\right)^{2/3} \sim 1$$

(6.185)

where a' is defined by $a(T) = a'(t-1)$, and in the last estimate we used $N(E_F) = \frac{1}{\pi v_F}$. Note that the width of Ginzburg critical region of the order of unity, in fact, corresponds to the absence of true phase transition (long-range order).

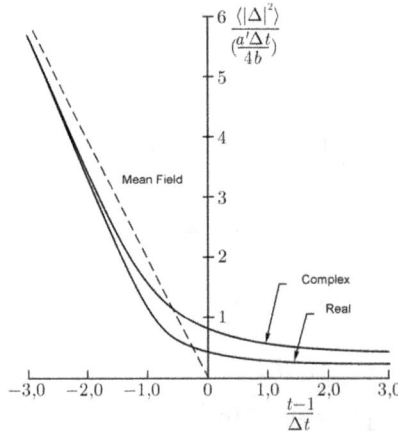

Fig. 6.17 Temperature behavior of mean-square fluctuation of the order-parameter in one-dimensional GL model. Shown are cases of real and complex order parameters. Dashed line — mean-field approximation for the square of the order parameter.

---

[12] Peierls transition with incommensurate CDW is described by complex order parameter, while that with commensurate CDW — by the real one.

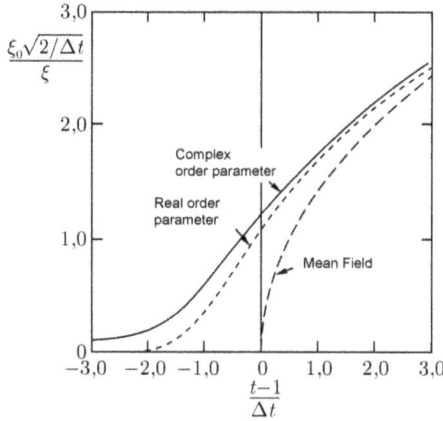

Fig. 6.18 Temperature behavior of inverse correlation length in one-dimensional GL model. Shown are cases of real and complex order parameters. Dashed line — mean-field approximation.

Basic qualitative conclusions, following from results shown in Figs. 6.17, 6.18, are as follows. Crudely we may estimate $< |\Delta|^2 > \sim \Delta(T) \sim T_{p0}$ in rather wide temperature region. Correlation length $\xi(T) \sim \xi_0 \sqrt{\frac{T_{p0}}{T-T_{p0}}}$ for $T \gg T_{p0}$, but $\xi(T) \to \infty$ only for $T \to 0$. At the same time, for temperatures $T < T_{p0}$, correlation length becomes very large — in our system appear rather large regions of *short-range order*, where we can speak about the existence of (fluctuating) Peierls CDW with the wave vector of the order of $Q \sim 2p_F$.

Real systems, where Peierls transition is observed experimentally, are always *quasi-one-dimensional*, with one-dimensional chains coupled by weak inter-chain interactions, which lead to stabilization of Peierls long-range order at some finite transition temperature. Interchain coupling may be due to electron tunnelling between chains, i.e. due to, in fact, three-dimensional (though strongly anisotropic) nature of electronic spectrum. Also of importance may be "direct" interaction of order-parameters on the nearby chains. For example, we may remember, that Peierls CDW creates real modulation of electronic charge density along the chain:

$$\rho(x) = ne \frac{\Delta}{\lambda E_F} \cos(Qx + \phi) \tag{6.186}$$

which, in turn, creates electrostatic potential around the chain, given by:

$$\varphi(r_\perp, x) = 2ne\frac{\Delta}{\lambda E_F}\cos(Qx + \phi)K_0(Qr_\perp) \qquad (6.187)$$

where $K_0(r)$ is Bessel function of imaginary argument (exponentially small for large distances). Thus, we obtain electrostatic interaction energy of CDW's (per chain length) of the following form:

$$U = U_0 \sum_n \sum_{<m>} \cos(\phi_n - \phi_m) \qquad (6.188)$$

where the second sum is taken over the nearest neighbors (chains) of the chain n. We conclude that there is an energy gain in case of CDW's on the neighboring chains being in antiphase, as shown in Fig. 6.19. Thus we obtain three-dimensional ordering and a real phase transition. Qualitatively, the temperature of such transition is determined by almost obvious condition $U_0\xi(T) \sim T$, which gives the following equation for the critical temperature:

$$1 \sim \frac{1}{T_c}U_0\xi(T_c) \qquad (6.189)$$

As temperature T lowers, the value of $\xi(T)$, as we have seen in purely one-dimensional model, grows ($\xi(T) \to \infty$ for $T \to 0$), so that solution of Eq. (6.189) exists even for arbitrarily small U_0. In real life everything depends on parameters. It can be that $T_c \sim T_{p0}$, or it may be that $T_c \ll T_{p0}$. Thus, in quasi-one-dimensional systems we may have wide enough temperature region, where $T_c \ll T \ll T_{p0}$, and long-range order is absent, though well developed fluctuations of short-range order exist and are characterized by correlation function of the type of (6.184). For high enough temperatures $T \sim T_{p0}$ these fluctuations may be considered

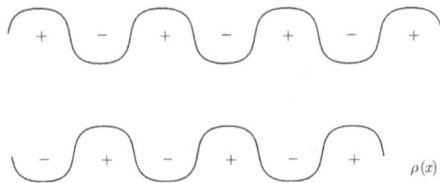

Fig. 6.19   CDW antiphase ordering on neighboring chains due to electrostatic interchain interaction.

as Gaussian, for $T \ll T_{p0}$ this is obviously wrong. Correlation length of these fluctuations is of order of $\xi(T)$ and may significantly greater than interatomic spacing.

Let us consider the problem of three-dimensional ordering in the system of interacting order parameters on the nearby chains in more details. For such a system (with chains enumerated by indices i, j), from purely phenomenological point of view we can write the following GL-expansion:

$$F\{\Delta_i\} = \int \frac{dx}{\xi_0} \left\{ \sum_i \left[a|\Delta_i(x)|^2 + b|\Delta_i(x)|^4 + c \left| \frac{d\Delta_i}{dx} \right|^2 \right] + \frac{1}{2} \sum_{<ij>} \lambda_{ij} \Delta_i(x) \Delta_j(x) \right\}$$

(6.190)

where we assume only nearest neighbor (chains) interaction.

The average value of the order parameter can be written as:

$$< \Delta_i >$$

$$= Z^{-1} \int \{\delta\Delta\} \Delta_i(x) \exp\left\{ -\frac{1}{T} \int \frac{dx'}{\xi_0} \left[\sum_i F_i(\Delta_i(x')) + \frac{1}{2} \sum_{ij} \lambda_{ij} \Delta_i(x') \Delta_j(x') \right] \right\}$$

$$\rightarrow Z^{-1} \int \{\delta\Delta\} \Delta_i(x) \exp\left\{ -\frac{1}{T} \int \frac{dx'}{\xi_0} \left[\sum_i F_i(\Delta_i(x')) + \sum_{ij} \lambda_{ij} \Delta_i(x') < \Delta_j > \right] \right\}$$

(6.191)

where in the second line we have made "mean-field" approximation over the interchain coupling. For $T \rightarrow T_c$ we have $< \Delta_i > \rightarrow 0$, then we can write:

$$< \Delta_i >$$

$$\approx Z^{-1} \int \{\delta\Delta\} \Delta_i(x) \exp\left\{ -\frac{1}{T} \int \frac{dx'}{\xi_0} \sum_i F_i \right\} \left\{ 1 + \frac{1}{T} \int \frac{dx'}{\xi_0} \sum_{ij} \Delta_i(x') < \Delta_j > \right\}$$

$$= \frac{1}{T} \int \frac{dx'}{\xi_0} \sum_j \lambda_{ij} < \Delta_i(x) \Delta_i(x') >< \Delta_j >$$

(6.192)

so that critical temperature T_c is determined by the equation:

$$1 = \frac{\lambda}{T} \int \frac{dx'}{\xi_0} < \Delta_i(x) \Delta_i(x') >, \qquad \lambda = \sum_{<j>} \lambda_{ij}$$

(6.193)

and, using

$$< \Delta_i(x) \Delta_i(x') > = < \Delta^2 > \exp\left\{ -|x - x'| \xi^{-1}(T) \right\}$$

(6.194)

we get equation similar to (6.189):

$$1 = \frac{\lambda}{T} < \Delta^2 > \frac{\xi(T)}{\xi_0}$$

(6.195)

and the same conclusions as above.

6.6.2   *Electron in a random field of fluctuations*

Consider an electron propagating in a random field of fluctuations $\Delta(x)$, which we assume to be Gaussian. Then we can use the usual "impurity" diagram technique, associating with interaction (scattering) lines correlation function of order parameter (6.184). In momentum representation we associate with interaction line the Fourier-transform of (6.184):

$$S(Q) = 2\Delta^2 \left\{ \frac{\kappa}{(Q - 2p_F)^2 + \kappa^2} + \frac{\kappa}{(Q + 2p_F)^2 + \kappa^2} \right\} \qquad (6.196)$$

where $\kappa = \xi^{-1}(T)$, and $< |\Delta|^2 >$ is denoted (for shortness) as Δ^2. In the following we shall first consider much oversimplified, but quite instructive, variant of this model, corresponding to the limit of $\xi \to \infty$ ($\kappa \to 0$), i.e. the asymptotics of very large correlation lengths of short-range order fluctuations.[13] This problem can be solved *exactly*, and we can sum *all* Feynman diagrams of perturbation theory for "interaction" (6.196), which in the limit of $\xi \to \infty$ ($\kappa \to 0$) becomes:[14]

$$S(Q) = 2\pi\Delta^2 \{\delta(Q - 2p_F) + \delta(Q + 2p_F)\} \qquad (6.197)$$

Consider the simplest contribution to electron self-energy, described by the diagram shown in Fig. 6.20, which we write in Matsubara representation:

$$\Sigma(\varepsilon_n p) = \int \frac{dQ}{2\pi} S(Q) \frac{1}{i\varepsilon_n - \xi_{p-Q}} \approx 2\Delta^2 \int_{-\infty}^{\infty} \frac{dx}{2\pi} \frac{\kappa}{x^2 + \kappa^2} \frac{1}{i\varepsilon_n + \xi_p - v_F x}$$

$$= 2\Delta^2 \int_{-\infty}^{\infty} \frac{dx}{2\pi} \frac{\kappa}{(x - i\kappa)(x + i\kappa)} \frac{1}{i\varepsilon_n + \xi_p - v_F x}$$

$$= \frac{\Delta^2}{i\varepsilon_n + \xi_p + iv_F\kappa} \qquad (6.198)$$

where, for definiteness we assume $p \sim +p_F$, $\varepsilon_n > 0$ and defined the new integration variable x via $Q = 2p_F + x$ (it is helpful to look once again at Fig. 6.3!).

---

[13]To avoid misunderstanding, note that this limit does not correspond to the appearance of long-range order! Electron is propagating in the Gaussian random field with specific pair correlator, not in the periodic system. In details this will be seen from the analysis which follows.

[14]Note the obvious analogy of this model with that of Keldysh, discussed above.

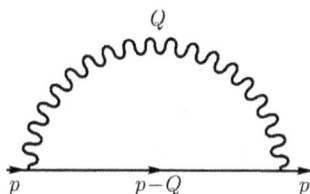

Fig. 6.20   Simplest diagram for electron self-energy. Wave-like line denotes correlator $S(Q)$.

The limit of $\xi(T) \to \infty$ ($\kappa \to 0$) should be understood as the requirement of:

$$v_F\kappa = v_F\xi^{-1} \ll Max\{2\pi T, \xi_p\} \qquad (6.199)$$

or

$$v_F\kappa = v_F\xi^{-1} \ll 2\pi T, \qquad \xi(T) \gg |p - p_F|^{-1} \qquad (6.200)$$

Then (6.198) gives just:

$$\Sigma(\varepsilon_n p) \approx \frac{\Delta^2}{i\varepsilon_n + \xi_p} \qquad (6.201)$$

Now, for "interaction" of the form of (6.197) there is no problem to write down the contribution of an arbitrary diagram for Green's function correction in any order, e.g. of the type shown in Fig. 6.21. In such a diagram of the n-th order in $S(Q)$ we have $2n$ vertices, connected by interaction lines in any possible combinations. These lines either "bring" or "take away" momenta $Q = 2p_F$.[15] As a result, in the analytic expression for the contribution of such a diagram we have a sequence of alternating Green's functions like $\frac{1}{i\varepsilon_n - \xi_p}$ (n times) and $\frac{1}{i\varepsilon_n + \xi_p}$ (also n times), plus one more (at the start of the sequence) $\frac{1}{i\varepsilon_n - \xi_p}$.[16] In

---

[15]These processes have to alternate with each other, so that the electron does not "leave" far from the Fermi level (Fermi points $\pm p_F$) in Fig. 6.3 or Fig. 6.7 (in opposite case large energies appear in denominators of Green's functions in higher orders). This requirement is absent in the case of commensurate fluctuations, e.g. like period-doubling, when we work with the spectrum shown in Fig. 6.8 and "bringing" or "taking away" any number of momenta $Q = (\pi/a, \pi/a)$ leave an electron close to the Fermi level. Accordingly, in this special case a different combinatorics of interaction lines (similar to that in Keldysh model) appears. We drop this special case for shortness of our presentation.

[16]Obviously, just in the same way we may solve for the case of arbitrary fixed scattering vector Q, when we have alternating $\frac{1}{i\varepsilon_n - \xi_p}$ and $\frac{1}{i\varepsilon_n - \xi_{p-Q}}$. Here we take $Q = 2p_F$ only for compactness of our expressions and because of the physics of Peierls transition.

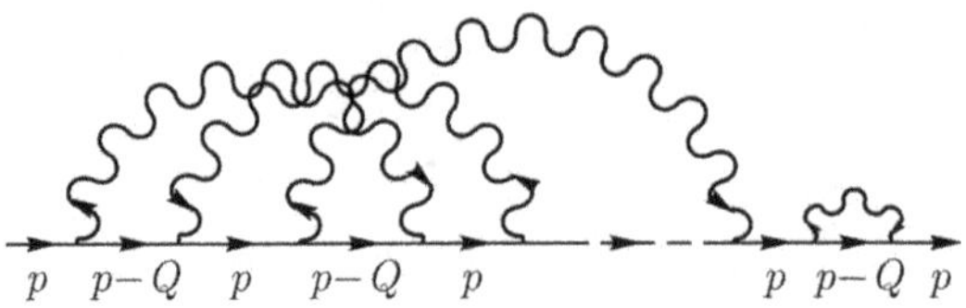

Fig. 6.21   Diagram of an arbitrary order for the single-electron Green's function.

addition we have the factor of Δ^{2n}. Finally, we see that contributions
of all diagrams in a given order just coincide and the total contribution
of this order can be obtained by multiplication by the total number of
these diagrams, which is easily calculated from combinatorics — it is
equal to $n!$. In fact, wee have $2n$ points (vertices), with "incoming"
or "outgoing" interaction lines. Of these, n points are connected with
"outgoing" line, which in any of $n!$ ways may "enter" into the remaining
"open" n vertices. Use now the identity:[17]

$$\sum_{n=0}^{\infty} n! z^n = \sum_{n=0}^{\infty} \int_0^\infty d\zeta e^{-\zeta}(\zeta z)^n = \int_0^\infty d\zeta e^{-\zeta}\frac{1}{1-\zeta z} \qquad (6.202)$$

Then we easily sum the *whole* series for Green's function and obtain an
exact solution of our problem (M.V. Sadovskii, 1974):[18]

$$G(\varepsilon_l p) = \sum_{n=0}^{\infty} \frac{\Delta^{2n} n!}{(i\varepsilon_l - \xi_p)^n (i\varepsilon_l + \xi_p)^n (i\varepsilon_l - \xi_p)} \equiv \sum_{n=0}^{\infty} n! z^n(\varepsilon_l, \xi_p) G_0(\varepsilon_l \xi_p)$$

$$= \int_0^\infty d\zeta e^{-\zeta}\frac{i\varepsilon_l + \xi_p}{(i\varepsilon_l)^2 - \xi_p^2 - \zeta\Delta^2} \equiv\; < G_{\zeta\Delta^2}(\varepsilon_l \xi_p) >, \quad \varepsilon_l = (2l+1)\pi T$$

$$(6.203)$$

where we have used the notation:

$$z(\varepsilon_l, \xi_p) = \Delta^2 G_0(\varepsilon_l, \xi_p) G_0(\varepsilon_l, -\xi_p) \qquad (6.204)$$

and we obtained the "normal" Green's function of the Peierls dielectric:

$$G_{\Delta^2}(\varepsilon_l p) = \frac{i\varepsilon_l + \xi_p}{(i\varepsilon_l)^2 - \xi_p^2 - \Delta^2} \qquad (6.205)$$

[17]As already noted above, in mathematics this procedure is called Borel summation.
[18]Let us stress once again the major difference of this problem from that of an electron,
propagating in coherent periodic field, analyzed above in connection with (6.57), (6.58).

"averaged" according to:

$$< \dots >_\zeta = \int_0^\infty d\zeta e^{-\zeta} \dots \tag{6.206}$$

It is easy to understand (formal proof is given in Appendix B), that
(6.203) represent the Green's function of an electron, which is propagating in an external periodic field of the form $2W\cos(2p_F x + \phi)$,
with amplitude W "fluctuating" according to so-called Rayleigh distribution:[19]

$$\mathcal{P}(W) = \frac{2W}{\Delta^2} e^{-\frac{W^2}{\Delta^2}} \tag{6.207}$$

while the phase ϕ is homogeneously distributed over the interval from
0 to 2π.

Performing analytic continuation $i\varepsilon_l \to \varepsilon \pm i\delta$, from (6.203) we obtain for $\varepsilon > 0$:

$$ImG^{R,A}(\varepsilon\xi_p) = \mp\pi(\varepsilon + \xi_p)\int_0^\infty d\zeta e^{-\zeta}\delta(\varepsilon^2 - \xi_p^2 - \zeta\Delta^2)$$

$$= \mp\frac{\pi}{\Delta^2}(\varepsilon + \xi_p)\theta(\varepsilon^2 - \xi_p^2)e^{-\frac{\varepsilon^2 - \xi_p^2}{\Delta^2}} \tag{6.208}$$

so that spectral density

$$A(\varepsilon\xi_p) = -\frac{1}{\pi}ImG^R(\varepsilon\xi_p) \tag{6.209}$$

acquires "non Fermi-liquid" form, shown in Fig. 6.22.

Let us write for completeness also the analytic expressions for $ReG^{R,A}(\varepsilon\xi_p)$:

$$ReG^{R,A}(\varepsilon\xi_p) = \begin{cases} \frac{\varepsilon+\xi_p}{\Delta^2} e^{-\frac{\varepsilon^2-\xi_p^2}{\Delta^2}} \overline{Ei}\left(\frac{\varepsilon^2-\xi_p^2}{\Delta^2}\right) & \text{for} \quad \varepsilon^2 - \xi_p^2 \geq 0 \\ \frac{\varepsilon+\xi_p}{\Delta^2} e^{\frac{|\varepsilon^2-\xi_p^2|}{\Delta^2}} Ei\left(-\frac{|\varepsilon^2-\xi_p^2|}{\Delta^2}\right) & \text{for} \quad \varepsilon^2 - \xi_p^2 < 0 \end{cases} \tag{6.210}$$

where $Ei(x)$ and $\overline{Ei}(x)$ are integral exponential functions. Let us stress, that our
Green's function (6.203) does nor possess poles in the vicinity of the Fermi level
and, in this sense, does not describe the spectrum of any "elementary excitations"
(quasiparticles), once again demonstrating "non-Fermi liquid" behavior.

---

[19] This distribution is widely used in statistical radiophysics: S.M. Rytov. Introduction
to statistical radiophysics. Part I. "Nauka", Moscow, 1976.

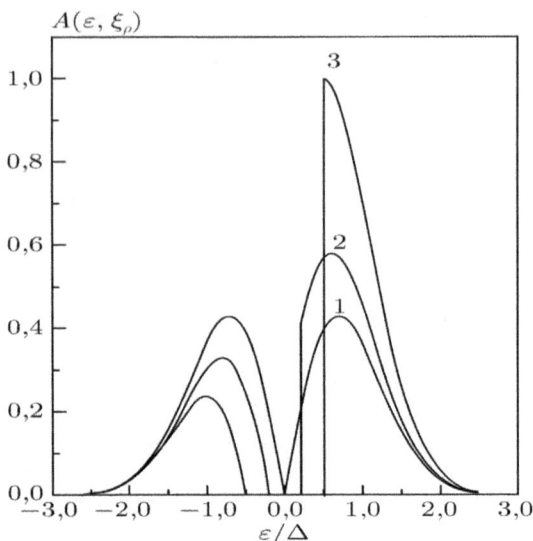

Fig. 6.22   Spectral density in the model of the pseudogap state: (1) $\xi_p = 0$; (2) $\xi_p = 0.1\Delta$; (3) $\xi_p = 0.5\Delta$.

Electron density of states of is now given by:

$$\frac{N(\varepsilon)}{N_0(E_F)} = \left|\frac{\varepsilon}{\Delta}\right| \int_0^{\frac{\varepsilon^2}{\Delta^2}} d\zeta \frac{e^{-\zeta}}{\sqrt{\frac{\varepsilon^2}{\Delta^2} - \zeta}} = 2\left|\frac{\varepsilon}{\Delta}\right| \exp\left(-\frac{\varepsilon^2}{\Delta^2}\right) Erfi\left(\frac{\varepsilon}{\Delta}\right)$$

$$= \begin{cases} 1 & \text{for} & |\varepsilon| \to \infty \\ \frac{2\varepsilon^2}{\Delta^2} & \text{for} & |\varepsilon| \to 0 \end{cases} \tag{6.211}$$

where $N_0(E_F)$ is the density of states of free electrons at the Fermi level, while $Erfi(x) = \int_0^x dx e^{x^2}$ is probability integral of an imaginary argument. Characteristic form of this density of states is shown in Fig. 6.23 and demonstrates the presence of "soft" *pseudogap* in the vicinity of the Fermi level. In fact, it is just the density of states of Peierls dielectric (6.77), shown above in Fig. 6.9, averaged over the gap fluctuations, determined by Rayleigh distribution (6.207).

Generalization of these results for the case of finite correlation lengths $\xi(T)$ (or finite κ) is much more difficult (M.V. Sadovskii, 1979). First of all, we have to learn how to calculate the contribution of an arbitrary diagram in any order. Unfortunately, this problem can not be solved exactly, as integrations become more and more cumbersome. However, we can formulate some very effective

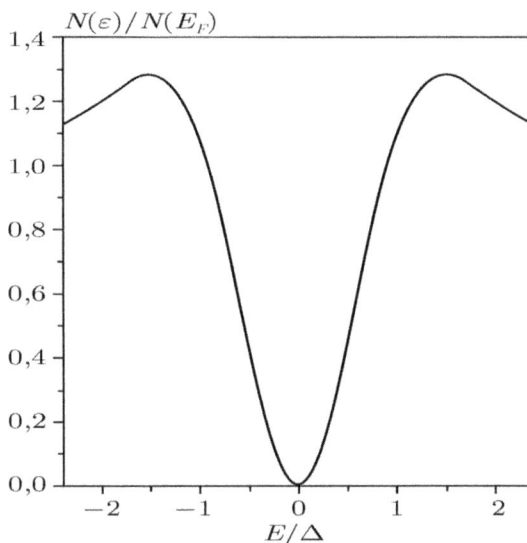

$$N(\varepsilon)/N(E_F)$$

Fig. 6.23 Density of states with pseudogap.

(as we shall see below) approximate *Ansatz*, allowing to write down an explicit expression for any diagram in any order. On Fig. 6.24 we show all essential diagrams of third order. Assume we are working with linearized spectrum of Fig. 6.3, and scattering vector $Q < p_F$, so that scattering takes place only within one branch ("right" or "left") of the spectrum. In this case we *can* calculate the contribution of *any* diagram of the type shown in Fig. 6.24, as it happens that we can guarantee, that only nonzero contributions to integrals come from the poles of the Lorentzian $S(Q)$ (6.196), like in (6.198). This is due to the fact, that electron velocity does not change sign, while we remain within a single branch of the spectrum. For example, after elementary calculations we find that the contribution of Fig. 6.24(d) is given by:

$$\Delta^6 \frac{1}{i\varepsilon_n - \xi_p} \frac{1}{i\varepsilon_n - \xi_{p-Q} + iv_F\kappa} \frac{1}{i\varepsilon_n - \xi_p + 2iv_F\kappa} \frac{1}{i\varepsilon_n - \xi_{p-Q} + 3iv_F\kappa}$$
$$\times \frac{1}{i\varepsilon_n - \xi_p + 2iv_F\kappa} \frac{1}{i\varepsilon_n - \xi_{p-Q} + iv_F\kappa} \frac{1}{i\varepsilon_n - \xi_p} \tag{6.212}$$

Assume now, that ξ_p and ξ_{p-Q} in (6.212) represent the real spectrum of an electron, which is, of course, a continuous function of momenta p. Then we can safely *continue* (6.212) to *any* value of Q, including $Q = 2p_F$. In this case, instead of (6.212) we immediately obtain (remember "nesting" condition (6.61)!):

$$\Delta^6 \frac{1}{i\varepsilon_n - \xi_p} \frac{1}{i\varepsilon_n + \xi_p + iv_F\kappa} \frac{1}{i\varepsilon_n - \xi_p + 2iv_F\kappa} \frac{1}{i\varepsilon_n + \xi_p + 3iv_F\kappa}$$

$$\times \frac{1}{i\varepsilon_n - \xi_p + 2iv_F\kappa} \frac{1}{i\varepsilon_n + \xi_p + iv_F\kappa} \frac{1}{i\varepsilon_n - \xi_p} \tag{6.213}$$

This is the essence of our *Ansatz*! In fact, it is exact in the limit of $\xi \to \infty$ (or $\kappa \to 0$), as is obvious from the direct comparison with the above discussion of this case. Thus, for $Q = 2p_F$ we actually take into account backward scattering (from one branch of the spectrum to the other) by Q exactly, the only approximation made concerns the account of small (at large ξ or small κ) "deviations" from scattering vector $Q = 2p_F$.

Now we can easily see, that contributions of all the other diagrams (calculated in the way just described) are entirely analogous: the numbers over the electron

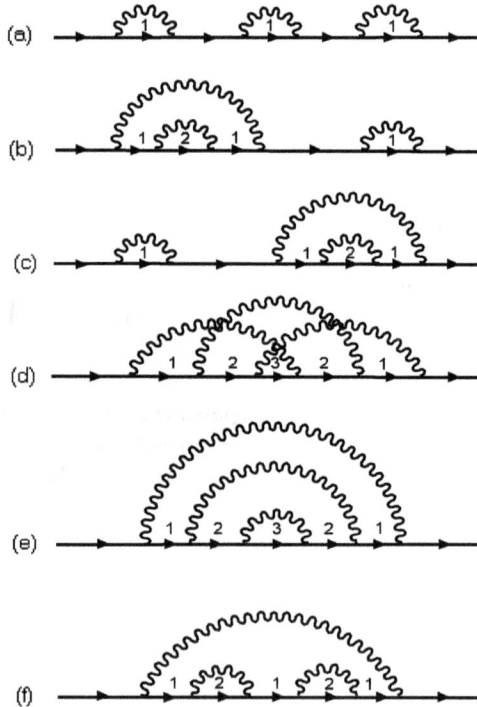

Fig. 6.24   All relevant diagrams of third order.

lines in Fig. 6.24 indicate how many times $iv_F\kappa$ occurs in the corresponding denominator. We note that the contribution of the diagram with crossing interaction lines in Fig. 6.24(d) is equal to that of diagram without crossings, shown in Fig. 6.24(e). We stress that simplicity of the expressions for the contributions of the various diagrams is due entirely to our *Ansatz*, but we shall see that this is, apparently, a very good approximation.

In fourth order there are $4! = 24$ relevant diagrams, all of the irreducible diagrams for self-energy are shown in Fig. 6.25. The corresponding contributions are now easily found and are analogous in form to those obtained in third order, with the use of the numbers over electron lines is as in Fig. 6.24. Furthermore, again there are quite a number of equalities among the diagrams: (a)=(b)=(c)=(d); (e)=(f)=(g)=(h); (i)=(j); (k)=(l).

The general rules for writing out the expression corresponding to an arbitrary diagram are now clear. The contribution of any diagram is determined by the arrangement of initial and final vertices (in Fig. 6.25 they are marked with letters i and f). In each electron line following a vertex of type i a term $iv_F\kappa$ is added in the denominator, and in an electron line following a vertex of type f, such a term is subtracted.

Thus, it is clear that the contribution of any diagram is determined by the arrangement of initial and final vertices. Furthermore any diagram with intersecting interaction lines can be uniquely represented by a diagram without any intersections. The recipe for the construction of the corresponding diagram without intersections (for a given arrangement of i and f vertices) is: Counting from the left, the first final vertex must be connected with an interaction line to the nearest initial vertex on its left, and so on for the remaining vertices not so far connected with interaction lines. Thus, for example, the diagrams of Fig. 6.25(b), (c), (d) reduce to the form of Fig. 6.25(a), the diagrams of Fig. 6.25(e), (f) reduce to the form of Fig. 6.25(g), and so on. For a fixed distribution of initial vertices, the final vertices can be chosen only from the points of opposite parity (as we limit ourselves to incommensurate case only). The numbers put on electron lines in Figs. 6.24 and 6.25 can be transferred to the vertices, by assigning to a vertex the number of terms $iv_F\kappa$ in the denominator corresponding to the line proceeding after that vertex. The general rule is: To an initial vertex is assigned the number $N_n = N_{n-1} + 1$, where N_{n-1} is the number assigned to the nearest vertex on the left. To a final vertex is assigned the number $N_n - 1$. Also $N_0 = 0$, and n is the order number of a vertex.

Let us define:

$$v(k) = \begin{cases} \frac{k+1}{2} & \text{for odd } k \\ \frac{k}{2} & \text{for even } k. \end{cases} \tag{6.214}$$

Then it can be verified that the number of irreducible self-energy diagrams which are equal to a given diagram without intersections of interaction lines is equal

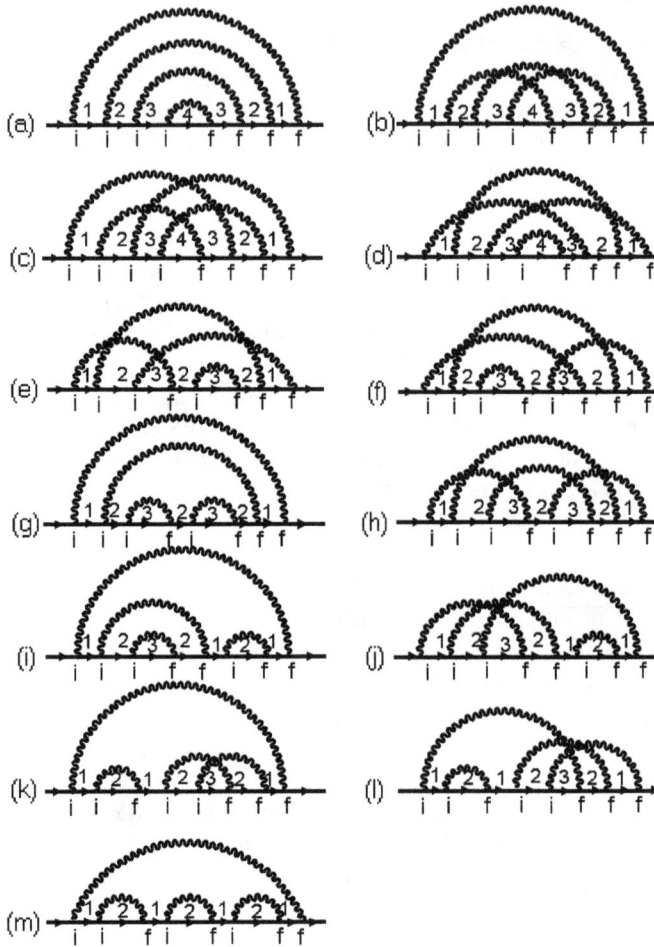

Fig. 6.25   All irreducible diagrams for self-energy in fourth order.

to the product of the factors of $v(N_n)$ for all initial vertices of that diagram (P.V. Elyuitin, 1977).[20] Accordingly, we can conduct all further discussion in terms of diagrams without intersections of interaction lines by applying to all initial vertices the appropriate factors $v(N_n)$.

Any diagram for an irreducible self-energy, when restructured according to the rules that have been formulated above, contains an all-surrounding interaction line, i.e. reduces to the form, shown in Fig. 6.26(a). This allows us to derive

_____

[20]The only change for commensurate case is that we define $v(k) = k$, for any k.

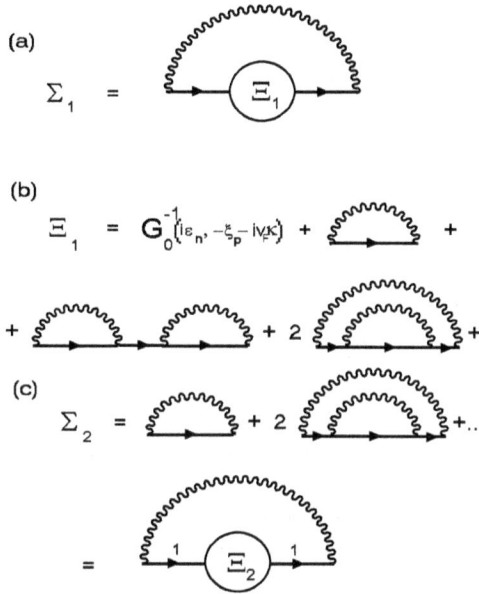

(a)

$$\Sigma_1 \;=\;$$

(b)

$$\Xi_1 \;=\; G_0^{-1}(i\varepsilon_n, -\xi_p - iv_F\kappa) \;+\; \cdots \;+$$

$$+ \quad\cdots\quad + \quad 2 \quad\cdots\quad +$$

(c)

$$\Sigma_2 \;=\; \quad + \; 2 \quad\cdots\quad +\ldots$$

$$=\; \Xi_2$$

Fig. 6.26 Representation of general irreducible self-energy via diagrams without intersecting interaction lines.

recursion equations determining the irreducible self-energy, which includes *all* diagrams of Feynman series. By definition of irreducible self-energy part, we write Dyson equation for the Green's function as:

$$G^{-1}(\varepsilon_n, \xi_p) = G_0^{-1}(\varepsilon_n, \xi_p) - \Sigma_1(\varepsilon_n, \xi_p) \tag{6.215}$$

where

$$\Sigma_1(\varepsilon_n, \xi_p) = \frac{\Delta^2}{(i\varepsilon_n + \xi_p - iv_F\kappa)^2}\Xi_1(\varepsilon_n, \xi_p) = \Delta^2 G_0^2(\varepsilon_n, -\xi_p - iv_F\kappa)\Xi_1(\varepsilon_n, \xi_p) \tag{6.216}$$

and for $\Xi_1(\varepsilon_n, \xi_p)$ we have an expansion shown in Fig. 6.26(b) in terms of diagrams without intersecting interaction lines, with factors $v(N_n)$ attributed to vertices. This expansion can be expressed in the standard way in terms of the corresponding irreducible diagrams:

$$\Xi_1(\varepsilon_n, \xi_p) = G_0^{-2}(\varepsilon_n, -\xi_p - iv_F\kappa)\{G_0^{-1}(\varepsilon_n, -\xi_p - iv_F\kappa) - \Sigma_2(\varepsilon_n, \xi_p)\}^{-1} \tag{6.217}$$

where $G_0(\varepsilon_n, \xi_p)$ denotes the free electron Green's function, and $\Sigma_2(\varepsilon_n, \xi_p)$ can be expressed as a sum of irreducible diagrams shown in Fig. 6.26(c):

$$\Sigma_2(\varepsilon_n, \xi_p) = \Delta^2 v(2) G_0^2(\varepsilon_n, \xi_p - 2iv_F\kappa)\Xi_2(\varepsilon_n, \xi_p) \tag{6.218}$$

$$\Xi_2(\varepsilon_n, \xi_p) = G_0^{-2}(\varepsilon_n, \xi_p - 2iv_F\kappa)\{G_0^{-1}(\varepsilon_n, \xi_p - 2iv_F\kappa) - \Sigma_3(\varepsilon_n, \xi_p)\}^{-1} \tag{6.219}$$

and so on. We have finally:

$$\Sigma_k(\varepsilon_n, \xi_p) = \Delta^2 v(k) G_0^2(\varepsilon_n, (-1)^k \xi_p - ikv_F\kappa)\Xi_k(\varepsilon_n, \xi_p) \tag{6.220}$$

$$\Xi_k(\varepsilon_n, \xi_p) = G_0^{-2}(\varepsilon_n, (-1)\xi_p - ikv_F\kappa)\{G_0^{-1}(\varepsilon_n, (-1)^k\xi_p - ikv_F\kappa) - \Sigma_{k+1}(\varepsilon_n, \xi_p)\}^{-1} \tag{6.221}$$

so that the fundamental recursion relation for self-energy takes the form (M.V. Sadovskii, 1979):

$$\Sigma_k(\varepsilon_n, \xi_p) = \frac{\Delta^2 v(k)}{G_0^{-1}(\varepsilon_n, (-1)^k\xi_p - ikv_F\kappa) - \Sigma_{k+1}(\varepsilon_n, \xi_p)} \tag{6.222}$$

Similarly we can write the recursion formula for the Green's function itself:

$$G_k(\varepsilon_n, \xi_p) = \{i\varepsilon_n - (-1)^k\xi_p + ikv_F\kappa - \Delta^2 v(k+1)G_{k+1}(\varepsilon_n, \xi_p)\}^{-1} \tag{6.223}$$

with *physical* Green's function being determined as $G(\varepsilon_n, \xi_p) \equiv G_{k=0}(\varepsilon_n, \xi_p)$, which represents the sum of all Feynman series for our problem. Actually, these recursion relations yield the representation of the single-electron Green's function in the form of the following *continuous fraction*:

$$G(\varepsilon_n, \xi_p)$$
$$= \cfrac{1}{i\varepsilon_n - \xi_p - \cfrac{\Delta^2}{i\varepsilon_n + \xi_p + iv_F\kappa - \cfrac{\Delta^2}{i\varepsilon_n - \xi_p + 2iv_F\kappa - \cfrac{2\Delta^2}{i\varepsilon_n + \xi_p + 3iv_F\kappa - \ldots}}}} \tag{6.224}$$

Symbolically, this recursion relation can be represented by Dyson-like equation shown diagrammatically in Fig. 6.27.

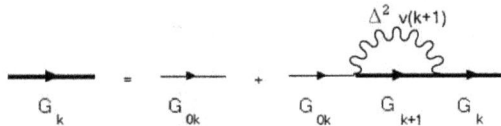

Fig. 6.27   Dyson-like representation of recursion relation for Green's function.

For $\kappa = 0$ we can use continuous fraction representation of the incomplete Γ-function:

$$\Gamma(\alpha, x) = \int_x^\infty dt e^{-t} t^{\alpha-1} = \cfrac{x^\alpha}{x + \cfrac{1-\alpha}{1 + \cfrac{1}{x + \cfrac{2-\alpha}{1+\ldots}}}} \tag{6.225}$$

and the relation $\Gamma(0, x) = -Ei(-x)$ to verify that (6.224), after the usual analytic continuation $i\varepsilon_n \to \varepsilon + i\delta$, reduces to (6.210), (6.208), thus reproducing our exact result (6.203), etc.

From the fundamental recursion relation (6.222), after analytic continuation $i\varepsilon_n \to \varepsilon + i\delta$, we obtain similar relations for real and imaginary parts of self-energy:

$$Re\Sigma_k(\varepsilon, \xi_p) = \frac{\Delta^2 v(k)[\varepsilon - (-1)^k \xi_p - Re\Sigma_{k+1}(\varepsilon, \xi_p)]}{[\varepsilon - (-1)^k \xi_p - Re\Sigma_{k+1}(\varepsilon, \xi_p)]^2 + [kv_F\kappa - Im\Sigma_{k+1}(\varepsilon, \xi_p)]^2}$$
(6.226)

$$Im\Sigma_k(\varepsilon, \xi_p) = \frac{-\Delta^2 v(k)[kv_F\kappa - Im\Sigma_{k+1}(\varepsilon, \xi_p)]}{[\varepsilon - (-1)^k \xi_p - Re\Sigma_{k+1}(\varepsilon, \xi_p)]^2 + [kv_F\kappa - Im\Sigma_{k+1}(\varepsilon, \xi_p)]^2}$$
(6.227)

Next we can use these relations for numerical calculations — start with some large enough (to guarantee convergence) value of k and e.g. $Re\Sigma_{k+1} = Im\Sigma_{k+1} = 0$, and perform calculations down to $k = 1$. In fact, convergence is pretty fast, and calculations take only seconds on any modern PC.

Let us start from spectral density:

$$A(\varepsilon, \xi_p) = -\frac{1}{\pi} Im G^R(\varepsilon, \xi_p) = \frac{Im\Sigma_1(\varepsilon, \xi_p)}{[\varepsilon - \xi_p - Re\Sigma_1(\varepsilon, \xi_p)]^2 + [Im\Sigma_1(\varepsilon, \xi_p)]^2}$$
(6.228)

Results of our calculations are shown in Fig. 6.28 for different values of dimensionless parameter $\Gamma = v_F\kappa/\Delta = v_F\xi^{-1}/\Delta$. As we know, in the case of well defined quasiparticles the spectral density is given by $\delta(\varepsilon - \xi_p)$ or similar narrow peak around the value of quasiparticle energy ξ_p. However, our results show that at small values of Γ, i.e. for large correlation lengths $\xi \gg v_F/\Delta$, our solution contains no contributions of quasiparticle type. Quite opposite, our spectral density shows rather wide double peak (pseudogap) structure due to strong renormalization by short-range order fluctuations (in the limit of $\xi \to \infty$ ($\kappa \to 0$) it reduces to that shown in Fig. 6.22), transforming continuously to a single peak, as we move far from the Fermi level (for large $\varepsilon, \xi_p \gg \Delta$). The second peak in spectral density is usually attributed to the so-called "shadow band". Similarly, at fairly large values of Γ (short correlation lengths $\xi \ll v_F/\Delta$), we also obtain a quasi-free single peak behavior at $\epsilon \sim \xi_p$, corresponding to weakly damped quasiparticles.

The physical reason for the free-like behavior ar small ξ (large κ) is clear. In the limit of $\xi = \kappa^{-1} \to 0$ our effective interaction (6.196) with fluctuations becomes short-ranged, but is not reduced to the common "white noise" limit. Although all momenta in the integral (for self-energy) over Q become important, the scattering amplitude itself becomes $\sim \Delta^2/\kappa$, so that scattering rate, estimated (via Fermi "golden rule") as $2\pi N_0(E_F)\Delta^2/\kappa = \Delta^2/v_F\kappa = \Delta/\Gamma \to 0$, as $\kappa \to \infty$ (where we used $N_0(E_F) = 1/2\pi v_F$ for electron density of states at the Fermi level for one-dimensional case). Correspondingly, in the limit of $\kappa \to \infty$ electrons become effectively "free" (as they do also if we move far from the Fermi level).

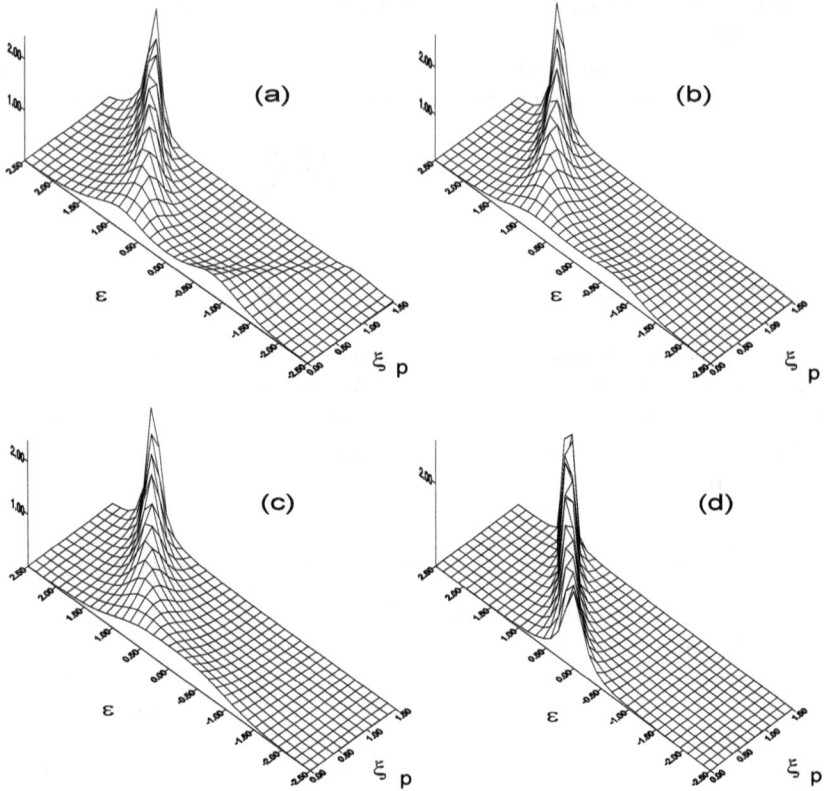

Fig. 6.28   The surfaces of spectral density $A(\varepsilon, \xi_p)$ for: (a) $\Gamma = 0.1$; (b) $\Gamma = 0.5$; (c) $\Gamma = 1.0$; (d) $\Gamma = 5.0$. All energies are in units of Δ.

Now let us calculate density of states:

$$\frac{N(\varepsilon)}{N_0(E_F)} = \int_{-\infty}^{\infty} d\xi_p A(\varepsilon, \xi_p) \tag{6.229}$$

Results for different value of $\Gamma = v_F \kappa / \Delta = v_F \xi^{-1}/\Delta$ are shown in Fig. 6.29 with full curves. We see that for finite $\kappa = \xi^{-1}$ density of states now is finite at the Fermi level (cf. Fig. 6.23). Pseudogap is gradually smeared (or "filled") by additional scattering due to finiteness of correlation length[21] and completely vanishes for $v_F \kappa \gg \Delta$.

---

[21]Physically, we are dealing now with an electron, propagating in the system of random one-dimensional "clusters" of length $\sim \xi$, with "periodic" field $2W \cos(Qx + \phi)$ with $Q \sim 2p_F$ within each of the "clusters" and random amplitude W, distributed "almost" according to (6.207).

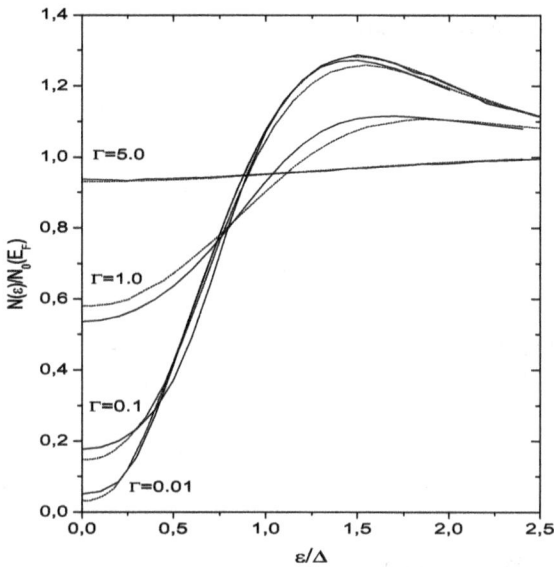

Fig. 6.29 Density of states with pseudogap for different values of $\Gamma = v_F \kappa / \Delta$. Full lines — our approximation. Dashed lines — results of exact numerical simulation (L. Bartosch, P. Kopietz, 1999).

Dashed curves in Fig. 6.29 show the results of *exact* numerical simulation of the density of states for our (one-dimensional) problem, obtained by "crude force", i.e. via direct solution of Schroedinger equation for many configurations of Gaussian random field (with correlator (6.196)) subsequent averaging (L. Bartosch, P. Kopietz, 1999). We can clearly see that our approximation (based on the *Ansatz* (6.213) for contribution of higher-order diagrams) is in fact very good *quantitatively*, probably except the close vicinity of the Fermi level (center of the pseudogap)[22] (cf. (6.199), (6.200)). Obviously, our method has many advantages in comparison with "direct" numerical approaches, it is much less time-consuming, and also it can be generalized to more complicated situations, e.g. to the study of pseudogaps in two-dimensions (very important in high-temperature copper oxide superconductors).[23]

---

[22]In case of commensurate fluctuations similar comparison shows, that our *Ansatz* is less accurate — it misses the formation of so-called Dyson singularity in the center of the pseudogap (L. Bartosch, P. Kopietz, 1999).

[23]We refer the reader for further discussion and references to our review paper: M.V. Sadovskii. Physics Uspekhi **44**, 515 (2001).

6.6.3   *Electromagnetic response*

Remarkable property of the model under consideration is the availability of an exact (in the limit of correlation length $\xi \to \infty$) solution (i.e. our ability to sum *all* diagrams) also for the response function, describing reaction to an external electromagnetic field (polarization operator) (M.V. Sadovskii, 1974).

First of all, let us write down some general relations in zero-temperature technique ($T = 0$). Apply to our system a small perturbing external vector-potential:

$$\delta H_{int} = -\frac{e}{mc} \int d^3 r \psi^+(\mathbf{r}) \mathbf{p} \cdot \delta \mathbf{A}(\mathbf{r}t) \psi(\mathbf{r}) \qquad (6.230)$$

where $\delta \mathbf{A}(\mathbf{r}t) = \delta \mathbf{A}_{\mathbf{q}\omega} e^{i\mathbf{q}\mathbf{r} - i\omega t}$. Appropriate variation of the single-electron Green's function can be written as [Abrikosov A.A., Gorkov L.P., Dzyaloshinskii I.E. (1963)]:

$$\delta G(\varepsilon\mathbf{p}) = -G(\varepsilon\mathbf{p}) \frac{e}{mc} (\mathbf{p} \cdot \delta \mathbf{A}_{\mathbf{q}\omega}) G(\varepsilon + \omega\mathbf{p} + \mathbf{q}) + iG(\varepsilon\mathbf{p}) G(\varepsilon + \omega\mathbf{p} + \mathbf{q})$$

$$\times \int \frac{d^3 p'}{(2\pi)^3} \int \frac{d\varepsilon'}{2\pi} \Gamma(\varepsilon\mathbf{p}, \varepsilon'\mathbf{p}'; \mathbf{q}\omega) G(\varepsilon'\mathbf{p}') \frac{e}{mc} (\mathbf{p}' \cdot \delta \mathbf{A}_{\mathbf{q}\omega}) G(\varepsilon' + \omega\mathbf{p}' + \mathbf{q})$$

$$(6.231)$$

or

$$\delta G(\varepsilon\mathbf{p}) = G(\varepsilon\mathbf{p}) \mathbf{J}(\varepsilon\mathbf{p}; \varepsilon + \omega\mathbf{p} + \mathbf{q}) G(\varepsilon + \omega\mathbf{p} + \mathbf{q}) \delta \mathbf{A}_{\mathbf{q}\omega} \qquad (6.232)$$

which is shown diagrammatically in Fig. 6.30. From here, by the way, it is clear that for the free-electron Green's function:

$$\delta G_0(\varepsilon\mathbf{p}) = -G_0(\varepsilon\mathbf{p}) \frac{e}{mc} \mathbf{p} G_0(\varepsilon + \omega\mathbf{p} + \mathbf{q}) \delta \mathbf{A}_{\mathbf{q}\omega}$$

$$\equiv G_0(\varepsilon\mathbf{p}) \mathbf{J}_0(\mathbf{p}; \mathbf{p} + \mathbf{q}) G_0(\varepsilon + \omega\mathbf{p} + \mathbf{q}) \delta \mathbf{A}_{\mathbf{q}\omega} \qquad (6.233)$$

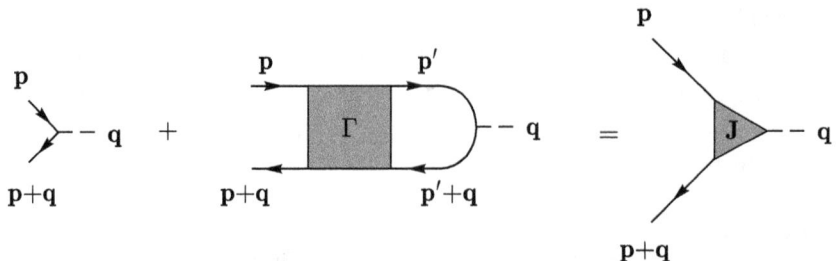

Fig. 6.30   Variation of Green's function due to a small external electromagnetic field.

where

$$\mathbf{J}_0(\mathbf{p};\mathbf{p}+\mathbf{q}) = -\frac{e}{mc}\mathbf{p} \tag{6.234}$$

is the "current" vertex for the free particle. The full vertex is defined from (6.232) as:

$$\mathbf{J}(\mathbf{p};\mathbf{p}+\mathbf{q}) = -\frac{\delta G^{-1}(\varepsilon\mathbf{p})}{\delta \mathbf{A}_{\mathbf{q}\omega}} \tag{6.235}$$

Quite similar expressions appear also for the case of response to an external scalar potential:

$$\delta H_{int} = e\int d^3 r \psi^+(\mathbf{r})\delta\varphi(\mathbf{r}t)\psi(\mathbf{r}) \tag{6.236}$$

where $\delta\varphi(\mathbf{r}t) = \delta\varphi_{\mathbf{q}\omega}e^{i\mathbf{q}\mathbf{r}-i\omega t}$. In particular, similarly to (6.231) we have:

$$\delta G(\varepsilon\mathbf{p}) = G(\varepsilon\mathbf{p})e\delta\varphi_{\mathbf{q}\omega}G(\varepsilon+\omega\mathbf{p}+\mathbf{q}) - iG(\varepsilon\mathbf{p})G(\varepsilon+\omega\mathbf{p}+\mathbf{q})$$
$$\times \int \frac{d^3 p'}{(2\pi)^3}\int\frac{d\varepsilon'}{2\pi}\Gamma(\varepsilon\mathbf{p},\varepsilon'\mathbf{p}';\mathbf{q}\omega)G(\varepsilon'\mathbf{p}')e\delta\varphi_{\mathbf{q}\omega}G(\varepsilon'+\omega\mathbf{p}'+\mathbf{q})$$
$$\tag{6.237}$$

or

$$\delta G(\varepsilon\mathbf{p}) = G(\varepsilon\mathbf{p})J_0(\varepsilon\mathbf{p};\varepsilon+\omega\mathbf{p}+\mathbf{q})G(\varepsilon+\omega\mathbf{p}+\mathbf{q})\delta\varphi_{\mathbf{q}\omega} \tag{6.238}$$

where we have defined "scalar" vertex $J_0(p;p+q)$:

$$J^0(p;p+q) = -\frac{\delta G^{-1}(\varepsilon\mathbf{p})}{\delta\varphi_{\mathbf{q}\omega}} \tag{6.239}$$

Diagrammatically (6.238) is again expressed by Fig. 6.30. Analogously to (6.233):

$$\delta G_0(\varepsilon\mathbf{p}) = G_0(\varepsilon\mathbf{p})G_0(\varepsilon+\omega\mathbf{p}+\mathbf{q})e\delta\varphi_{\mathbf{q}\omega}$$
$$\equiv G_0(\varepsilon\mathbf{p})J_0^0(\mathbf{p};\mathbf{p}+\mathbf{q})G_0(\varepsilon+\omega\mathbf{p}+\mathbf{q})\delta\varphi_{\mathbf{q}\omega} \tag{6.240}$$

where $J_0^0(p;p+q) = e$ is the "free" vertex.

It is convenient to introduce a general definition of the vertex:

$$J^\mu(p;p+q) = -\frac{\delta G^{-1}(\varepsilon\mathbf{p})}{\delta A_\mu(\mathbf{q}\omega)} \tag{6.241}$$

where $A_\mu(\mathbf{q}\omega) = \{\varphi_{\mathbf{q}\omega}, \mathbf{A}_{\mathbf{q}\omega}\}$, so that the "free" vertex is:

$$J_0^\mu(p; p+q) = \begin{cases} -\frac{e}{mc}\mathbf{p} & \mu = 1, 2, 3 \\ e & \mu = 0 \end{cases} \qquad (6.242)$$

or

$$\frac{\delta G_0(\varepsilon\mathbf{p})}{\delta A_\mu(\mathbf{q}\omega)} = G_0(\varepsilon\mathbf{p}) J_0^\mu(p; p+q) G_0(\varepsilon + \omega\mathbf{p} + \mathbf{q}) \qquad (6.243)$$

while for the "full" Green's function:

$$\frac{\delta G(\varepsilon\mathbf{p})}{\delta A_\mu(\mathbf{q}\omega)} = G(\varepsilon\mathbf{p}) J^\mu(p; p+q) G(\varepsilon + \omega\mathbf{p} + \mathbf{q}) \qquad (6.244)$$

Now it is time to start! Let us consider our model in the asymptotic limit of large correlation lengths $\xi \to \infty$. From previous discussion it is clear, that an arbitrary diagram, describing response to an external field, can be obtained from the arbitrary diagram for the single-electron Green's function (of the type shown in Fig. 6.21) by "insertion" of the line of external field into any of electron lines of this diagram, as it is shown in Fig. 6.31 (and we have to do *all* possible insertions!). Performing such "differentiation" of the whole series (6.203),[24] we obtain (m-number of the "block" z, to which we insert the line of $A^\mu(\mathbf{q}\omega)$):

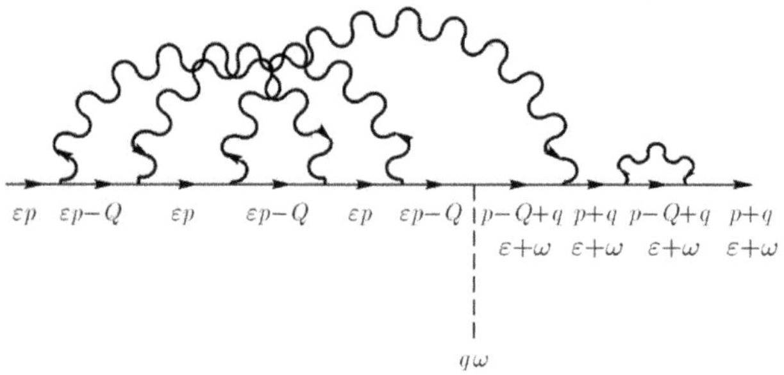

$$\varepsilon p \quad \varepsilon p - Q \quad \varepsilon p \quad \varepsilon p - Q \quad \varepsilon p \quad \varepsilon p - Q \underset{|}{\vert} p - Q + q \quad p + q \quad p - Q + q \quad p + q$$
$$\varepsilon + \omega \quad \varepsilon + \omega \quad \varepsilon + \omega \quad \varepsilon + \omega$$
$$q\omega$$

Fig. 6.31 Diagram of an arbitrary order for the vertex correction, describing interaction with external electromagnetic field.

---

[24]In fact, we are explicitly calculating functional derivative of the whole perturbation series for the single-particle Green's function and "generating" all diagrams for the appropriate vertex-part.

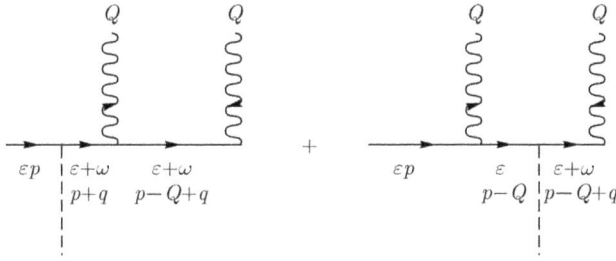

Fig. 6.32 Diagrams for functional derivative of "block" $z(\varepsilon \mathbf{p})$.

$$\frac{\delta G(\varepsilon \mathbf{p})}{\delta A_\mu(\mathbf{q}\omega)} = \left\langle \sum_{n=1}^{\infty} \sum_{m=1}^{n} (\zeta z(\varepsilon \mathbf{p}))^{m-1} \zeta \frac{\delta z}{\delta A_\mu(\mathbf{q}\omega)} (\zeta z(\varepsilon+\omega \mathbf{p}+\mathbf{q}))^{n-m} G_0(\varepsilon+\omega \mathbf{p}+\mathbf{q}) \right.$$

$$\left. + \sum_{n=0}^{\infty} (\zeta z(\varepsilon \mathbf{p}))^n \frac{\delta G_0(\varepsilon \mathbf{p})}{\delta A_\mu(\mathbf{q}\omega)} \right\rangle_\zeta \qquad (6.245)$$

Here $\frac{\delta G_0(\varepsilon \mathbf{p})}{\delta A_\mu(\mathbf{q}\omega)}$ is defined by (6.243), and derivative of the "block" $z(\varepsilon \mathbf{p})$ is determined by Fig. 6.32 and is equal to:

$$\frac{\delta z(\varepsilon \mathbf{p})}{\delta A_\mu(\mathbf{q}\omega)} = \Delta^2 G_0(\varepsilon \mathbf{p}) J_0^\mu(\mathbf{p};\mathbf{p}+\mathbf{q}) G_0(\varepsilon+\omega \mathbf{p}+\mathbf{q}) G_0(\varepsilon+\omega \mathbf{p}-\mathbf{Q}+\mathbf{q})$$

$$+\Delta^2 G_0(\varepsilon \mathbf{p}) G_0(\varepsilon \mathbf{p}-\mathbf{Q}) J_0^\mu(\mathbf{p}-\mathbf{Q};\mathbf{p}-\mathbf{Q}+\mathbf{q}) G_0(\varepsilon+\omega \mathbf{p}-\mathbf{Q}+\mathbf{q})$$

$$= G_0(\varepsilon \mathbf{p}) J_0^\mu(\mathbf{p};\mathbf{p}+\mathbf{q}) z(\varepsilon+\omega \mathbf{p}+\mathbf{q}) + z(\varepsilon \mathbf{p}) J_0^\mu(\mathbf{p}-\mathbf{q};\mathbf{p}-\mathbf{Q}+\mathbf{q})$$

$$G_0(\varepsilon+\omega \mathbf{p}-\mathbf{Q}+\mathbf{q}) \qquad (6.246)$$

Substituting (6.246) into (6.245), and taking into account (6.243), we obtain:

$$\frac{\delta G(\varepsilon \mathbf{p})}{\delta A_\mu(\mathbf{q}\omega)} = \left\langle \underbrace{\sum_{n=1}^{\infty} \sum_{m=1}^{n} (\zeta z(\varepsilon \mathbf{p}))^{m-1} (\zeta z(\varepsilon+\omega \mathbf{p}+\mathbf{q}))^{n-m+1} G_0(\varepsilon \mathbf{p}) J_0^\mu(\mathbf{p};\mathbf{p}+\mathbf{q}) G_0(\varepsilon+\omega \mathbf{p}+\mathbf{q})}_{I} \right.$$

$$+ \underbrace{\sum_{n=1}^{\infty} \sum_{m=1}^{n} (\zeta z(\varepsilon \mathbf{p}))^{m} (\zeta z(\varepsilon+\omega \mathbf{p}+\mathbf{q}))^{n-m} G_0(\varepsilon+\omega \mathbf{p}-\mathbf{Q}+\mathbf{q}) J_0^\mu(\mathbf{p}-\mathbf{Q};\mathbf{p}-\mathbf{Q}+\mathbf{q}) G_0(\varepsilon+\omega \mathbf{p}+\mathbf{q})}_{II}$$

$$\left. + \underbrace{\sum_{n=0}^{\infty} (\zeta z(\varepsilon \mathbf{p}))^{n} G_0(\varepsilon \mathbf{p}) J_0^\mu(\mathbf{p};\mathbf{p}+\mathbf{q}) G_0(\varepsilon+\omega \mathbf{p}+\mathbf{q})}_{III} \right\rangle_\zeta \qquad (6.247)$$

The first and third terms of this expression together give:

$$< I + III >_\zeta = \left\langle J_0^\mu(\mathbf{p}; \mathbf{p} + \mathbf{q}) G_0(\varepsilon \mathbf{p}) G_0(\varepsilon + \omega \mathbf{p} + \mathbf{q}) \left\{ \sum_{n=0}^{\infty} \zeta^n {}_z(\varepsilon \mathbf{p}) \right.\right.$$

$$\left.\left. + \sum_{n=1}^{\infty} \sum_{m=1}^{n} \zeta^{m-1} {}_z{}^{m-1}(\varepsilon \mathbf{p}) \zeta^{n-m+1} {}_z{}^{n-m+1}(\varepsilon + \omega \mathbf{p} + \mathbf{q}) \right\} \right\rangle_\zeta = (m - 1 \to m)$$

$$= J_0^\mu(\mathbf{p}; \mathbf{p} + \mathbf{q}) G_0(\varepsilon \mathbf{p}) G_0(\varepsilon + \omega \mathbf{p} + \mathbf{q}) \left\langle \sum_{n=0}^{\infty} \sum_{m=0}^{n} \zeta^m {}_z{}^m(\varepsilon \mathbf{p}) \zeta^{n-m} {}_z{}^{n-m}(\varepsilon + \omega \mathbf{p} + \mathbf{q}) \right\rangle_\zeta$$

$$= \left\langle J_0^\mu(\mathbf{p}; \mathbf{p} + \mathbf{q}) G_0(\varepsilon \mathbf{p}) G_0(\varepsilon + \omega \mathbf{p} + \mathbf{q}) \sum_{n=0}^{\infty} \zeta^n {}_z{}^n(\varepsilon \mathbf{p}) \sum_{m=0}^{\infty} \zeta^m {}_z{}^m(\varepsilon + \omega \mathbf{p} + \mathbf{q}) \right\rangle_\zeta \qquad (6.248)$$

where we have used the standard rule for multiplication of series: $\left(\sum_{n=0}^{\infty} a_n\right)\left(\sum_{m=0}^{\infty} b_m\right) = \sum_{n=0}^{\infty} \sum_{m=0}^{n} a_n b_{n-m}$.

Then, using (6.203), we get:

$$< I + III >_\zeta = J_0^\mu(\mathbf{p}; \mathbf{p} + \mathbf{q}) < G_{\zeta\Delta^2}(\varepsilon \mathbf{p}) G_{\zeta\Delta^2}(\varepsilon + \omega \mathbf{p} + \mathbf{q}) >_\zeta \qquad (6.249)$$

where

$$G_{\Delta^2}(\varepsilon \mathbf{p}) = \frac{\varepsilon + \xi_p}{\varepsilon^2 - \xi_p^2 - \Delta^2} \qquad (\varepsilon \to \varepsilon \pm i\delta) \qquad (6.250)$$

is the normal Green's function of Peierls dielectric.

Analogous calculations for the second term in (6.247) give:

$$< II >_\zeta = J_0^\mu(\mathbf{p} - \mathbf{Q}; \mathbf{p} - \mathbf{Q} + \mathbf{q}) G_0(\varepsilon + \omega \mathbf{p} - \mathbf{Q} + \mathbf{q}) G_0(\varepsilon + \omega \mathbf{p} + \mathbf{q})$$

$$\times \left\langle \sum_{n=1}^{\infty} \sum_{m=1}^{n} \zeta^m z^m(\varepsilon \mathbf{p}) \zeta^{n-m} z^{n-m}(\varepsilon + \omega \mathbf{p} + \mathbf{q}) \right\rangle_\zeta$$

$$= J_0^\mu(\mathbf{p} - \mathbf{Q}; \mathbf{p} - \mathbf{Q} + \mathbf{q}) \left\langle \sum_{n=1}^{\infty} \sum_{m=1}^{n} \zeta^m z^m(\varepsilon \mathbf{p}) \zeta^{n-m} z^{n-m+1}(\varepsilon + \omega \mathbf{p} + \mathbf{q}) \frac{1}{\zeta \Delta^2} \right\rangle_\zeta$$

$$= J_0^\mu(\mathbf{p} - \mathbf{Q}; \mathbf{p} - \mathbf{Q} + \mathbf{q}) \left\langle \frac{1}{\zeta \Delta^2} \sum_{n=1}^{\infty} \zeta^n z^n(\varepsilon \mathbf{p}) \sum_{m=1}^{\infty} \zeta^m z^m(\varepsilon + \omega \mathbf{p} + \mathbf{q}) \right\rangle_\zeta$$

$$= J_0^\mu(\mathbf{p} - \mathbf{Q}; \mathbf{p} - \mathbf{Q} + \mathbf{q}) < F_{\zeta\Delta^2}(\varepsilon \mathbf{p}) F_{\zeta\Delta^2}^+(\varepsilon + \omega \mathbf{p} + \mathbf{q}) >_\zeta \qquad (6.251)$$

where appeared "ζ-average" of the product of two *anomalous* Green's functions of Peierls dielectric:

$$F_{\Delta^2}^+(\varepsilon) = \frac{\Delta}{\varepsilon^2 - \xi_p^2 - \Delta^2} \qquad (\varepsilon \to \varepsilon \pm i\delta) \qquad (6.252)$$

despite the obvious absence of Peierls long-range order in the problem we are discussing! Here we used $\left(\sum_{n=1}^{\infty} a_n\right)\left(\sum_{m=1}^{\infty} b_m\right) =$

$\sum_{n=1}^{\infty} \sum_{m=1}^{n} a_n b_{n-m+1}$ and summed progressions in the term before the last one in (6.251).

Thus, finally we obtain:

$$\frac{\delta G(\varepsilon \mathbf{p})}{\delta A_\mu(\mathbf{q}\omega)} = G(\varepsilon \mathbf{p}) J^\mu G(\varepsilon + \omega \mathbf{p} + \mathbf{q})$$

$$= \int_0^\infty d\zeta e^{-\zeta} \{ G_{\zeta \Delta^2}(\varepsilon \mathbf{p}) J_0^\mu(\mathbf{p}; \mathbf{p} + \mathbf{q}) G_{\zeta \Delta^2}(\varepsilon + \omega \mathbf{p} + \mathbf{q})$$

$$+ F_{\zeta \Delta^2}(\varepsilon \mathbf{p}) J_0^\mu(\mathbf{p} - \mathbf{Q}; \mathbf{p} - \mathbf{Q} + \mathbf{q}) F_{\zeta \Delta^2}^+(\varepsilon + \omega \mathbf{p} + \mathbf{q}) \} \quad (6.253)$$

which can be expressed by diagrams shown in Fig. 6.33. Let us stress that this result was obtained by summation of *all* diagrams of perturbation theory for the vertex part. This answer is "almost obvious", if we remember the nature of the random field, scattering an electron in our problem (cf. remarks after Eq. (6.206)) — we have to obtain Peierls "dielectric" response, averaged over gap fluctuations.

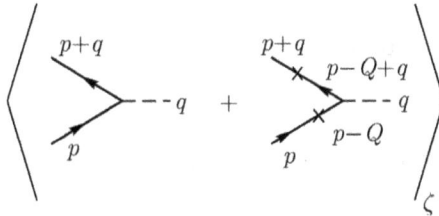

Fig. 6.33 Diagrammatic representation of functional derivative of $G(\varepsilon \mathbf{p})$.

Now, the appropriate polarization operator (which we write down in Matsubara technique) is:

$$\Pi(\mathbf{q}\omega_m) = \int_0^\infty d\zeta e^{-\zeta} 2T \sum_n \int_{-\infty}^\infty \frac{dp}{2\pi} \{ G_{\zeta \Delta^2}(\varepsilon_n \mathbf{p}) G_{\zeta \Delta^2}(\varepsilon_n + \omega_m \mathbf{p} + \mathbf{q})$$

$$+ F_{\zeta \Delta^2}(\varepsilon_n \mathbf{p}) F_{\zeta \Delta^2}^+(\varepsilon_n + \omega_m \mathbf{p} + \mathbf{q}) \} = < \Pi_{\zeta \Delta^2}(\mathbf{q}\omega_m) >_\zeta$$

$$(6.254)$$

which is represented by diagrams shown in Fig. 6.34. We see that under the averaging procedure over gap fluctuations we have here just the polarization operator of Peierls dielectric. Accordingly, the structure of our (exact!) solution for electromagnetic response is clear — we

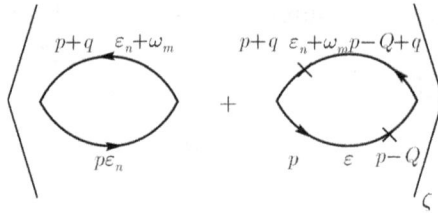

Fig. 6.34   Polarization operator for the model of pseudogap state.

have to calculate the response of Peierls insulator (with fixed gap in the spectrum) and then average over gap fluctuations (with Rayleigh distribution (6.207)). Thus, during calculations which follow we, in fact, are analyzing two physical problems — that of Peierls dielectric response to electromagnetic field, and that of the response in exactly solvable model of the pseudogap state.

Let us now perform detailed calculations of $\Pi_{\Delta^2}(\mathbf{q}\omega_m)$ — polarization operator of Peierls insulator with fixed gap. Substituting into the relevant expression (directly following from (6.254) both normal and anomalous Green's functions of Peierls insulator:[25]

$$G_{\Delta^2}(\varepsilon_n \mathbf{p}) = \frac{u_p^2}{i\varepsilon_n - E_p} + \frac{v_p^2}{i\varepsilon_n + E_p} \qquad (6.255)$$

$$F_{\Delta^2}^+(\varepsilon_n \mathbf{p}) = \frac{\Delta}{(i\varepsilon_n - E_p)(i\varepsilon_n + E_p)} \qquad (6.256)$$

where

$$u_p^2 = \frac{1}{2}\left\{1 + \frac{\xi_p}{E_p}\right\}, \qquad v_p^2 = \frac{1}{2}\left\{1 - \frac{\xi_p}{E_p}\right\} \qquad (6.257)$$

with $E_p = \sqrt{\xi_p^2 + \Delta^2}$, we write the sum over Fermion frequencies via the contour integral (3.38) and obtain:

$$\Pi_{\Delta^2}(\mathbf{q}\omega_m) = -2\int_{-\infty}^{\infty}\frac{dp}{2\pi i}\int_C \frac{d\varepsilon}{2\pi} n(\varepsilon)\left\{\frac{u_p^2 u_{p+q}^2}{(\varepsilon - E_p)(\varepsilon + i\omega_m - E_{p+q})}\right.$$
$$+ \frac{v_p^2 v_{p+q}^2}{(\varepsilon + E_p)(\varepsilon + i\omega_m + E_{p+q})} + \frac{u_p^2 v_{p+q}^2}{(\varepsilon - E_p)(\varepsilon + i\omega_m + E_{p+q})}$$
$$\left. + \frac{v_p^2 u_{p+q}^2}{(\varepsilon + E_p)(\varepsilon + i\omega_m - E_{p+q})} + \frac{\Delta^2}{(\varepsilon + E_p)(\varepsilon - E_p)(\varepsilon + i\omega_m - E_{p+q})(\varepsilon + i\omega_m + E_{p+q})}\right\} \qquad (6.258)$$

[25]In the following we may assume Δ to be real.

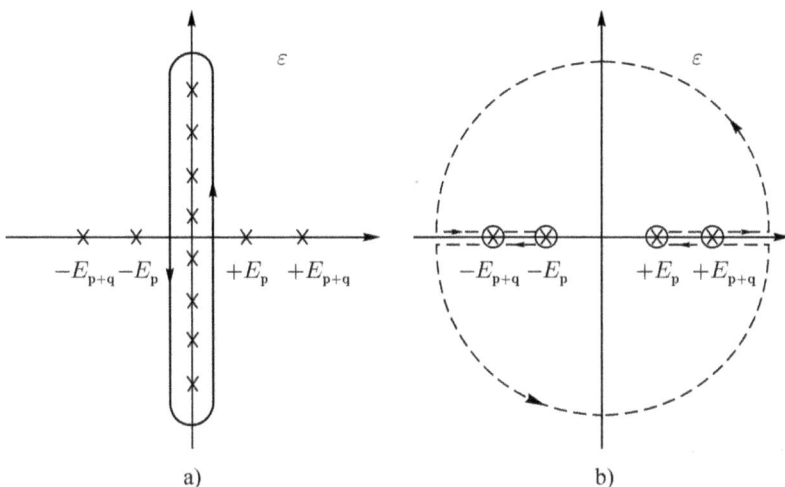

Fig. 6.35 Integration contours used during the calculation of the sum over frequencies in polarization operator.

where integration contour C is shown in Fig. 6.35(a). This contour may be deformed as shown in Fig. 6.35(b), and then "stretched" to infinity. Then our integral is determined by contributions of four poles $\varepsilon = \pm E_p$ and $\varepsilon = \pm E_{p+q}$. Calculating appropriate residues and using the property of Fermi function: $n(\varepsilon + i\omega_m) = n(\varepsilon)$, where $\omega_m = 2\pi m T$, as well as $n(-\varepsilon) = 1 - n(\varepsilon)$, changing integration variable from p to ξ_p (taking into account both "ends" of the Fermi surface (line), giving an additional factor of 2), we obtain:

$$
\Pi_{\Delta^2}(\mathbf{q}\omega_m)
$$
$$
= 2N_0(E_F)\int_{-\infty}^{\infty} d\xi_p \frac{1}{E_p - E_{p+q} + i\omega_m}[n(E_p) - n(E_{p+q})]\left\{u_p^2 u_{p+q}^2 + \frac{\Delta^2}{4E_p E_{p+q}}\right\}
$$
$$
+ 2N_0(E_F)\int_{-\infty}^{\infty} d\xi_p \frac{1}{E_p - E_{p+q} - i\omega_m}[n(E_p) - n(E_{p+q})]\left\{v_p^2 v_{p+q}^2 + \frac{\Delta^2}{4E_p E_{p+q}}\right\}
$$
$$
+ 2N_0(E_F)\int_{-\infty}^{\infty} d\xi_p \frac{1}{E_p + E_{p+q} + i\omega_m}[n(E_p) + n(E_{p+q}) - 1]\left\{u_p^2 v_{p+q}^2 - \frac{\Delta^2}{4E_p E_{p+q}}\right\}
$$
$$
+ 2N_0(E_F)\int_{-\infty}^{\infty} d\xi_p \frac{1}{E_p + E_{p+q} - i\omega_m}[n(E_p) + n(E_{p+q}) - 1]\left\{v_p^2 u_{p+q}^2 - \frac{\Delta^2}{4E_p E_{p+q}}\right\}
$$
$$
(6.259)
$$

where $N_0(E_F)$ is free-electron density of states at the Fermi level (for both spin projections). Now use:

$$u_p^2 u_{p+q}^2 = \frac{1}{4}\left\{1 + \frac{\xi_p \xi_{p+q}}{E_p E_{p+q}} + \frac{\xi_p}{E_p} + \frac{\xi_{p+q}}{E_{p+q}}\right\},$$

$$v_p^2 v_{p+q}^2 = \frac{1}{4}\left\{1 + \frac{\xi_p \xi_{p+q}}{E_p E_{p+q}} - \frac{\xi_p}{E_p} - \frac{\xi_{p+q}}{E_{p+q}}\right\}$$

$$u_p^2 v_{p+q}^2 = \frac{1}{4}\left\{1 - \frac{\xi_p \xi_{p+q}}{E_p E_{p+q}} + \frac{\xi_p}{E_p} - \frac{\xi_{p+q}}{E_{p+q}}\right\},$$

$$v_p^2 u_{p+q}^2 = \frac{1}{4}\left\{1 - \frac{\xi_p \xi_{p+q}}{E_p E_{p+q}} - \frac{\xi_p}{E_p} + \frac{\xi_{p+q}}{E_{p+q}}\right\} \qquad (6.260)$$

Terms linear over ξ_p and ξ_{p+q} drop after the integration due to odd parity of the integrand.

After the analytic continuation $i\omega_m \to \omega + i\delta$ we finally get:[26]

$$\Pi_{\Delta^2}(\mathbf{q}\omega)$$

$$= \frac{1}{2}N_0(E_F)\int_{-\infty}^{\infty} d\xi_p \left\{\frac{E_p E_{p+q} + \xi_p \xi_{p+q} + \Delta^2}{E_p E_{p+q}}\right\} [n(E_p) - n(E_{p+q})]$$

$$\times \left\{\frac{1}{E_p - E_{p+q} + \omega + i\delta} + \frac{1}{E_p - E_{p+q} - \omega - i\delta}\right\}$$

$$- \frac{1}{2}N_0(E_F)\int_{-\infty}^{\infty} \left\{\frac{E_p E_{p+q} - \xi_p \xi_{p+q} - \Delta^2}{E_p E_{p+q}}\right\} [1 - n(E_p) - n(E_{p+q})]$$

$$\times \left\{\frac{1}{E_p + E_{p+q} + \omega + i\delta} + \frac{1}{E_p + E_{p+q} - \omega - i\delta}\right\} \qquad (6.261)$$

This is the general expression for polarization operator of Peierls dielectric with fixed gap Δ^2. For $\Delta^2 \to 0$ the second term in (6.261) goes to zero, while the first one reduces to the usual (retarded) polarization

---

[26]Similar calculations for a superconductor give the same results, differing only by the sign before Δ^2 in the numerator of the integrand, which is due to antisymmetry of Gorkov's function $F_{\alpha\beta}$ over spin indices (5.53). Thus, in the expression for polarization operator of a superconductor, in comparison to (6.254), we have a change of $FF^+ \to -FF^+$. This expression determines e.g. ultrasound absorption in superconductors [Mahan G.D. (1981)], but it is insufficient for calculations of conductivity $\sigma(\omega)$ via Eq. (2.111). To obtain the correct expression for polarization operator (and also for dielectric permeability) of a superconductor we have to take into account also contributions from collective excitations (R. Prange, 1963). In FF^+-loop with "current" vertices for a superconductor this change of signs is compensated by the change of the relative sign of these vertices, as Gorkov's F-functions describe $\pm\mathbf{p} \to \mp\mathbf{p}$ transitions. As a result, we obtain a combination of signs written above in (5.301).

operator of electron gas. On the other hand, for $T \to 0$, but $\Delta^2 \neq 0$, the first term in (6.261) becomes zero, so that for polarization operator at $T = 0$ we get:

$$\Pi_{\Delta^2}(\mathbf{q}\omega) = -\frac{1}{2}N_0(E_F) \int_{-\infty}^{\infty} d\xi_p \left\{ \frac{E_p E_{p+q} - \xi_p \xi_{p+q} - \Delta^2}{E_p E_{p+q}} \right\}$$

$$\times \left\{ \frac{1}{E_p + E_{p+q} + \omega + i\delta} \frac{1}{E_p + E_{p+q} - \omega - i\delta} \right\} \quad (6.262)$$

Performing the simple expansions[27] in powers of q, in the limit of $v_F q \ll \Delta$ we obtain:

$$\Pi_{\Delta^2}(\mathbf{q}\omega) = -\frac{1}{2}N_0(E_F)v_F^2 q^2 \int_0^{\infty} d\xi_p \left\{ \frac{1}{2E_p + \omega + i\delta} + \frac{1}{2E_p - \omega - i\delta} \right\} \frac{\Delta^2}{E_p^4}$$

$$(6.263)$$

Then:

$$Re\Pi_{\Delta^2}(q\omega) = -\frac{1}{2}N_0(E_F)v_F^2 q^2 \int_0^{\infty} d\xi_p \left\{ \frac{1}{2E_p + \omega} + \frac{1}{2E_p - \omega} \right\} \frac{\Delta^2}{E_p^4}$$

$$(6.264)$$

$$Im\Pi_{\Delta^2}(q\omega) = \frac{\pi}{2}N_0(E_F)v_F^2 q^2 \int_0^{\infty} d\xi_p \frac{\Delta^2}{E_p^4} \left\{ \delta(2E_p + \omega) - \delta(2E_p - \omega) \right\}$$

$$(6.265)$$

Let us calculate now dielectric permeability and conductivity. Using the standard definition (2.8), we have:

$$Re\epsilon_{\Delta^2}(q\omega) = 1 - \frac{4\pi e^2}{q^2} Re\Pi_{\Delta^2}(q\omega) \quad (6.266)$$

$$Im\epsilon_{\Delta^2}(q\omega) = -\frac{4\pi e^2}{q^2} Im\Pi_{\Delta^2}(q\omega) \quad (6.267)$$

Then from (6.264) and (6.265) we obtain:

$$Re\epsilon_{\Delta^2}(q\omega) = 1 + \frac{\omega_p^2}{4} \int_0^{\infty} d\xi_p \frac{1}{E_p^2 - \frac{\omega^2}{4}} \frac{\Delta^2}{E_p^3} \quad (6.268)$$

---

[27] For $p \sim +p_F$ and $q > 0$, we have: $\xi_{p+q} \approx \xi_p + v_F q$,

$E_{p+q} = \sqrt{\xi_{p+q}^2 + \Delta^2} \approx E_p + v_F q \frac{\xi_p}{E_p} + \frac{1}{2}v_F^2 q^2 \frac{\Delta^2}{E_p^3}$

Accordingly:

$\xi_p \xi_{p+q} \approx \xi_p^2 + v_F q \xi_p$

$E_p E_{p+q} \approx E_p^2 + v_F q \xi_p + \frac{1}{2}v_F^2 q^2 \frac{\Delta^2}{E_p^2}$

so that:

$E_p E_{p+q} - \xi_p \xi_{p+q} - \Delta^2 \approx \frac{1}{2}v_F^2 q^2 \frac{\Delta^2}{\xi_p^2 + \Delta^2}$.

$$Im\epsilon_{\Delta^2}(q\omega) = \frac{\pi}{4}\omega_p^2 \int_0^\infty d\xi_p \frac{\Delta^2}{(\xi_p^2 + \Delta^2)^2} \{\delta(2E_p - \omega) - \delta(2E_p + \omega)\}$$

(6.269)

where we have introduced:

$$\omega_p^2 = v_F\kappa_D^2, \qquad \kappa_D^2 = 8\pi e^2 N_0(E_F)$$

(6.270)

— squares of plasma frequency and inverse of screening length.[28]

For $\omega = 0$ (6.262) reduces to:

$$\Pi_{\Delta^2}(q0) = -2N_0(E_F)\int_0^\infty d\xi_p \frac{E_p E_{p+q} - \xi_p\xi_{p+q} - \Delta^2}{E_p E_{p+q}} \frac{1}{E_p + E_{p+q}}$$

(6.271)

For $v_F q \ll \Delta$ from this expression (or from (6.264)) we obtain:

$$\Pi_{\Delta^2}(q0) = -\frac{1}{2}N_0(E_F)v_F^2 q^2 \int_0^\infty d\xi_p \frac{\Delta^2}{E_p^5} = -\frac{1}{3}N_0(E_F)\frac{v_F^2 q^2}{\Delta^2} \quad (6.272)$$

which gives (use also (6.270)):

$$\epsilon_{\Delta^2}(q0) = 1 + \frac{4\pi e^2 N_0(E_F)v_F^2}{3\Delta^2} = 1 + \frac{\omega_p^2}{6\Delta^2}$$

(6.273)

— the static dielectric permeability of Peierls insulator.

For $v_F q \gg \Delta$ from (6.261), dropping the details of calculations, we obtain:

$$\Pi_{\Delta^2}(q0) = 2N_0(E_F) = \frac{\kappa_D^2}{4\pi e^2}$$

(6.274)

so that:

$$\epsilon(q0) = 1 + \frac{\kappa_D^2}{q^2}$$

(6.275)

where we again took into account (6.270). Eq. (6.275) obviously corresponds to Debye screening in a *metal* — for $v_F q \gg \Delta$ Peierls gap is insignificant!

---

[28] For $d = 1$ we have $n = \frac{2p_F}{\pi}$, $p_F = \frac{\pi}{2}n$, and $v_F^2\kappa_D^2 = 8\pi e^2 v_F^2 \frac{1}{\pi v_F} = 4e^2 v_F = 4e^2 \frac{p_F}{m} = \frac{4\pi ne^2}{m}$, which coincides with the usual definition of plasma frequency. If we are dealing with three-dimensional system, consisting of one-dimensional chains of atoms, our expressions for polarization operator has to be multiplied by the number of chain per unit square of specimen cross-section, i.e. by $1/a^2$, where a is the lattice constant of two-dimensional (for simplicity square) lattice, which is formed by chains in the orthogonal plane. Then all expressions remain valid, only n denotes electron density in three-dimensional system.

Returning to the case of $\omega \neq 0$ and $v_F q \ll \Delta$, we can write (6.264) as:

$$Re\epsilon_{\Delta^2} = 1 + \frac{\omega_p^2}{4} \int_0^\infty d\xi_p \frac{\Delta^2}{(\xi_p^2 + \Delta^2)^{3/2}} \frac{1}{\xi_p^2 + \Delta^2 - \frac{\omega^2}{4}} \qquad (6.276)$$

For $\omega \to 0$ ($\omega \ll 2\Delta$) it naturally leads to (6.273), while for $\omega \gg 2\Delta$ we obtain the usual plasma limit:

$$Re\epsilon_{\Delta^2}(\omega \gg 2\Delta) = 1 - \frac{\omega_p^2}{\omega^2} \qquad (6.277)$$

The full expression (6.276) describes continuous crossover from (6.273) to (6.277), taking place at $\omega \sim 2\Delta$. In more details we can proceed as follows. Using in (6.276) the variable change $\xi_p = \Delta sh(z)$, after simple transformations and taking (tabular) integrals, we obtain:

$$Re\epsilon_{\Delta^2}(\omega) = 1 - \frac{\omega_p^2}{\omega^2}\left\{ 1 + \frac{2\Delta}{\omega}\frac{1}{\sqrt{1 - \frac{\omega^2}{4\Delta^2}}}\left[arctg\left(\sqrt{\frac{4\Delta^2}{\omega^2} - 1} \right) - \frac{\pi}{2} \right] \right\}, \omega^2 < 4\Delta^2$$

$$(6.278)$$

$$Re\epsilon_{\Delta^2}(\omega) = 1 - \frac{\omega_p^2}{\omega^2}\left\{ 1 - \frac{2\Delta}{\omega}\frac{1}{\sqrt{\frac{\omega^2}{4\Delta^2} - 1}}\left[arcth\left(\sqrt{1 - \frac{4\Delta^2}{\omega^2}} \right) - 1 \right] \right\}, \omega^2 > 4\Delta^2$$

$$(6.279)$$

which gives us the quoted asymptotic behavior.

Consider now $Im\epsilon_{\Delta^2}(\omega)$. Eq. (6.269) is written as:

$$Im\epsilon_{\Delta^2}(\omega) = \frac{\pi}{4}\omega_p^2\int_0^\infty d\xi_p \frac{\Delta^2}{(\xi_p^2 + \Delta^2)^2}\left\{ \delta\left(2\sqrt{\xi_p^2 + \Delta^2} - \omega \right) - \delta\left(2\sqrt{\xi_p^2 + \Delta^2} + \omega \right) \right\}$$

$$(6.280)$$

Calculating integral with the use of well known expressions $\delta(ax) = \frac{1}{a}\delta(x)$ and $\int_y^\infty dx \delta(x - a) = \theta(a - y)$, we get:

$$Im\epsilon_{\Delta^2}(\omega) = \pi\Delta\frac{\omega_p^2}{\omega^3}\frac{\theta(|\omega| - 2\Delta)}{\sqrt{\frac{\omega^2}{4\Delta^2} - 1}} \qquad (6.281)$$

Then, for the real part of conductivity we obtain:

$$Re\sigma_{\Delta^2}(\omega) = \frac{\omega}{4\pi}Im\epsilon_{\Delta^2}(\omega) = \begin{cases} \frac{ne^2}{m\omega}\frac{\pi}{\sqrt{\frac{\omega^2}{4\Delta^2} - 1}}\frac{\Delta}{\omega} & \text{for} \quad |\omega| > 2\Delta \\ 0 & \text{for} \quad |\omega| < 2\Delta \end{cases}$$

$$(6.282)$$

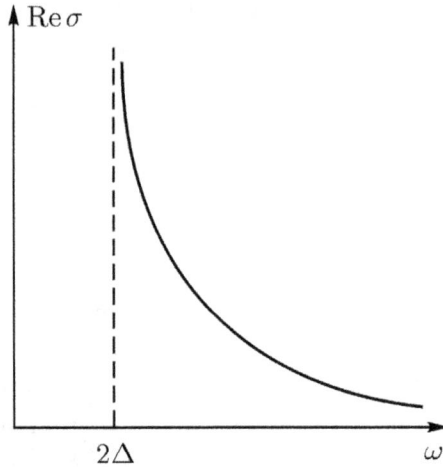

Fig. 6.36 Frequency dependence of the real part of conductivity in Peierls dielectric.

The frequency dependence given by (6.282) is shown in Fig. 6.36. We see that electromagnetic absorption takes place via quasiparticle excitation through Peierls gap 2Δ, i.e. different from zero only for $\omega > 2\Delta$. This is typical insulating (semiconductor) behavior.

For $\omega \gg 2\Delta$ we have:

$$Im\epsilon_{\Delta^2}(\omega) \approx 2\pi \left(\frac{\Delta}{\omega}\right)^2 \left(\frac{\omega_p}{\omega}\right)^2, \qquad Re\sigma_{\Delta^2}(\omega) \approx \frac{ne^2}{m\omega}2\pi \left(\frac{\Delta}{\omega}\right)^2 \tag{6.283}$$

In our model of the pseudogap state (with asymptotically large correlation length of Gaussian short-range order fluctuations $\xi \to \infty$) all these expressions have to be averaged over fluctuations of Δ, distributed according to (6.206) or (6.207). Thus, from (6.281) and (6.282) we obtain:

$$Im\epsilon(\omega) = \pi\Delta\frac{\omega_p^2}{\omega^3} \int_0^{\frac{\omega^2}{4\Delta^2}} d\zeta e^{-\zeta} \frac{\zeta}{\sqrt{\frac{\omega^2}{4\Delta^2} - \zeta}} \tag{6.284}$$

$$Re\sigma(\omega) = \frac{\omega_p^2}{4}\frac{\Delta}{\omega^2} \int_0^{\frac{\omega^2}{4\Delta^2}} d\zeta e^{-\zeta} \frac{\zeta}{\sqrt{\frac{\omega^2}{4\Delta^2} - \zeta}} \tag{6.285}$$

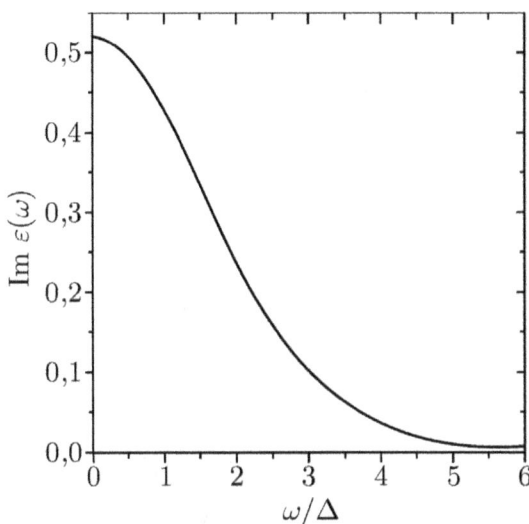

Fig. 6.37 Frequency dependence of imaginary part of dielectric permeability in the model of pseudogap state. The values of $Im\epsilon(\omega)$ are given in units of $\frac{\omega_p^2}{\Delta^2}$.

Characteristic behavior following from these expressions is shown in Figs. 6.37, 6.38. Analytically, from (6.284) and (6.285) it is easy to show that for $\omega \ll 2\Delta$ the following asymptotic behavior is valid:[29]

$$Im\epsilon(\omega) \approx \frac{\pi}{6}\frac{\omega_p^2}{\Delta^2}, \qquad Re\sigma(\omega) \approx \frac{1}{24}\left(\frac{\omega_p}{\Delta}\right)^2 \omega \to 0 \quad \text{for} \quad \omega \to 0$$
$$(6.286)$$

Of course, we can perform numerical calculations of $Re\epsilon(\omega)$ using expressions following from (6.279), (6.278) with further averaging (6.206):

$$Re\epsilon_{\Delta^2}(\omega) = 1 - \frac{\omega_p^2}{\omega^2}\int_0^\infty d\zeta e^{-\zeta}\left\{1 + \frac{4\zeta\Delta^2}{\omega}\frac{1}{\sqrt{4\zeta\Delta^2 - \omega^2}}\left[arctg\left(\sqrt{\frac{4\zeta\Delta^2}{\omega^2} - 1}\right) - \frac{\pi}{2}\right]\right\}$$
$$(6.287)$$

for $\omega^2 < 4\Delta^2$, and

---

[29] This immediately follows with the account of asymptotic behavior of the integral for $a \to 0$: $\int_0^a dx e^{-x}\frac{x}{\sqrt{a-x}} \to \int_0^a dx \frac{x}{\sqrt{a-x}} = \frac{4}{3}a^{3/2}$.

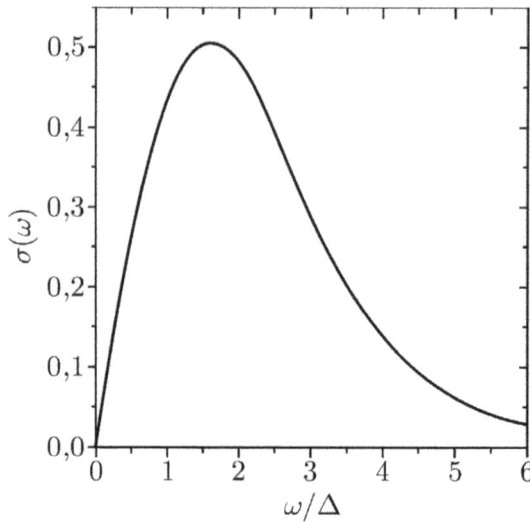

Fig. 6.38   Frequency dependence of the real part of conductivity in the pseudogap state. Conductivity is given in units of $\frac{\omega_p^2}{4\pi\Delta}$.

$$Re\epsilon_{\Delta^2}(\omega) = v1 - \frac{\omega_p^2}{\omega^2}\int_0^\infty d\zeta e^{-\zeta}\left\{1 - \frac{4\zeta\Delta^2}{\omega}\frac{1}{\sqrt{\omega^2 - 4\zeta\Delta^2}}\left[arcth\left(\sqrt{1 - \frac{4\zeta\Delta^2}{\omega^2}}\right) - 1\right]\right\}$$

$$(6.288)$$

for $\omega^2 > 4\Delta^2$. However, it is much simpler to use, instead of (6.279) and (6.278), a simplest interpolation:

$$Re\epsilon_{\Delta^2}(\omega) = 1 - \frac{\omega_p^2}{\omega^2 - 6\Delta^2} \qquad (6.289)$$

which correctly reproduces the limiting behavior for $\omega \ll 2\Delta$ and $\omega \gg 2\Delta$. Then we get:

$$Re\epsilon(\omega) = 1 - \omega_p^2\int_0^\infty d\zeta e^{-\zeta}\frac{1}{\omega^2 - 6\zeta\Delta^2} = 1 - \frac{\omega_p^2}{6\Delta^2}e^{-\frac{\omega^2}{6\Delta^2}}\overline{Ei}\left(\frac{\omega^2}{6\Delta^2}\right)$$

$$(6.290)$$

Direct numerical calculations show, that (6.287), (6.288) and (6.290) give (quantitatively) very close results, as is seen from Fig. 6.39. Using the asymptotic behavior:

$$\overline{Ei}(x) = \begin{cases} \frac{e^x}{x} & \text{for} \quad x \gg 1 \\ C + \ln x + ... & \text{for} \quad x \to 0, \quad C = \ln\gamma \end{cases} \qquad (6.291)$$

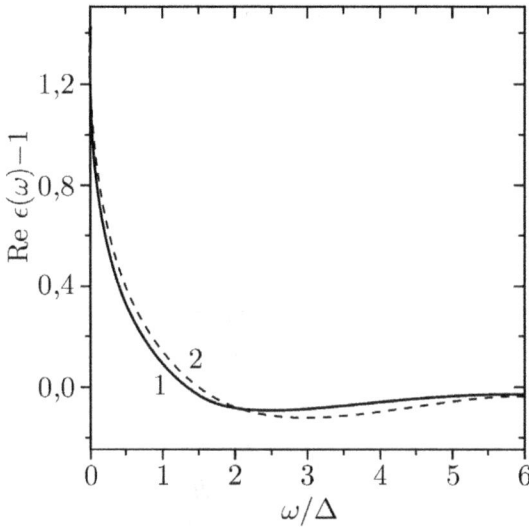

Fig. 6.39 Real part of dielectric permeability as a function of frequency. (1) Dependence obtained by numerical calculations from exact expressions, (2) dependence obtained from interpolation formula. Dielectric permeability is given in units of $\frac{\omega_p^2}{\Delta^2}$.

we can find from (6.290):

$$Re\epsilon(\omega) \to 1 - \frac{\omega_p^2}{\omega^2} \qquad \text{for} \qquad \omega \gg 2\Delta \qquad (6.292)$$

$$Re\epsilon(\omega) \approx 1 - \frac{\omega_p^2}{6\Delta^2} \ln \gamma \frac{\omega^2}{6\Delta^2} \qquad (6.293)$$

Logarithmic divergence of $Re\epsilon(\omega)$ at small frequencies demonstrates intermediate (between metallic and insulating) behavior, characteristic of our (oversimplified!) model of the pseudogap state.

In a similar way we can analyze q-dependence of dielectric permeability in the static limit of $\omega = 0$. Let us write down again a simple interpolation, connecting the limiting cases of (6.272) and (6.274):

$$\Pi_{\Delta^2}(q0) = -2N_0(E_F)\frac{v_F^2 q^2}{v_F^2 q^2 + 6\Delta^2} \qquad (6.294)$$

so that

$$\epsilon_{\Delta^2}(q0) = 1 + \frac{\kappa_D^2}{q^2 + \frac{6\Delta^2}{v_F^2}} \qquad (6.295)$$

interpolating between (6.273) and (6.275). Then in our model of the pseudogap state:

$$\epsilon(q0) = \int_0^\infty d\zeta e^{-\zeta} \frac{q^2 + \kappa_D^2 + 6\zeta \frac{\Delta^2}{v_F^2}}{q^2 + 6\zeta \frac{\Delta^2}{v_F^2}} = 1 - \frac{v_F^2 \kappa_D^2}{6\Delta^2} e^{\frac{v_F^2 q^2}{6\Delta^2}} Ei\left(-\frac{v_F^2 q^2}{6\Delta^2}\right)$$

$$(6.296)$$

This behavior is shown in Fig. 6.40, from which we again can see a close agreement of our interpolation with the results of numerical calculations, using exact expressions (which we drop for brevity). For $v_F q \gg \Delta$ we can use asymptotic behavior for $x \gg 1$: $Ei(-x) \to -\frac{e^{-x}}{x}$. Then, as expected, we obtain:

$$\epsilon(q0) = 1 + \frac{\kappa_D^2}{q^2} \qquad (6.297)$$

i.e. "metallic" (Debye) behavior. However, for $v_F q \ll \Delta$, using asymptotics $Ei(-x) \approx \ln \gamma x \quad (x \to 0)$, we get:

$$\epsilon(q0) \approx 1 - \frac{v_F^2 \kappa_D^2}{6\Delta^2} \ln \gamma \frac{v_F^2 q^2}{6\Delta^2} \qquad (6.298)$$

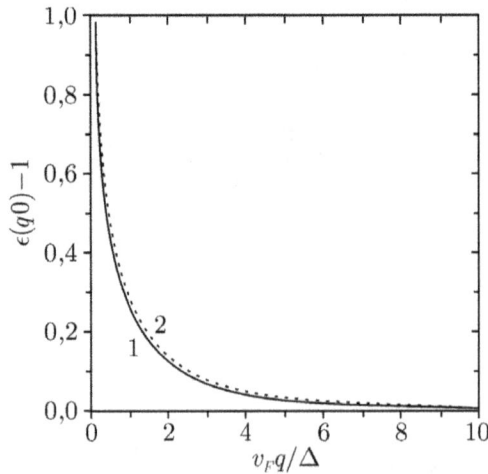

Fig. 6.40   Dielectric permeability as a function of wave vector. (1) Dependence obtained from exact expressions, (2) dependence obtain from interpolation. Dielectric permeability is given in units of $\frac{\omega_p^2}{\Delta^2}$.

With such behavior of $\epsilon(q0)$ in the limit of small q, effective Coulomb interaction acquires the following form:

$$\mathcal{V}(q) = \frac{4\pi\tilde{e}^2(q)}{q^2} \tag{6.299}$$

where

$$\tilde{e}^2(q) = \frac{e^2}{1 - \frac{v_F^2\kappa_D^2}{6\Delta^2}\ln\gamma\frac{v_F^2 q^2}{6\Delta^2}} \to 0 \qquad \text{for} \qquad q \to 0 \tag{6.300}$$

which is analogous to the well known "zero-charge" behavior in quantum electrodynamics [Sadovskii M.V. (2019b)]. Thus, gap fluctuations in the spectrum of our model lead to complete charge screening, though of very peculiar form. Again we observe intermediate behavior, which is between typical "metallic" one and "insulating".[30]

The basic conclusion from our analysis of this simplified and rather artificial model is rather important. Results of our exact solution (complete summation of diagrammatic series) are quite different from what we can obtain (or expect) by approximate methods, such as partial summation. Unfortunately, complete summation is usually possible only in simplified and unrealistic models.

Generalization of these results for the case of finite correlation lengths ξ (or finite κ) can be done if we formulate recursion relation for the vertex part, describing electromagnetic response, along the lines of our derivation of recursion relations for electron self-energy (or single-electron Green's function), described above.

Arbitrary diagram for the vertex part, as we have seen above, can be obtained by an insertion of an external field line to the appropriate diagram for the self-energy. The basic idea now is that in our model we can limit ourselves only to diagrams with non-intersecting interaction lines with additional combinatorial factors $v(k)$ at "initial" interaction vertices. It is clear then that to calculate vertex corrections we have to consider only diagrams of the type shown in Fig. 6.41. Then we immediately obtain the system of recurrence equations for the vertex parts shown by diagrams of Fig. 6.42. To find appropriate analytic expressions consider the simplest vertex correction shown in Fig. 6.43(a). Performing explicit

---

[30]Of course, these anomalies are mainly due to our artificial assumptions, used in our model of the pseudogap state, and mostly disappear, when we go to a more realistic situation, e.g. take into account the finite values of correlation length of short-range order.

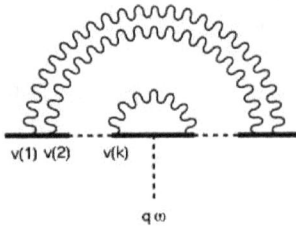

Fig. 6.41   General diagram for vertex correction.

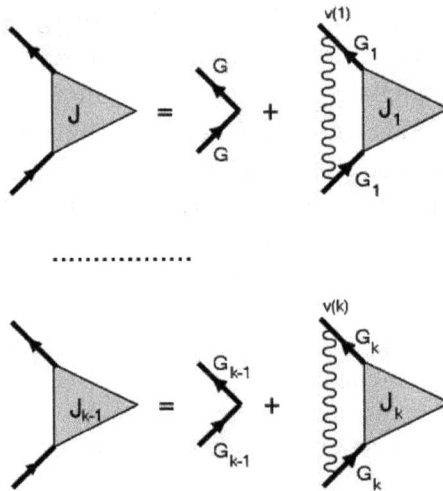

Fig. 6.42   Recursion relations for vertex-part.

calculations for $T = 0$ in RA-channel we find its contribution to be:

$$\mathcal{J}_1^{(1)RA}(\varepsilon, \xi_p; \varepsilon + \omega, \xi_{p+q}) = \Delta^2 \int \frac{dQ}{2\pi} G_0^A(\varepsilon, \xi_{p-Q}) G_0^R(\varepsilon + \omega, \xi_{p-Q+q})$$

$$= \Delta^2 \left\{ G_0^A(\varepsilon, -\xi_p + iv_F\kappa) - G_0^R(\varepsilon + \omega, \xi_{p+q} - iv_F\kappa) \right\} \frac{1}{\omega + v_Fq}$$

$$= \Delta^2 G_0^A(\varepsilon, -\xi_p + iv_F\kappa) G_0^R(\varepsilon + \omega, -\xi_{p+q} - iv_F\kappa) \left\{ 1 + \frac{2iv_F\kappa}{\omega + v_Fq} \right\}$$

$$\tag{6.301}$$

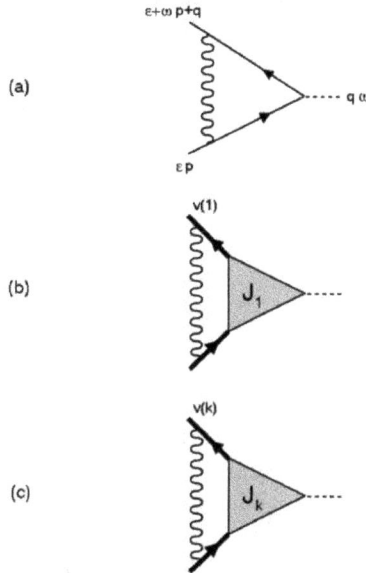

Fig. 6.43   Simplest corrections for vertex-part.

where during the integral calculations we have used the following identity, valid for the free-electron Green's functions:

$$G_0^A(\varepsilon, \xi_p) G_0^R(\varepsilon + \omega, \xi_{p+q}) = \left\{ G_0^A(\varepsilon, \xi_p) - G_0^R(\varepsilon + \omega, \xi_{p+q}) \right\} \frac{1}{\omega - v_F q} \quad (6.302)$$

"Dressing" the internal electronic lines by fluctuations we obtain the diagram shown in Fig. 6.43(b), so that using the identity:

$$G^A(\varepsilon, \xi_p) G^R(\varepsilon + \omega, \xi_{p+q}) = \left\{ G^A(\varepsilon, \xi_p) - G^R(\varepsilon + \omega, \xi_{p+q}) \right\}$$

$$\times \frac{1}{\omega - v_F q - \Sigma_1^R(\varepsilon + \omega, \xi_{p+q}) + \Sigma_1^A(\varepsilon, \xi_p)} \quad (6.303)$$

valid for exact Green's functions (6.215), we can write the contribution of this diagram as:

$$\mathcal{J}_1^{RA}(\varepsilon, \xi_p; \varepsilon + \omega, \xi_{p+q}) = \Delta^2 v(1) G_1^A(\varepsilon, \xi_p) G_1^R(\varepsilon + \omega, \xi_{p+q})$$

$$\times \left\{ 1 + \frac{2i v_F \kappa}{\omega + v_F q - \Sigma_2^R(\varepsilon + \omega, \xi_{p+q}) + \Sigma_2^A(\varepsilon, \xi_p)} \right\} J_1^{RA}(\varepsilon, \xi_p; \varepsilon + \omega, \xi_{p+q}) \quad (6.304)$$

Here we have assumed that interaction line in the vertex correction of Fig. 6.43(b) "transforms" self-energies $\Sigma_1^{R,A}$ of internal lines into $\Sigma_2^{R,A}$, in accordance with the

main idea of our *Ansatz* for the self-energy.[31] Now we can write down the similar
expression for the general diagram, shown in Fig. 6.43(c):

$$\mathcal{J}_k^{RA}(\varepsilon, \xi_p; \varepsilon + \omega, \xi_{p+q}) = \Delta^2 v(k) G_k^A(\varepsilon, \xi_p) G_k^R(\varepsilon + \omega, \xi_{p+q})$$

$$\times \left\{ 1 + \frac{2 i v_F \kappa k}{\omega - (-1)^k v_F q - \Sigma_{k+1}^R(\varepsilon + \omega, \xi_{p+q}) + \Sigma_{k+1}^A(\varepsilon, \xi_p)} \right\} J_k^{RA}(\varepsilon, \xi_p; \varepsilon + \omega, \xi_{p+q})$$

$$(6.305)$$

Then we can write recurrence relation for the vertex-part, shown diagrammatically in Fig. 6.42, in the following form (M.V. Sadovskii, A.A. Timofeev,
1991):

$$J_{k-1}^{RA}(\varepsilon, \xi_p; \varepsilon + \omega, \xi_{p+q}) = 1 + \Delta^2 v(k) G_k^A(\varepsilon, \xi_p) G_k^R(\varepsilon + \omega, \xi_{p+q})$$

$$\times \left\{ 1 + \frac{2 i v_F \kappa k}{\omega - (-1)^k v_F q - \Sigma_{k+1}^R(\varepsilon + \omega, \xi_{p+q}) + \Sigma_{k+1}^A(\varepsilon, \xi_p)} \right\} J_k^{RA}(\varepsilon, \xi_p; \varepsilon + \omega, \xi_{p+q})$$

$$(6.306)$$

where all self-energies and Green's functions are determined from appropriately
analytically continued recursion relations of the type of (6.222), (6.223). The
"physical" vertex $J^{RA}(\varepsilon, \xi_p; \varepsilon + \omega, \xi_{p+q})$ is determined as $J_{k=0}^{RA}(\varepsilon, \xi_p; \varepsilon + \omega, \xi_{p+q})$.
Recurrence procedure (6.306) takes into account *all* perturbation theory diagrams
for the vertex-part. In case of RR and AA-type of vertices we have the same type
of recursion procedure, with obvious replacements $G^R \leftrightarrow G^A$ and expression in
large brackets in the r.h.s. replaced by 1. For $\kappa \to 0$ ($\xi \to \infty$) these procedures
are equivalent to perturbation series studied above, which was summed exactly
in analytic form. Standard "ladder" approximation corresponds in our scheme to
the case of combinatorial factors in (6.306) $v(k) = 1$.

According to (2.116), (4.78) conductivity of our system can be expressed via
retarded density–density response function $\chi(q, \omega)$ as:

$$\sigma(\omega) = e^2 \lim_{q \to 0} \left(-\frac{i\omega}{q^2} \right) \chi(q\omega) \tag{6.307}$$

To simplify numerical calculations it is tempting to use small ω expression (4.105):

$$\chi(q\omega) = \omega \left\{ \Phi^{RA}(q\omega) - \Phi^{RA}(0\omega) \right\} \tag{6.308}$$

where two-particle Green's function $\Phi^{RA}(q, \omega)$ was defined (4.103) (cf. general
discussion of Chapter IV and definitions (4.88), (4.94) etc.).[32] However, due to

---

[31] One of the main motivations for this trick is that it guarantees the fulfillment of an
exact Ward identity (6.309).

[32] Direct numerical computations confirm that the recursion procedure (6.306) satisfies
an exact (in the limit of $\omega \to 0$) Ward identity (4.104):

$$\Phi^{RA}(0\omega) = -\frac{N(E_F)}{\omega} \tag{6.309}$$

where $N(E_F)$ is the density of states at the Fermi level, which can be independently
calculated via (6.226)–(6.229). Actually, this is probably the main argument for the
validity of an *Ansatz* used to derive Eqs. (6.304), (6.305) and (6.306).

existence in our problem of an additional energy scale $\Delta \ll E_F$ (the width of the pseudogap) the use of (6.308) leads to certain (quantitative, not qualitative!) inaccuracy, especially notable in the limit of small κ. Thus, it is much better to use complete (integral) representation for $\chi(q, \omega)$, given by (4.87), (4.95). This allows us to reproduce exact results for conductivity obtained above in the limit of $\kappa \to 0$ via recursion relations for the vertex part, used here. However, due to additional integration this procedure obviously leads to more time-consuming numerical calculations. Below we present results of calculations using full expression (4.95). Convergence of numerical procedure for the vertex part itself is rather good (except the limit of very small frequencies and small $\kappa = \xi^{-1}$), though conductivity calculations are obviously much more time-consuming, than e.g. calculations of the density of states.

Typical dependences of the real part of conductivity on frequency are shown in Fig. 6.44 (for the case of incommensurate short-range order fluctuations).[33] One

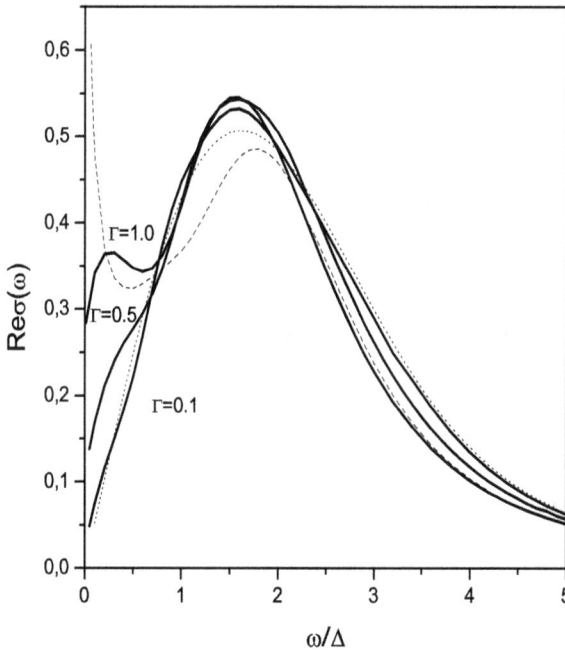

Fig. 6.44 Frequency dependence of the real part of conductivity in the case of incommensurate pseudogap fluctuations for different values of $\Gamma = v_F \kappa / \Delta$. Dotted curve — $\Gamma = 0$. Dashed curve — results of the "ladder" approximation for $\Gamma = 1.0$. Conductivity is given in units of $\frac{\omega_p^2}{4\pi\Delta}$.

---

[33]I am grateful to Dr. E.Z. Kuchinskii for making full calculations of conductivity for these lectures.

can see the gradual growth of absorption within the pseudogap with decrease of the correlation length $\xi = \kappa^{-1}$. Most striking anomaly is the appearance of additional shallow maximum (or non monotonic behavior) in the frequency dependence of conductivity within the pseudogap, which we attribute to Anderson localization of carriers, ever present in one-dimensional system. Localization nature of this anomaly is directly confirmed by comparison of "exact" (i.e. taking into account all diagrams) calculations with that of "ladder" approximation, obtained by putting combinatorial factor $v(k) = 1$ in all relations. Typical dependence of conductivity, obtained in this approximation, is shown in Fig. 6.44 by dashed curve. It is clearly seen that localization behavior is transformed into narrow Drude-like "metallic" peak at small frequencies, with no signs of localization behavior. It is quite natural, as we seen above in Chapter IV, that localization is intimately related with diagrams with intersecting interaction lines, absent in "ladder" approximation. Direct check shows that all our results for conductivity satisfy the exact sum rule (5.336).

This approach can also be generalized to studies of conductivity in two-dimensional models of pseudogap behavior, relevant to high-temperature super-conductors.[34]

6.7   Tomonaga–Luttinger model and non Fermi-liquid behavior

Practically in all problems analyzed above, the starting point was Landau Fermi-liquid and the single-particle Green's function with a pole:

$$G(p) = \frac{Z}{\varepsilon - v_F(p - p_F) + i\delta} + \cdots \qquad (6.310)$$

where $0 < Z < 1$ is some constant, determining the discontinuity of distribution function of particles at the Fermi surface $p = p_F$. At the same time, in the previous sections we have shown, that an exact solution of one-dimensional problem leads to quite different form of the Green's function, which does not possess poles and is in no way similar to that assumed in Fermi-liquid theory. In fact, this is rather the general property of interacting Fermions in one dimension — Fermi liquid behavior is *always* absent. As probably most striking (and general) example, in this section we shall briefly consider so-called Tomonaga–Luttinger model (S. Tomonaga, 1950; J.M. Luttinger, 1963).

---

[34]More details can be found in reviews: M.V. Sadovskii. Physics Uspekhi **44**, 515 (2001) and ArXiV: cond-mat/0408489.

This model describes a gas of Fermions with density n (Fermi momentum $p_F = \pi n/2$), mass m (Fermi velocity $v_F = p_F/m$) and interaction potential $\lambda(|x|)$, with Fourier components $\lambda(k)$ being different from zero only in very narrow interval of momenta $|k| \leq \Lambda \ll p_F$. The Hamiltonian of this model is written as:

$$H = \sum_p \frac{p^2}{2m} a_p^+ a_p + \frac{1}{2} \sum_{pp'k} \lambda(k) a_p^+ a_{p'}^+ a_{p'-k} a_{p+k} \qquad (6.311)$$

Tomonaga has shown, that in case of very long-range interaction, i.e. neglecting all contributions of the order of $\Lambda/p_F \to 0$, the spectrum of (6.311) coincides with the spectrum of (free) *Bosons*, described by Boson operators b_k, b_k^+:[35]

$$H = \sum_k v(k) k b_k^+ b_k, \qquad v^2(k) = v_F^2 + \frac{2v_F}{\pi} \lambda(k) \qquad (6.312)$$

Below we shall prove this by diagram technique (I.E. Dzyaloshinskii, A.I. Larkin, 1973). We shall also show that single-particle Green's function coincides with (6.310) in the region of $|p - p_F| \gg \Lambda$, but has completely different form close to the Fermi surface, i.e. for $|p - p_F| \ll \Lambda$.

Having in mind one-dimensional system of free electrons with the spectrum shown in Fig. 6.3, we shall calculate Green's functions close to the "right" and "left" Fermi points $\pm p_F$, denoting these $G_+(p)$ and $G_-(p)$ ($\pm$-Fermions). For the gas of free particles:

$$G_+^{(0)} = \frac{1}{\varepsilon - p + p_F + i\delta}, \qquad G_-^{(0)} = \frac{1}{\varepsilon + p + p_F + i\delta} \qquad (6.313)$$

where, for brevity, we are using the units with $v_F = 1$.

Particles from the vicinity of right or left Fermi point, can be considered as different Fermions also in the interacting system, and even for $|p - p_F| \gg p_F$, as in the limit of $\Lambda/p_F \to 0$ our interaction can not transform one sort of particles into the other. This means that with the same accuracy, the values of $p - p_F$ for "+"-particles and $p + p_F$ for "−"-particles may change on the interval from $-\infty$ to $+\infty$.

First of all, we have to calculate effective interaction, which we denote $D(k)$ and express by wave-like line diagrammatically, as well as

---

[35]In other words, there are no Fermion excitations at all, the spectrum consists only of "sound-like" collective excitations.

"triangular" vertex $\Gamma(p, k)$. Due to our condition $\Lambda/p_F \to 0$, only momenta $k \ll p_F$ are relevant in all vertices, thus both Fermion Green's functions, entering Γ belong to the same Fermi point $(+p_F$ or $-p_F)$, so that we can introduce $\Gamma_+(p, k)$ and $\Gamma_-(p, k)$. We shall consider a certain generalization of Tomonaga–Luttinger model assuming different interactions of particles of the same "sign" $(+$ or $-)$ and of different "signs", as is shown in Fig. 6.45. Accordingly we introduce notations:

$$\lambda_{++} = \lambda_{--} = \lambda_1; \quad \lambda_{+-} = \lambda_2; \qquad D_{++} = D_{--}; \quad D_{+-} = D_{-+} \tag{6.314}$$

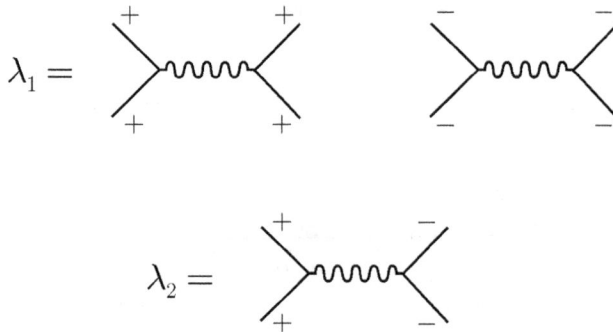

Fig. 6.45   Interactions of particles in Tomonaga–Luttinger models.

Dyson equations for D has the form:

$$D_{++} = \lambda_1 + \lambda_1\Pi_+ D_{++} + \lambda_2\Pi_- D_{-+} \tag{6.315}$$

$$D_{-+} = \lambda_2 + \lambda_2\Pi_+ D_{++} + \lambda_1\Pi_- D_{-+} \tag{6.316}$$

Polarization operators entering here are given by diagrams shown in Fig. 6.46. Dyson equation for Green's function G has the standard

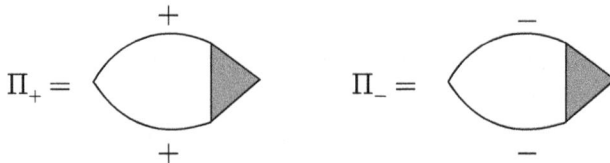

Fig. 6.46   Polarization operators in Tomonaga–Luttinger model.

form:

$$G_{\pm}^{-1} = \varepsilon \mp p + p_F - \Sigma_{\pm} \tag{6.317}$$

where Σ_+ and Σ_- are shown in Fig. 6.47.

Fig. 6.47 Electron self-energies in Tomonaga–Luttinger model.

Usually, as we have seen on different examples above, vertex part Γ is determined by an infinite series of diagrams and can be found only approximately. However, in this model the problem can be solved exactly in the limit of $\Lambda/p_F \to 0$, using the Ward identity, connecting Γ and G and taking the following simple form:

$$\Gamma_+(p, k) = \frac{G_+^{-1}(p) - G_+^{-1}(p - k)}{\omega - k} \tag{6.318}$$

$$\Gamma_-(p, k) = \frac{G_-^{-1}(p) - G_-^{-1}(p - k)}{\omega + k} \tag{6.319}$$

Here, as usual, we understand that k in Green's functions and vertices denote the pair (k, ω). Eqs. (6.318) and (6.319) can be derived directly, analyzing diagrams of different orders and using the identity:

$$G_{\pm}^{(0)}(p)G_{\pm}^{(0)}(p + k) = \frac{1}{\omega \mp k} \left(G_{\pm}^{(0)}(p) - G_{\pm}^{(0)}(p + k) \right) \tag{6.320}$$

following directly from (6.313). The thing is that interaction (wave-like line) transfers (almost) zero momentum ($\leq \Lambda$) $\to 0$. Then in all diagrams for G or Σ we have the continuous line of particles of the "same sign", carrying an "external" momentum p, as can be seen analyzing typical diagrams shown in Fig. 6.48. Thus, all diagrams for Γ can be obtained by arbitrary insertions of external interaction lines into diagrams for self-energy. This is shown in Fig. 6.49, where we show diagrams for the vertex, obtained from diagrams for self-energy, shown in Fig. 6.48. Using (6.320) at any insertion point of this type we immediately obtain Ward identities (6.318), (6.319).

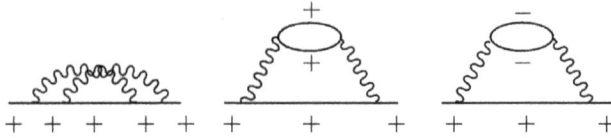

Fig. 6.48 Examples of typical diagrams for electron self-energy in Tomonaga–Luttinger model.

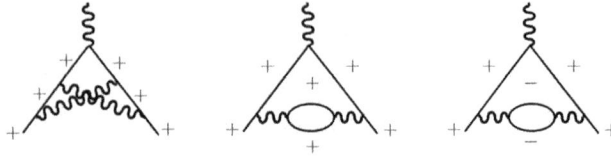

Fig. 6.49 Examples of diagrams for the vertex-part in Tomonaga–Luttinger model.

It is quite important that in this model with very long-range interaction and "bare" Green's functions (6.313) with linear spectrum we have another remarkable property — all diagrams, containing closed loops with more than two Fermion lines, are just zero. Or, more precisely, appropriately symmetrized sum of such diagrams gives zero contribution. Thus, we can drop contributions of the type shown in Fig. 6.50, so that really *all* diagrams for the vertex are generated as shown in Fig. 6.49. The proof is based on the particle number conservation for $\pm$-particles (separately) and is similar to the case of quantum electrodynamics [Sadovskii M.V. (2019b)], where such diagrams also drop out from Ward identity.

Let us write down equations of motion for free operators of $+$-particles:

$$i\frac{\partial \psi_+}{\partial t} + i\frac{\partial \psi_+}{\partial x} = 0 \qquad (6.321)$$

from which we have the particle number (charge) conservation law as:

$$\frac{\partial \rho_+}{\partial t} + \frac{\partial j_+}{\partial x} = 0, \qquad \rho_+ = j_+ = \psi_+^+ \psi_+ \qquad (6.322)$$

For $-$-particles similarly:

$$i\frac{\partial \psi_-}{\partial t} + i\frac{\partial \psi_-}{\partial x} = 0 \qquad (6.323)$$

$$\frac{\partial \rho_-}{\partial t} + \frac{\partial j_-}{\partial x} = 0, \qquad \rho_- = -j_+ = \psi_-^+ \psi_- \qquad (6.324)$$

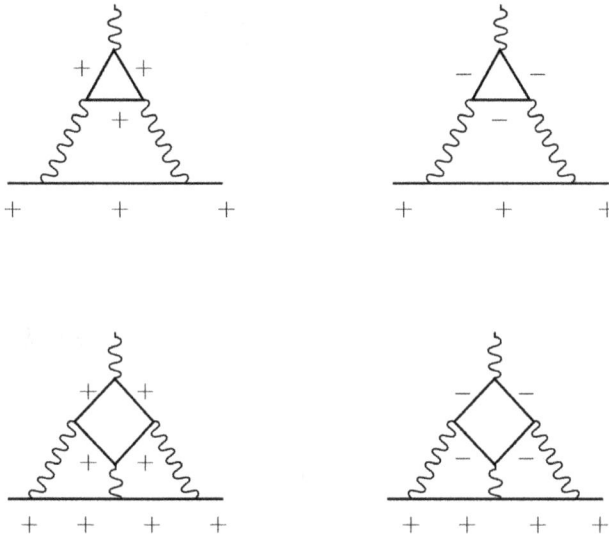

Fig. 6.50 Examples of diagrams for the vertex-part in Tomonaga–Luttinger model giving zero contribution.

As an arbitrary closed loop represents the Fourier component of the ground state average of the product of appropriate number of density operators $< T\rho_+(1)\rho_+(2)...\rho_+(n) >$ (and similarly for $-$-particles), it follows from conservation laws (6.322), (6.324) that:

$$(\omega_1 - k_1)(\omega_2 - k_2)...(\omega_n - k_n) < \rho_+(k_1)\rho_+(k_2)...\rho_+(k_n) >= 0 \qquad (6.325)$$

$$(\omega_1 + k_1)(\omega_2 + k_2)...(\omega_n + k_n) < \rho_-(k_1)\rho_-(k_2)...\rho_-(k_n) >= 0 \qquad (6.326)$$

Then our statement concerning loops follows from (6.325), (6.326), if all momenta integrals converge. It is easily seen that this is so for all loops, containing three and more electron lines.

The loop containing two lines is proportional (for +-particles) to integral:

$$\int d^2p \frac{1}{\varepsilon - p + p_F + i\delta} \frac{1}{\varepsilon - \omega - p + k + p_F + i\delta} \qquad (6.327)$$

which formally diverges. Physically, the finite value of this integral is guaranteed by square dependence of particle energy on momentum far from the Fermi surface (which is neglected in our model). Technically this may be achieved if we first integrate over ε, and only afterwards over p. Result of such integration is finite and proportional to $\frac{k}{\omega - k}$.

Now we can write down the *closed* system of equations for G and D, expressing vertices Γ in Π and Σ, defined by Figs. 6.46, 6.47 via G using (6.318), (6.319). In this way we obtain:

$$\Pi_+(k) = -\frac{i}{2\pi^2}\frac{1}{\omega - k}\int d^2p\,[G_+(p-k) - G_+(p)] \qquad (6.328)$$

$$\Pi_-(k) = -\frac{i}{2\pi^2}\frac{1}{\omega + k}\int d^2p\,[G_-(p-k) - G_-(p)] \qquad (6.329)$$

and equations for $G_\pm$ take the form:

$$(\varepsilon - p + p_F)G_+(p) = 1 + \frac{i}{4\pi^2}\int d^2k\,\frac{D_{++}(k)}{\omega - k}G_+(p-k) \qquad (6.330)$$

$$(\varepsilon + p + p_F)G_+(p) = 1 + \frac{i}{4\pi^2}\int d^2k\,\frac{D_{--}(k)}{\omega + k}G_-(p-k) \qquad (6.331)$$

Let us now calculate Π_+ and Π_-. Introduce momentum cut-off A, so that $|p - p_F| \leq A \ll p_F$. Then:

$$\Pi_+(k) = -\frac{i}{2\pi^2}\frac{1}{\omega - k}\int_{p_F - A}^{p_F + A} dp \int_{-\infty}^{\infty} d\varepsilon\,[G_+(p-k, \varepsilon - \omega) - G_+(p, \varepsilon)] \qquad (6.332)$$

Contribution of the region of $|p - p_F| \geq A$ can not be found from (6.328), (6.329) as Ward identities (6.318), (6.319) are valid only close to Fermi surface. We can convince ourselves that this contribution is zero if we calculate it with "free electron" Green's functions and vertices. Calculating the integral over ε in (6.332) we have:

$$\Pi_+(k) = \frac{1}{\pi(\omega - k)}\int_{p_F - A}^{p_F + A} dp\,[n_+(p - k) - n_+(p)] \qquad (6.333)$$

where we used the general expression for distribution function:

$$n_+(p) = -i \lim_{t \to -0} \int_{-\infty}^{\infty} \frac{d\varepsilon}{2\pi}e^{-i\varepsilon t}G_+(\varepsilon p) \qquad (6.334)$$

Now, the integral over p in (6.333) can be rewritten as:

$$\int_{p_F - A}^{p_F + A} dp... = \int_{p_F - A - k}^{p_F - A} dp\,n_+(p) - \int_{p_F + A - k}^{p_F + A} dp\,n_+(p) \qquad (6.335)$$

where, up to terms of the order of $\Lambda/A \to 0$, we can take $n_+(p) = n_F(p)$, where n_F is the usual Fermi-step function at $T = 0$. Thus we obtain:

$$\Pi_+(k) = \frac{k}{\pi(\omega - k)} \qquad (6.336)$$

Similarly we get:

$$\Pi_-(k) = -\frac{k}{\pi(\omega + k)} \tag{6.337}$$

Using (6.336), (6.337) in (6.315), (6.316) and solving these equations we find:

$$D_{++}(k) = (\omega - k)\frac{\lambda_1(\omega + k) + (\lambda_1^2 - \lambda_2^2)k/\pi}{\omega^2 - u^2 k^2 + i\delta} \tag{6.338}$$

where

$$u = \left(1 + \frac{2\lambda_1}{\pi} + \frac{\lambda_1^2 - \lambda_2^2}{\pi^2}\right)^{1/2} \tag{6.339}$$

Thus we have obtained collective (Boson) excitations with spectrum determined by the pole of (6.338). For $\lambda_1 = \lambda_2 = \lambda$ these expressions, in fact, give Tomonaga result (6.312) (remember that we use units with $v_F = 1$).

To find the single-electron Green's function we still have to solve the linear integral equation (6.330) with D_{++}, determined above. We shall not do it in details, considering only some simplest cases and just quoting the general results.

Let $\lambda_2 = 0$, then particles on one $(+)$ side of Fermi surface (line) do not interact at all with particles on the other $(-)$ side, so that $D_{-+} = 0$, and

$$D_{++}(k) = \frac{\pi(w - 1)(\omega - k)}{\omega - wk + i\delta} \tag{6.340}$$

where

$$w = 1 + \frac{\lambda_1}{\pi} \tag{6.341}$$

Equation for G_+ takes now the form:

$$(\varepsilon - p)G(p) = 1 + \frac{i}{4\pi}\int d^2 k\, G(p - k)\frac{w - 1}{\omega - wk + i\delta} \tag{6.342}$$

Here and below we drop index $+$ at G_+ and put the origin for p at $+p_F$. By direct substitution we can check that Eq. (6.342) is satisfied by:

$$G(p) = \{(\varepsilon - p + i\delta)(\varepsilon - wp + i\delta)\}^{-1/2} \tag{6.343}$$

where the cut in the complex plane of ε is a line, connecting $p - i\delta$ and $wp - i\delta$ ($sign\delta = sign(p)$). This Green's function does not possess

poles, corresponding to single-particle excitations, but simple calculation using (6.334) shows, that momentum distribution remains Fermi-
like:

$$n(p) = n_F(p) \qquad (6.344)$$

Then we can analyze the case of small λ_1 and λ_2, when u given by
(6.339) is of the order of unity (v_F!). Analysis of (6.330), (6.338) and
(6.339) shows, that in this case electron Green's function has the form:

$$G(p) = \{(\varepsilon - p)(\varepsilon - up)\}^{-1/2} \exp\left\{-\frac{\lambda_2^2}{4\pi^2} \ln \frac{\Lambda}{|p|}\right\} \qquad (6.345)$$

where

$$u = 1 + \frac{\lambda_1}{\pi} + \frac{\lambda_1^2 - \lambda_2^2}{2\pi^2} \qquad (6.346)$$

Calculation of distribution function $n(p)$ using (6.334) and (6.345)
shows that for $\lambda_2 \neq 0$ there is no discontinuity at $p = 0$ and we have
the following behavior instead (E. Lieb, D. Mattis, 1965):

$$n(p) = \frac{1}{2} - \frac{1}{2}\left(\frac{|p|}{\Lambda}\right)^{\frac{\lambda_2^2}{4\pi^2}} sign(p) \qquad (6.347)$$

Consider at last the physically "realistic" case of $\lambda_1 = \lambda_2 = \lambda$. Then
we have:

$$D_{++}(k) = \frac{\lambda(\omega^2 - k^2)}{\omega^2 - v^2 k^2 + i\delta} \qquad (6.348)$$

where

$$v = \left(1 + \frac{2\lambda}{\pi}\right)^{1/2} \qquad (6.349)$$

and for G we have an integral equation:

$$(\varepsilon - p)G(p) = 1 + \frac{i}{4\pi^2} \int d^2k\, G(p - k)\frac{\lambda(\omega + k)}{\omega^2 - v^2 k^2 + i\delta} \qquad (6.350)$$

This equation may be solved after transformation to time-coordinate
representation x, t. We shall not do that and only quote the result
for momentum distribution function $n(p)$ close to Fermi point $p = 0$.
For small interaction, when $\lambda \to 0$, $v \to 1$ we obtain (6.347) with

$\lambda_1 = \lambda_2 = \lambda$. This expression is conserved until interaction is not too strong:

$$n(p) = \frac{1}{2} - const|p|^{2\alpha} sign(p) \tag{6.351}$$

where

$$\alpha = \frac{(v-1)^2}{8v}, \qquad \text{for} \qquad \alpha < 1/2 \tag{6.352}$$

However, when $\alpha > 1/2$, the leading term in the expansion of $n(p)$ near the Fermi point is linear:

$$n(p) = \frac{1}{2} - const \cdot p \tag{6.353}$$

In any case there is no discontinuity at the Fermi point!

Let us quote the results for asymptotic behavior of $G(\varepsilon p)$ in the region of $p \sim \varepsilon \ll \Lambda$. For $\alpha < 1/2$:

$$G(\varepsilon \sim p) \sim \frac{1}{\varepsilon^{1-2\alpha}} \tag{6.354}$$

For $\alpha > 1/2$:

$$G(\varepsilon \sim p) \sim A + B\varepsilon^{2\alpha-1} \tag{6.355}$$

For $3/2 > \alpha > 1$:

$$G(\varepsilon \sim p) \sim A + B\varepsilon + C\varepsilon^{2\alpha-1}, \qquad \text{etc.} \tag{6.356}$$

Thus, in Tomonaga–Luttinger model basic assumptions of Landau Fermi-liquid theory are violated. Already for arbitrarily weak interaction singularity of Green's function at Fermi surface is weaker than a simple pole (cf. (6.354)), while for strong enough interaction Green's function remains finite at the Fermi point (cf. (6.354), (6.356)). In this case, Fermi point manifests itself only in the derivatives of high enough order.

These anomalies are connected with a kind of "infrared catastrophe" taking place in one-dimensional systems. Any particle from close vicinity of one of the Fermi points can emit (satisfying all conservation laws) any number of real particle-hole pairs, which also are in close vicinity of this Fermi point. Mathematically this is expressed by the presence in perturbation series for self-energy $\Sigma(p)$ of singular contributions, containing poles of higher orders, like:

$$\frac{\lambda^n}{(\varepsilon - p)^{n-1}} \qquad (6.357)$$

Consider e.g. an expression for Σ, corresponding to (6.343):

$$\Sigma(p) = \varepsilon - p - [(\varepsilon - p)(\varepsilon - p - \lambda p/\pi)]^{1/2} \qquad (6.358)$$

Expanding it in powers of λ, we obtain:

$$\Sigma(p) = \frac{\lambda p}{2\pi} + \frac{\lambda^2 p^2}{8\pi^2(\varepsilon - p)} + \cdots \qquad (6.359)$$

All terms of this expansion (besides the first one) has a structure given by (6.357).

Conclusions from this study of Tomonaga–Luttinger model are very important and instructive. In fact, most of these results are qualitatively valid also for more general models of interactions in one dimension. Fermi-liquid behavior is always absent, and we observe *bosonization* of spectrum of elementary excitations. In this sense, one-dimensional systems present a picture, alternative to that of Fermi-liquid. Usually it is called "Luttinger-liquid". In recent years major interest is attracted to situation, realizing in two-dimensional case, where in case of strong correlations (typical for high-temperature copper oxide superconductors) scenario of "Luttinger-liquid" behavior competes with that of traditional Fermi-liquid (or "marginal" Fermi-liquid mentioned previously) [Varma C.N., Nussinov Z., Wim van Saarloos (2002)].

Appendix A

Fermi surface as topological object

During our discussion of the basics of Fermi-liquid theory we assumed that Fermi surface is conserved after adiabatic "switching" of arbitrary strong interaction of Fermions. Below we present an elegant proof of this assumption, based on topological arguments (G.E. Volovik, 1991).

In an ideal Fermi-gas the Fermi surface represents a natural border dividing (in momentum space) the regions of occupied ($n(\mathbf{p}) = 1$) and unoccupied ($n(\mathbf{p}) = 0$) states [Sadovskii M.V. (2019a)]. It is clear that in such a gas the Fermi surface is a stable object — small changes of particle energies only slightly deform the border between occupied and unoccupied states, leading to small deformation of the Fermi surface.

If we "switch on" interaction between particles, distribution function $n(\mathbf{p})$ in the ground state (as we have seen above) is no more just 1 or 0. However, the Fermi surface is conserved and is reflected in the singularity (discontinuity) of $n(\mathbf{p})$. Such stability of the Fermi surface follows from certain topological property of Fermion Green's function. Let us write down this function in an ideal gas for a given momentum $\mathbf{p}$ and *imaginary*[1] frequency $z = ip_0$:

$$G(p_0, \mathbf{p}) = \frac{1}{ip_0 - v_F(p - p_F)} \tag{A.1}$$

It is obvious that this Green's function still contains singularity at the hypersurface ($p_0 = 0$, $p = p_F$) in four-dimensional space of (p_0, $\mathbf{p}$), where this function is undefined. This singularity is stable, i.e. it can not be destroyed by small perturbations. The reason is, that the phase

---

[1]Imaginary frequency is introduced here to avoid the usual singularity at $z = \xi(\mathbf{p})$, and is not connected, in general, with Matsubara formalism.

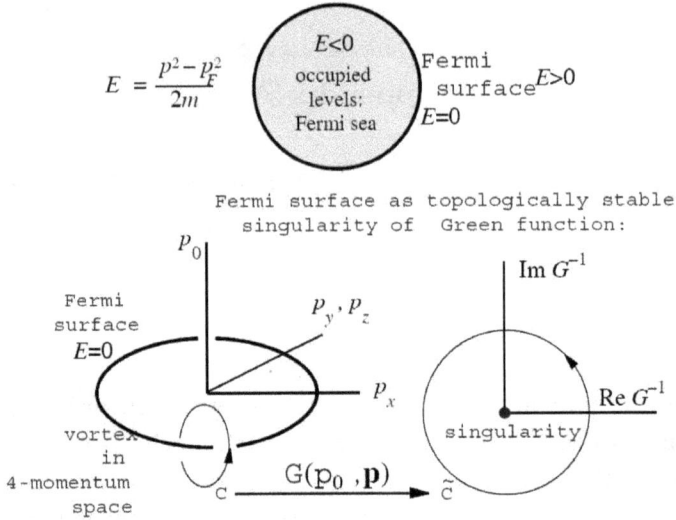

$$E = \frac{p^2 - p_F^2}{2m}$$

Fig. A.1   Fermi surface as topological object in momentum space. Above: In ideal Fermi-gas, Fermi surface surrounds the sphere of occupied states with negative energy. Below: Fermi surface is conserved after "switching on" interaction. The reason is that it is a topologically stable object — the "vortex" in four-dimensional space of $(p_0, \mathbf{p})$.

Φ of Green's function, considered as a complex number, i.e. $G = |G|e^{i\Phi}$, changes by 2π as we move around any contour C, encircling arbitrary element of this (singular) hypersurface.

To convince ourselves, let us drop one spatial dimension, so that Fermi surface becomes a closed line in two-dimensional space of (p_x, p_y). Singularities of the Green's function (A.1) then lie on a closed line in three-dimensional space of (p_0, p_x, p_y), shown in the lower part of Fig. A.1. The phase of the Green's function changes by 2π during each walk around the arbitrary contour C, encircling an arbitrary element of this "vortex line"[2] in three-dimensional space of (p_0, p_x, p_y). Appropriate "circulation number" $N_1 = 1$ can not change continuously and is stable towards adiabatic "switching" of arbitrary interaction. Thus, singularity of the Green's function and the presence of zero excitation energies in the system of Fermions is also conserved.

---

[2]This is in direct analogy with topological stability of Abrikosov vortices in type II superconductors, where the order parameter has the form of $\Psi = |\Psi|e^{i\Phi}$ [De Gennes P.G. (1966)].

In general case, Green's function of a Fermion is a matrix with spin indices. In periodic crystal it is characterized by an additional band index, etc. In such cases the notion of the phase of the Green's function looses its meaning, though topological properties, described above, survive. It can be shown that in such a general case we may define a topological invariant, given by:

$$N_1 = Sp \oint_C \frac{dl}{2\pi i} G(p_0, \mathbf{p}) \frac{\partial G^{-1}(p_0, \mathbf{p})}{\partial l} \qquad (A.2)$$

where Green's function is represented by appropriate matrix, while integral is taken around an arbitrary contour C in the space of $(p_0, \mathbf{p})$, surrounding the hypersurface of singularities in this space (as shown in lower part of Fig. A.1). In (A.2) Sp is taken over spin, band and other (possible) indices.

In Landau Fermi-liquid the single-particle Green's function, as we know, takes the following form:

$$G(p_0, \mathbf{p}) = \frac{Z}{ip_0 - v_F(p - p_F)} + \cdots \qquad (A.3)$$

The difference with the case of an ideal Fermi gas is that Fermi velocity v_F is no more simply p_F/m, but is an additional "fundamental constant" of the theory. It defines an effective mass of a quasiparticle $m^* = p_F/v_F$. The change of v_F and of residue at the pole Z does not change the value of topological invariant (A.2). This justifies Landau assumption of direct correspondence between low energy quasiparticles in Fermi-liquid and particles in an ideal Fermi-gas.

Thus (if there are no "infrared" singularities of the type appearing in Tomonaga–Luttinger model), in isotropic Fermi-liquid the spectrum of Fermion quasiparticles is described by universal dependence:

$$\varepsilon(\mathbf{p}) \to v_F(|\mathbf{p}| - p_F) \qquad (A.4)$$

with two "fundamental constants" v_F and p_F. Their values are determined by "microscopic" interactions, but in Fermi-liquid they are just phenomenological constants.

Topological stability of the Fermi surface means that any continuous change of the system does not change topological invariant. In particular, such a change may be due to adiabatic "switching on" interactions between particles and (or) adiabatic deformation of the Fermi surface. During such adiabatic perturbations, energy levels of the system do not cross the Fermi surface. The state without excited quasiparticles is transformed to another such state, i.e. vacuum is transformed into another vacuum (ground state). This leads to validity of Luttinger theorem, which is equivalent to the statement that the volume of Fermi surface is an "adiabatic invariant" (if the total number of particles does not change).

For isotropic Fermi-liquid with spherical Fermi surface, Luttinger theorem reduces to the validity of usual (gas-like!) relation between particle density and Fermi momentum:

$$n = \frac{p_F^3}{3\pi^2 \hbar^3} \tag{A.5}$$

Fermi surface characterized by topological invariant N_1 exists for any space dimensionality. In two-dimensions it is represented by the closed line in two-dimensional momentum space, corresponding to "vortex loop" in three-dimensional momentum-frequency space. In one-dimensional systems Fermi surface is represented by point-like "vortex".

We have seen above that in one-dimensional system Green's function may have the form quite different from canonical Fermi-liquid like, given by (A.3). In particular, for Tomonaga–Luttinger model it has no poles due to "infrared" divergences. However, Fermi surface survives, as well as the existence of excitations with arbitrary small energies, due to conservation of topological invariant $N_1 = 1$. We may see it if in explicit expressions for Green's function obtained above, like Eq. (6.345), we make transformation $\varepsilon \to ip_0$. Then again we have singularity at ($p_0 = 0$, $p = 0$). Thus, the Fermi surface survives, despite basic assumptions of Landau theory are broken.

Let us consider from this point of view our model of the pseudogap state, in the exactly solvable limit of large correlation lengths ($\xi \to \infty$). We may rewrite (6.203) as:

$$G(p_0, p) = \int_0^\infty d\zeta e^{-\zeta} \frac{ip_0 + \xi_p}{(ip_0)^2 - \xi_p^2 - \zeta \Delta^2}$$

$$= \frac{ip_0 + \xi_p}{\Delta^2} \exp\left(\frac{p_0^2 + \xi_p^2}{\Delta^2}\right) Ei\left(-\frac{p_0^2 + \xi_p^2}{\Delta^2}\right) \tag{A.6}$$

Then for $p_0 \to 0$ and $\xi_p \to 0$ (i.e. $p \to p_F$), we have:

$$G(p_0, p) \approx \frac{ip_0 + \xi_p}{\Delta^2} \ln\left(\gamma \frac{p_0^2 + \xi_p^2}{\Delta^2}\right) \equiv \frac{Z(p_0, \xi_p)}{ip_0 - \xi_p} \tag{A.7}$$

where

$$Z(p_0, \xi_p) = -\frac{p_0^2 + \xi_p^2}{\Delta^2} \ln \left(\gamma \frac{p_0^2 + \xi_p^2}{\Delta^2} \right) \to 0 \quad \text{for} \quad p_0 \to 0, \xi_p \to 0$$

$$(\text{A.8})$$

Now the effect of the "residue" is so strong, that it transforms the pole in the Green's function to zero of the Green's function. But the singularity of the Green's function at the Fermi surface is not destroyed: the zero is also the singularity and it has the same topological invariant as pole.[3] So in this sense our model is similar to some kind of Luttinger or "marginal" Fermi-liquid with very strong renormalization of singularity at the Fermi surface.

The difference with Landau Fermi-liquid in one-dimensional systems is clearly seen when we analyze *real* frequencies: quasiparticle poles in Green's function are absent, instead we have a cut in the complex plane of frequency, so that single-particle excitations are not defined. However, distribution function, as we have seen e.g. in (6.351), though not possessing discontinuity itself, may still contain singular behavior in its derivatives.

---

[3] I am grateful to G.E.Volovik for the clarification of this point.

Appendix B

Electron in a random field and Feynman path integrals

Returning to the problem of an electron propagating in a field of scatterers randomly distributed in space, let us show how it is possible to obtain formally exact expression for the averaged single-particle Green's function via Feynman path integral [Sadovskii M.V. (2019b)], equivalent to the sum of all diagrams of perturbation theory. For electron propagating in a potential field (4.1), in time-coordinate representation we can write down the Green's function as a standard path integral of the following form (N is total number of scatterers):

$$G(\mathbf{rr'};t) = \int_{\mathbf{r}(0)=\mathbf{r'}}^{\mathbf{r}(t)=\mathbf{r}} \mathcal{D}\mathbf{r}(\tau) \exp\left\{ \frac{i}{\hbar} \int_0^t d\tau \left[\frac{m\dot{\mathbf{r}}^2}{2} - \sum_{j=1}^{N} v(\mathbf{r} - \mathbf{R}_j) \right] \right\}$$

$$(\text{B.1})$$

where $\int_{\mathbf{r}(0)=\mathbf{r'}}^{\mathbf{r}(t)=\mathbf{r}} \mathcal{D}\mathbf{r}(\tau)$ denotes Feynman–Wiener functional (path) integration [Sadovskii M.V. (2019b)], and $\mathbf{r}$ and $\mathbf{r'}$ are final and initial points for electron propagation during the time-interval t.

Consider general enough case, when we know N-particle distribution functions [Sadovskii M.V. (2019a)] of scatterers $F_N(\mathbf{R}_1, ..., \mathbf{R}_N)$. Then, performing averaging of the part of (B.1), depending on scatterers, we obtain:

$$\left\langle \exp\left\{ -\frac{i}{\hbar}\sum_j \int_0^t d\tau v(\mathbf{r}(\tau) - \mathbf{R}_j) \right\} \right\rangle$$

$$= \frac{1}{V^N}\int d\mathbf{R}_1 ... \int d\mathbf{R}_N \exp\left\{ -\frac{i}{\hbar}\sum_j \int_0^t d\tau v(\mathbf{r}(\tau) - \mathbf{R}_j) \right\} F_N(\mathbf{R}_1, ..., \mathbf{R}_N)$$

$$= \exp\left\{ \sum_{n=0}^{\infty}\left(-\frac{i}{\hbar} \right)^n \frac{1}{n!}\int_0^t d\tau_1 ... \int_0^t d\tau_n \left\langle \sum_i v(\mathbf{r}(\tau_1) - \mathbf{R}_i)... \sum_j v(\mathbf{r}(\tau_n) - \mathbf{R}_j) \right\rangle_c \right\}$$

$$\equiv \exp\left\{ \sum_n \left(-\frac{i}{\hbar} \right)^n K_n \right\} \qquad (\text{B.2})$$

where $< ... >_c$ are cumulant averages, defined in (4.13), V is the volume of the system. Then, the averaged Green's function is given by:

$$< G(\mathbf{rr}'; t) > = \int_{\mathbf{r}(0)=\mathbf{r}'}^{\mathbf{r}(t)=\mathbf{r}} \mathcal{D}\mathbf{r}(\tau) \exp\left\{ \frac{i}{\hbar}\int_0^t d\tau \frac{m\dot{\mathbf{r}}^2}{2} + \sum_n \left(-\frac{i}{\hbar} \right)^n K_n \right\} \qquad (\text{B.3})$$

Limiting ourselves with $n = 2$ ($n = 1$ contribution gives trivial phase factor) i.e. in Gaussian approximation for the statistics of the random field, we get:

$$< G(\mathbf{rr}'; t) >$$
$$= \int_{\mathbf{r}(0)=\mathbf{r}'}^{\mathbf{r}(t)=\mathbf{r}} \mathcal{D}\mathbf{r}(\tau) \exp\left\{ \frac{i}{\hbar}\int_0^t d\tau \frac{m\dot{\mathbf{r}}^2}{2} - \frac{1}{2\hbar^2}\int_0^t d\tau' \int_0^t d\tau W[\mathbf{r}(\tau) - \mathbf{r}'(\tau')] \right\} \qquad (\text{B.4})$$

where

$$W(\mathbf{r} - \mathbf{r}') = < V(\mathbf{r})V(\mathbf{r}') >_c \qquad (\text{B.5})$$

which for the case of randomly distributed (in space) "impurities" is determined via (4.12), (4.16) and reduces to "white noise" correlator (4.20).

As an example of application of Eq. (B.4), consider an electron propagating in one-dimensional system with Gaussian random field with correlator defined in (6.184), which was interest to us in the model of pseudogap state (ξ-correlation length of short-range order fluctuations):

$$W(r - r') = 2 < |\Delta|^2 > \exp\left\{ -\frac{|r - r'|}{\xi} \right\} \cos 2p_F(r - r') \qquad (\text{B.6})$$

Fourier-transform of this correlator is given by (6.196):

$$S(Q) = 2\Delta^2 \left\{ \frac{\kappa}{(Q - 2p_F)^2 + \kappa^2} + \frac{\kappa}{(Q + 2p_F)^2 + \kappa^2} \right\} \tag{B.7}$$

where $\kappa = \xi^{-1}$, and is represented by two Lorentzians of the width $\sim \xi^{-1}$, centered at $Q = \pm K = \pm 2p_F$. Introducing the variable q (deviation from the center of the peak) via $Q = \pm K + q$, we can write:

$$\int_0^t d\tau \int_0^t d\tau' W[r(\tau) - r'(\tau')] = \int \frac{dQ}{2\pi} S(Q) \int_0^t d\tau \int_0^t d\tau' e^{iQr(\tau)} e^{-iQr(\tau')}$$

$$= \Delta^2 \int \frac{dq}{\pi} \frac{\kappa}{q^2 + \kappa^2} \int_0^t d\tau \int_0^t d\tau' e^{iKr(\tau)} e^{iqr(\tau)} e^{-iKr(\tau')} e^{-iqr(\tau')}$$

$$+ \Delta^2 \int \frac{dq}{\pi} \frac{\kappa}{q^2 + \kappa^2} \int_0^t d\tau \int_0^t d\tau' e^{-iKr(\tau)} e^{iqr(\tau)} e^{iKr(\tau')} e^{-iqr(\tau')}$$

$$= 2\Delta^2 \int \frac{dq}{\pi} \frac{\kappa}{q^2 + \kappa^2} \int_0^t d\tau \int_0^t d\tau' e^{iqr(\tau)} e^{-iqr(\tau')} \cos K[r(\tau) - r(\tau')]$$

$$= 2\Delta^2 \int \frac{dq}{\pi} \frac{\kappa}{q^2 + \kappa^2} \int_0^t d\tau \int_0^t d\tau' e^{iqr(\tau)} e^{-iqr(\tau')} \left\{ \cos Kr(\tau) \cos Kr(\tau') \right.$$

$$+ \sin Kr(\tau) \sin Kr(\tau') \Big\} \tag{B.8}$$

Above we have considered the asymptotic behavior for large correlation lengths $\xi \to \infty$ (or $\kappa \to 0$). In this limit we have:

$$\int_0^t d\tau \int_0^t d\tau' W[r(\tau) - r'(\tau')] \approx 2\Delta^2 \int_0^t d\tau \int_0^t d\tau' \left\{ \cos Kr(\tau) \cos Kr(\tau') \right.$$

$$+ \sin Kr(\tau) \sin Kr(\tau') \Big\} = 2\Delta^2 \left\{ \int_0^t d\tau \cos Kr(\tau) \right\}^2 + 2\Delta^2 \left\{ \int_0^t d\tau \sin Kr(\tau) \right\}^2 \tag{B.9}$$

Then we easily obtain the following representation for nontrivial part of the exponential in (B.4):

$$\exp\left\{ -\frac{1}{2\hbar^2} \int_0^t d\tau \int_0^t d\tau' W[r(\tau) - r(\tau')] \right\}$$

$$= \exp\left\{ -\frac{\Delta^2}{\hbar^2} \left[\int_0^t \cos Kr(\tau) \right]^2 \right\} \exp\left\{ -\frac{\Delta^2}{\hbar^2} \left[\int_0^t \sin Kr(\tau) \right]^2 \right\}$$

$$= \int_{-\infty}^\infty \frac{dx}{\sqrt{\pi}} e^{-x^2 + 2ix\frac{\Delta}{\hbar} \int_0^t \cos Kr(\tau)} \int_{-\infty}^\infty \frac{dy}{\sqrt{\pi}} e^{-y^2 + 2iy\frac{\Delta}{\hbar} \int_0^t \sin Kr(\tau)} \tag{B.10}$$

where in the last equality we have used the well known Hubbard–Stratonovich trick. As a result, after the obvious changes of variables, we obtain the averaged Green's function as:

$$
< G(rr';t) >= \int_{-\infty}^{\infty} \frac{dV_x}{\sqrt{\pi}} e^{-\frac{V_x^2}{\Delta^2}} \int_{-\infty}^{\infty} \frac{dV_y}{\sqrt{\pi}} e^{-\frac{V_y^2}{\Delta^2}} \int_{r(0)=r'}^{r(t)=r} \mathcal{D}r(\tau) \exp \frac{i}{\hbar} \left\{ \int_0^t d\tau \frac{m\dot{r}^2(\tau)}{2} \right.
$$

$$
\left. - 2 \int_0^t d\tau V_x \cos Kr(\tau) - 2 \int_0^t d\tau V_y \sin Kr(\tau) \right\} \tag{B.11}
$$

Transforming to polar coordinates in (V_x, V_y) plane, i.e. introducing $W = \sqrt{V_x^2 + V_y^2}$ and $\phi = arctg \frac{V_y}{V_x}$, we obtain:

$$
< G(rr';t) >= \int_0^{\infty} dW \frac{2W}{\Delta^2} e^{-\frac{W^2}{\Delta^2}} \int_0^{2\pi} \frac{d\phi}{2\pi}
$$

$$
\times \int_{r(0)=r'}^{r(t)=r} \mathcal{D}r(\tau) \exp \left\{ \frac{i}{\hbar} \int_0^t d\tau \left[\frac{m\dot{r}^2(\tau)}{2} - 2W \cos(Kr(\tau) + \phi) \right] \right\}
$$

$$
= \int_0^{\infty} dW \mathcal{P}\{W\} \int_0^{2\pi} \mathcal{G}_{2W \cos(Kr+\phi)}(rr';t) \tag{B.12}
$$

where $\mathcal{P}(W)$ is Rayleigh distribution (6.207), and $\mathcal{G}_{2W \cos(Kr+\phi)}(rr';t)$ is the single-electron Green's function in periodic field (potential) $2W \cos(Kr + \phi)$. If we make transformation to momentum space and find the Green's function in this field in the simplest (two-wave) approximation [Ziman J.M. (1972)], we get (6.57), (6.60) and (in Matsubara technique, under "nesting" conditions, valid for $K = 2p_F$ (6.71) (cf. Fig. 6.7)). Then it is clear that (B.12) reduces to the result (6.203) obtained above by diagram summation. Appropriate anomalous Green's function (6.72) gives zero after we average over the phase in (B.12), which corresponds to the absence of long-range order in our system. Thus, our simplified model of the pseudogap state, analyzed in Chapter VI, is really equivalent to the model of an electron propagating in potential field $2W \cos(Kr + \phi)$, with amplitude W independent of coordinate and distributed according to Rayleigh, while phase ϕ is distributed homogeneously on interval $(0, 2\pi)$. The appearance of Rayleigh distribution here is intimately connected with our assumption of the Gaussian nature of the random field.

The use of asymptotics of $\xi \to \infty$, i.e. neglecting the width of Lorentzian peaks in $S(Q)$, is obviously corresponding to neglect of large scale fluctuations of our random field on distances of the order (or less than) $\xi \sim \kappa^{-1}$. Physically it is clear that such fluctuations lead to additional scattering with characteristic times of the order of $\tau \sim \frac{\xi}{v_F} \sim (v_F \kappa)^{-1}$ (cf. (6.198)). It is clear that this scattering can be neglected for energies, satisfying the inequality $v_F \kappa \ll \xi_p$. It is also unimportant for $v_F \kappa \ll T$. These conditions were written above in (6.199). Our analysis of the problem with finite values of κ confirms these qualitative expectations. The main effect of additional large scale scattering is the "filling" of the pseudogap, which completely disappears for $v_F \kappa \sim \Delta$.

Bibliography

Abrikosov A.A., Gorkov L.P., Dzyaloshinskii I.E. (1963) Methods of Quantum
Field Theory in Statistical Physics. Pergamon Press, Oxford
Allen P.B., Mitrović B. Solid State Physics, Vol. 37 (Eds. F. Seitz, D. Turnbull,
H. Ehrenreich), Academic Press, NY, 1982, p. 1
Altshuler B.L., Aronov A.G. (1985) Electron-Electron Interaction in Disordered
Conductors. In "Electron-Electron Interactions in Disordered Systems", Ed.
by A.L. Efros and M. Pollak. Elsevier Science
Altshuler B.L., Aronov A.G., Khmelnitskii D.E., Larkin A.I. (1982) Coherent
Effects in Disordered Conductors. In "Quantum Theory of Solids", Ed. by
I.M. Lifshits. Mir Publishers, Moscow
Bogoliubov N.N. (1991a) Lectures on Quantum Statistics. Selected Works, Part
II: Quantum and Statistical Mechanics. Gordon and Breach, NY
Bogoliubov N.N. (1991b) Quasi-Averages in the Problems of Statistical Mechan-
ics. Selected Works, Part II: Quantum and Statistical Mechanics. Gordon
and Breach, NY
De Gennes P.G. (1966) Superconductivity of Metals and Alloys. Benjamin, NY
Ginzburg V.L., Kirzhnits D.A., editors. (1982) High-Temperature Superconduc-
tivity. Consultants Bureau, NY
Ginzburg S.L. (1989) Irreversible Phenomena in Spin Glasses, Nauka, Moscow
(in Russian)
Khomskii D.I. (2010) Basic Aspects of the Quantum Theory of Solids, Cambridge
University Press, Cambridge
Lee P.A., Ramakrishnan T.V. (1985) Disordered Electronic Systems. Rev. Mod.
Phys. **57**, 287
Levitov L.S., Shitov A.V. (2003) Green's Functions. Problems and Solutions.
Fizmatlit, Moscow (in Russian)
Lifshits E.M., Pitaevskii L.P. (1980) Statistical Physics. Part 2. Pergamon Press,
Oxford
Lifshits I.M., Gredeskul S.A., Pastur L.A. (1988) Introduction to the Theory of
Disordered Systems. Wiley, NY
Mahan G.D. (1981) Many-Particle Physics. Plenum Press, NY

Mattuck R.D. (1968) Quantum Theory of Phase Transitions in Fermi Systems. Adv. Phys. **17**, 509

Migdal A.B. (1967) Theory of Finite Fermi Systems and Applications to Atomic Nuclei. Interscience Publishers, NY

Mott N.F. (1974) Metal-Insulator Transitions. Taylor and Francis, London

Nozieres P. (1964) Theory of Interacting Fermi Systems. Benjamin, NY 1964

Pines D., Nozieres P. (1966) The Theory of Quantum Liquids. Benjamin, NY

Sadovskii M.V. (2000) Superconductivity and Localization. World Scientific, Singapore

Sadovskii M.V. (2019a) Statistical Physics. Walter De Gruyter GmbH, Berlin-Boston

Sadovskii M.V. (2019b) Quantum Field Theory. Walter de Gruyter GmbH, Berlin-Boston

Schrieffer J.R. (1964) Theory of Superconductivity. Benjamin, Reading, Mass.

Shklovskii B.I., Efros A.L. (1984) Electronic Properties of Doped Semiconductors. Springer Verlag, Berlin-Heidelberg-NY

Varma C.M., Nussinov Z., Wim van Saarloos (2002) Singular or non-Fermi liquids. Physics Reports **361**, 267

Vonsovsky S.V., Izyumov Yu.A., Kurmaev E.Z. (1977) Superconductivity of Transition Metals, their Alloys and Compunds. Springer Verlag, Berlin-Heidelberg-NY

White R.M., Geballe T.H. (1979) Long Range Order in Solids. Academic Press, NY

Ziman J.M. (1972) Principles of the Theory of Solids. Cambridge University Press, Cambridge

Zubarev D.N. (1974) Nonequilibrium Statistical Thermodynamics. Plenum, NY

Index

Abrikosov–Gorkov equation, 224
Anderson insulator, 164, 168
Anderson localization, 166
Anderson theorem, 223, 235
Anderson transition, 164, 165
anomalous average, 271
anomalous Green's functions, 199, 322

Bethe–Salpeter equation, 137, 143
Bogoliubov's coefficients, 202
Born approximation for impurity scattering, 113
Bose-condensation of phonons, 271
bosonization, 350

charge density wave, 261, 277, 291
coherence length, 228, 239, 266
commensurate Peierls transition, 274
compressibility, 67
condensate of Cooper pairs, 198
conductivity, 44, 128, 134, 155
conductivity in the pseudogap state, 332
conductivity of superconducting condensate, 252
Cooper instability, 193
Cooperon, 143
Coulomb interaction, 19
cumulant averages, 110

density matrix, 11
density of states, 133, 151, 204
density of states with pseudogap, 309, 317
density of superconducting electrons, 240
density–density response function, 43, 129
diagram rules for electrons and phonons, 72
dielectric function, 20
dielectric permeability of Peierls insulator, 328
diffusion pole, 141
diffuson, 141
dimensionless electron–phonon coupling constant, 91
dirty superconductors, 233, 236
discontinuity in particle distribution, 49
dispersion of the soft mode, 266
Dyson equation, 12, 113

effective mass, 30, 67
electric conductivity, 43
electron density of states, 308
electron-hole pairs, 290
electronic anomalous averages, 277
Eliashberg equations, 212, 215, 216
Eliashberg function, 90
energy gap, 202